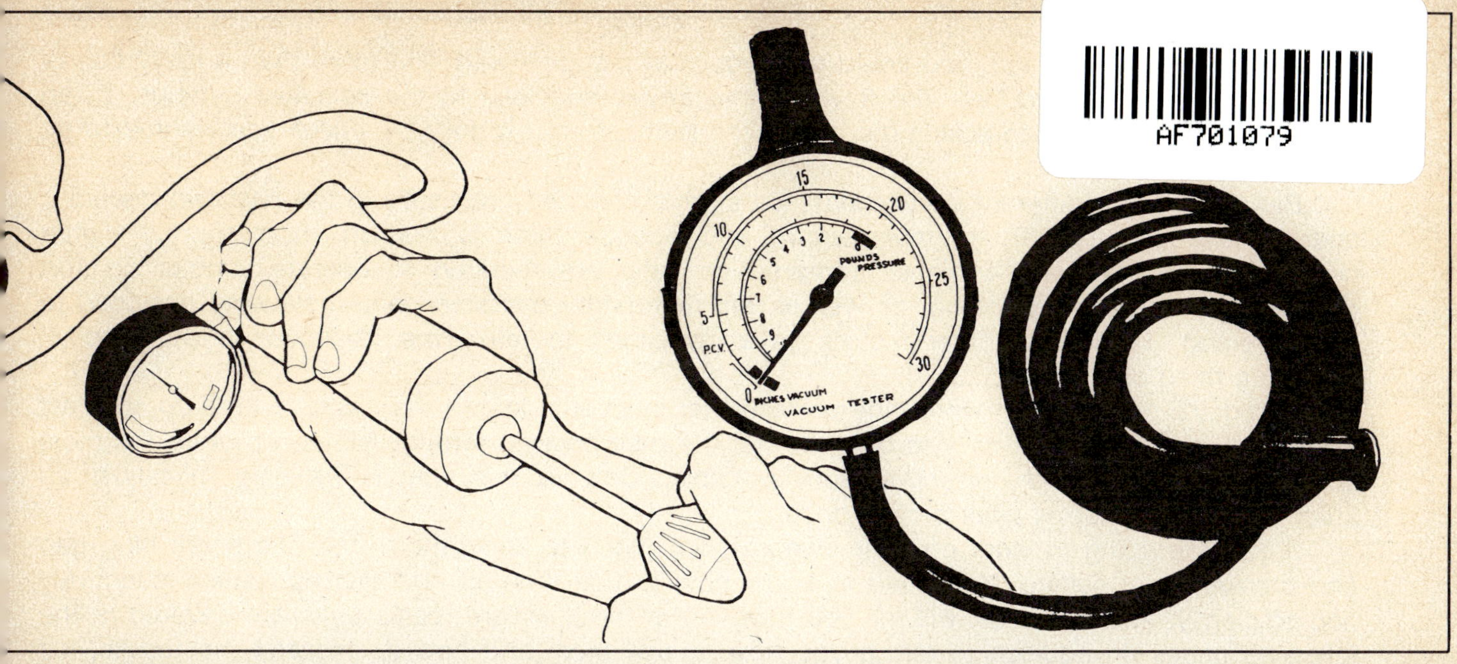

Basic Automotive TUNE-UP & TEST EQUIPMENT

INTRODUCTION

Not so many years ago, you could tune your car's engine with little more than a rusty pair of pliers and a bent screwdriver, and if you came anywhere close to the specified settings, it ran satisfactorily. Oh, it burned a little more gasoline perhaps, but it got you where you wanted to go without difficulty.

But we live in a different world today. Cars have changed drastically in the past two decades, and the energy necessary to manufacture and operate them is now in short supply. Higher horsepower engines with a higher compression ratio are required to operate various comfort options such as air conditioning, as well as the mandatory emission control systems designed to reduce smog. These and other factors have combined to make the "by gosh and by golly" tune-up a thing of the past.

In today's energy-short world, the professional mechanic knows the value of tune-up and test equipment, and he uses it daily to be sure that the cars passing through his stall receive the most accurate diagnostic and corrective work possible. Only in this way is he able to adjust an automotive engine for its peak performance.

If you are going to work on your own car, you need to know more than the spark plug gap and breaker point setting. You need to know how to use tune-up and test equipment effectively and efficiently, and this book will tell you how. All of the standard, basic units are covered in the following pages. In addition to telling you how to use them and how to interpret the results, we explain the system or component to be tested enough so that the beginner can relate the important HOW to the equally important WHY. In short, the following pages constitute a basic course in what makes the car work, what can go wrong, how to locate and pinpoint difficulties with tune-up and test equipment and how to correct the problems. Use it along with your equipment as a reference and see how much faster and more efficiently your next tune-up goes.

Don Whitt/Editor, Special Projects

ISBN 0-8227-0055-7

Library of Congress Catalog Card No. 74-78228

COVER

Even though a piece of test equipment may *look* complicated, it is usually very *easy* to use when properly understood. This book is designed to take away all the mystery surrounding tune-up and test equipment, such as this new Automotive Analyzer by Sears, Roebuck & Co. Cover design by Pat Taketa.

BASIC AUTOMOTIVE TUNE-UP & TEST EQUIPMENT

CONTENTS

CONTENTS

BEFORE YOU BUY

Tune-up and test equipment exists for a single reason—to help you diagnose and correct automotive malfunctions more accurately and quickly than you could without it. Professional mechanics recognize that the use of basic tune-up and test equipment is necessary to keep today's engines running correctly. Since you are reading this book, so do you.

Acquiring the tune-up and test equipment you need to help keep your car in good operating condition can cost you a substantial amount of money, though. Also, unless you shop carefully, you may find that a certain piece of test equipment does not do everything you thought it would. Or perhaps the equipment performs all the tests you expect it to but is so complicated that it now takes three times as long to work on your car as it did without its "help."

There is a great variety of tune-up and test equipment available, most of which is good quality. However, shoddy construction, inaccurate calibration and exaggerated claims for what it will do are not uncommon. To make sure that you get reliable equipment that does what it should without spending more money than necessary, there are several considerations you should keep in mind while shopping. Several manufacturers sell their tune-up and test equipment under a variety of brand names as well as private labels. The electronics are the same, but different housing designs, functional controls and meter scales are used so that the brand names will not compete with others made by the same company. Therefore the price varies from one name to another.

Unfortunately, the external appearance of tune-up and test equipment is a strong factor in its sale; people tend to buy the complicated-looking units with colorful meter scales and lots of dials, switches or buttons. This fact of life accounts for the combination of two or more pieces of equipment into what is popularly called an engine analyzer. To help you select your tune-up and test equipment, here are a few things to consider before you buy.

USE

While one well-known manufacturer sells a very handy compression tester for $57, it is hardly a wise purchase if you only expect to test engine compression two or three times a year. Plan your selection of equipment according to how much use you will make of it. A $7 compression tester will not look as impressive hanging over your workbench, but it is just as functional for the intended use and leaves $50 to spend on other equipment.

FUNCTION

Will that unit you've been thinking about buying really do the job you expect it to? For example, not all alternator-generator-regulator testers can be used to test alternator diodes, nor will all ignition testers allow you to check the complete ignition system. If you are considering an engine analyzer instead of separate units, the possibility of equipment limitation becomes even greater, because the variety and combination of units vary widely.

If you are not completely familiar with what tests you want or need to perform or exactly what a particular piece of equipment should be capable of doing, read the rest of this book first. Then compare the various units to determine which one comes closest to meeting your needs. Whenever possible, ask to see the instruction manual that comes with the equipment to make certain that it does everything you want it to. Don't buy by price alone, either, as the less expensive equipment is sometimes the more versatile.

CONSTRUCTION

This is what separates the men from the boys: How durable is the equipment? Is a metal housing really better than a plastic one? What happens to the meter if you drop it? Are the connecting cables well made, properly insulated and attached permanently to the unit to prevent losing them? These and many other questions should be answered when considering a particular unit.

Generally speaking, high-impact plastic housings will take just as rough usage as metal ones, but some manufacturers still use Bakelite and other brittle plastics for their housings—one good drop on the floor and suddenly you need a new unit. It should be possible to take the housing apart if repairs become necessary. Some plastic units are molded in such a way that the case must be broken to get inside.

Knob and switch functions should be stamped, printed or engraved directly on the case. Those paper labels used on some units have an annoying habit of falling off and getting lost. Knobs and switches should be large enough to use easily and should click into place when set.

Where equipment is likely to come into contact with chemicals (such as antifreeze and battery electrolyte), the materials used in its construction should be capable of withstanding the effect of the chemicals. Several brands of hydrometers use bulbs made of a rubber that is literally eaten away by the solutions after being used a few times. Chemical-resistant vinyl seems to last longer.

Connecting cables should be at least 36 inches long, well insulated and color-coded (red for positive, black for negative) to help you attach them correctly. Check to see that the alligator or spring return scissor clamp is securely fastened to each cable. If the cable is not permanently attached to the unit, a compartment should be provided in the unit for storing it or it is likely to get mislaid or lost.

METER SCALES

The main value of test equipment is in the information displayed on its meter scale. To be most useful, the scale should be simple in design, with a range of measurement sufficient for test requirements and marked with easy-to-read divisions and subdivisions. Don't be misled by multi-colored bands, lots of tiny lettering and dual-purpose scales. If color-coding is used for quick visual identification of test results, look for contrasting colors like black and white, red and green, etc. Some meters use a combination such as orange, yellow and red all on the same band; the colors tend to confuse the eye rather than clarify.

Lettering and numerals on each band should be easy to read at a distance of 14 inches. Where dual-purpose bands are provided on the meter scale, they should be legibly marked and readily identified as to function. Meter needles should have a zero adjusting screw as well as a marked adjustment setting on the meter face.

OVERALL DESIGN

If the equipment is meant to be hand-held during use, it should fit easily into your hand. If it is to stand upright, its weight and stability must be sufficient to keep it in place despite normal vibrations encountered when working on a car. For those units using accessories such as battery post adapters or spark plug adapters, look for compartments in the housing in which they can be stored.

Some of the larger units are designed for use on rollaway stands. These stands can prove useful for storage as well as providing a good display position while the equipment is in use. Check the stand before purchasing it to make sure that it is sturdy enough to hold the unit properly. Make sure that it rolls easily from one place to another and provides for locking the rollers or wheels in place.

If you stop to consider the points already mentioned before buying tune-up and test equipment, you will find it much easier to purchase the best equipment for your needs at a reasonable price. Remember too that the best piece of tune-up and test equipment ever manufactured will do you no good if it is not used. Equipment purchase is generally a once-in-a-lifetime proposition—buy wisely and use it often.

FACTORY SHOP MANUALS

One last word is in order before you read further. Every automotive manufacturer produces a factory shop manual for each model year. This rather large book contains detailed testing, teardown, repair and reassembly procedures for each system that makes up that particular car. Also included are specification charts and data that provide exact settings and/or allowable tolerances for testing and repair sequences. In many cases, such specifications will not be found elsewhere.

To obtain the factory shop manual for your particular automobile, look in the owner's handbook that you received from the dealer when the car was purchased. At the back of the book, you'll find a coupon to be used as an order blank. Fill it out completely, enclose payment as specified and mail it to the address shown on the coupon.

If you've lost your owner's handbook or bought the car secondhand, visit a dealer's service department and ask the parts manager to order one for your car. Factory shop manuals all have parts numbers and can be obtained in this manner.

DWELL TACHOMETER

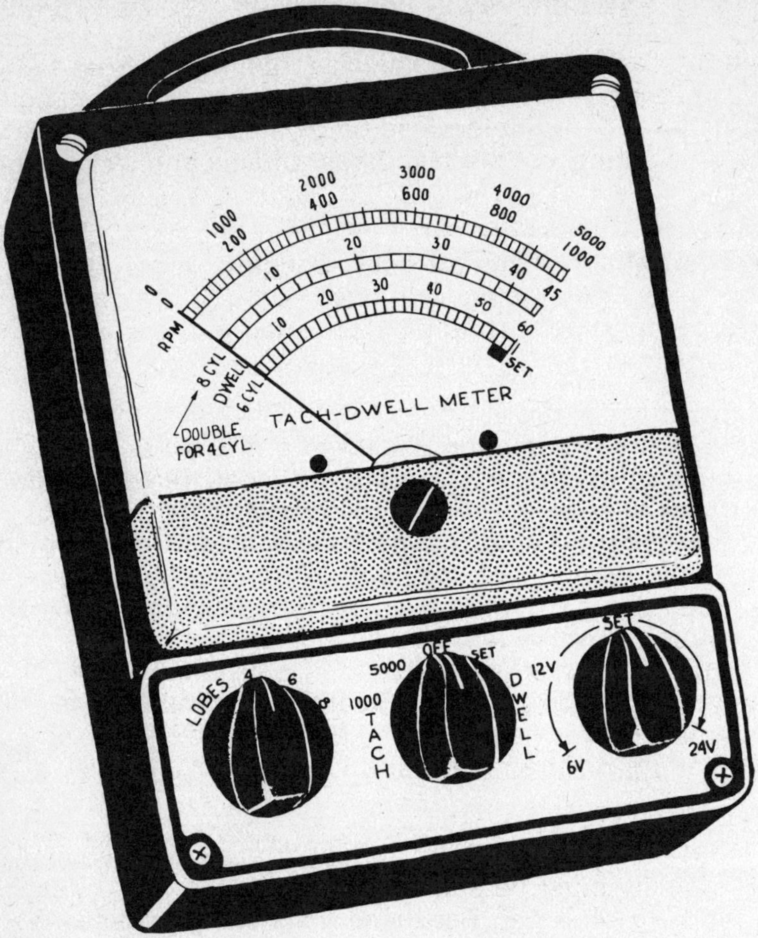

Do you actually need a dwell meter to set the breaker points correctly? Not really. Whether you set the point gap with a feeler gauge or set the cam angle with a dwell meter is immaterial, as long as you do it *right.* Breaker point gap and distributor cam angle are inversely proportional; that is, a wide gap gives less dwell and a narrow gap more dwell. Setting the breaker point gap is done with the distributor cap off and the engine dead, while dwell or cam angle is set with the engine turning at cranking speed. But depending upon how breaker point adjustment is made with your distributor, a dwell meter *can* be a handy instrument. Let's see why.

Internal-adjustment distributors use a solid distributor cap(A) which must be removed to make adjustments. If you set the breaker point gap correctly, the job is done once the distributor cap is replaced. But you may have to do it several times if a dwell meter is used, since dwell is different when the engine is running than when it is only cranking.

With the Delco-Remy external-adjustment or window-type distributor (B) used on General Motors cars, breaker points are set from outside the cap, using an Allen wrench (C). Working with a dwell meter is the fastest and easiest way to adjust this type of distributor, as its cap does not have to be removed.

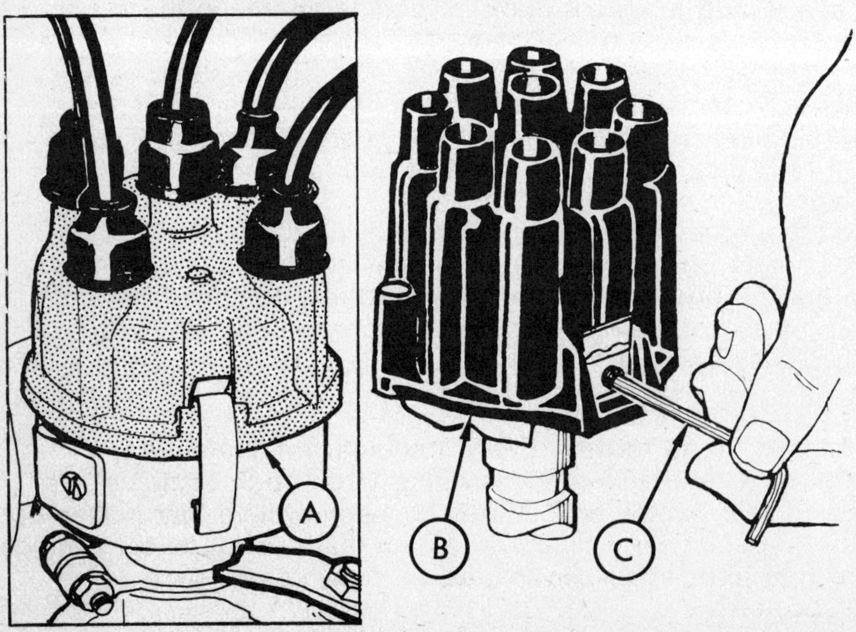

Dwell meters are often combined with a tachometer in a single instrument housing. With late-model cars, the tachometer is a necessity for setting idle speeds correctly. Many tune-ups were done in the past by setting engine idle speed according to the sound of the engine, but today's emission control settings demand that idle speed be "right on" as specified by the factory. Using a tachometer is the only accurate method of setting idle speed.

DWELL TACHOMETER CONTROLS

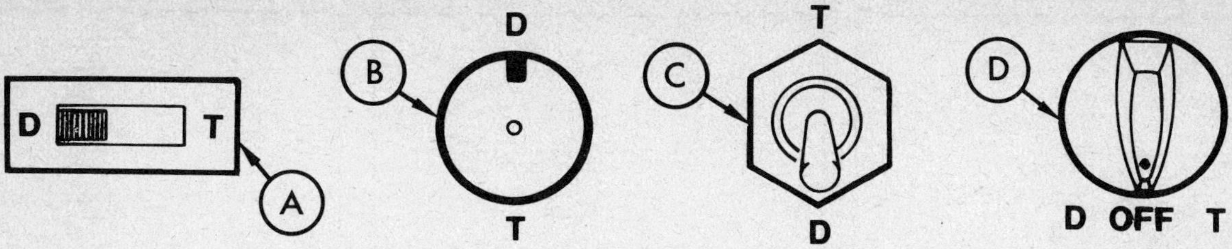

Dwell tachometers are equipped with a control switch for choosing the function of the unit. This control switch may be a positive slide switch (A), a rotary selector knob (B) or a toggle switch (C). On battery-powered units, it can be combined with an OFF switch (D).

Each unit has a cylinder or lobe switch (E). Older dwell tachometers are marked for use with six- or eight-cylinder cars only, while the newer ones also include a position for testing four-cylinder engines. If the dwell tachometer has no four-cylinder setting, it can still be used to adjust four-cylinder engines, but you must double the eight-cylinder readings on both the dwell and rpm (revolutions per minute) scales.

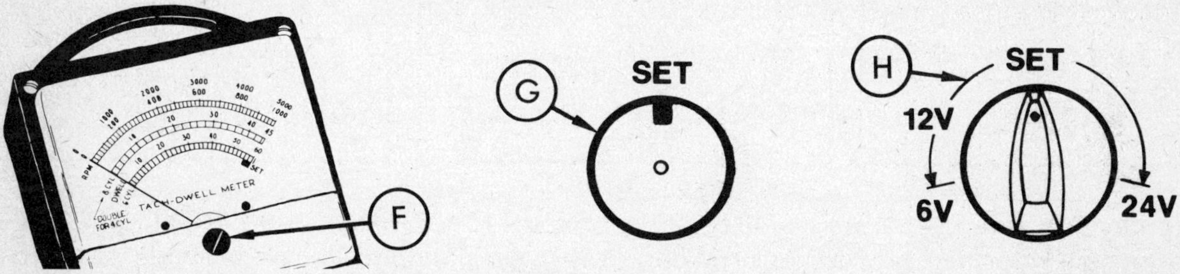

Dwell tachometers are often equipped with a zero adjusting screw. This is used to preset the meter needle to zero when necessary and is usually located on the face of the meter near the base of the needle (F). Some have a SET switch used with a voltage control (G) to calibrate the meter needle before testing. Battery-operated units may be equipped with a converter switch (H) and adjustable for six, 12 or 24 volts or any voltage setting in between.

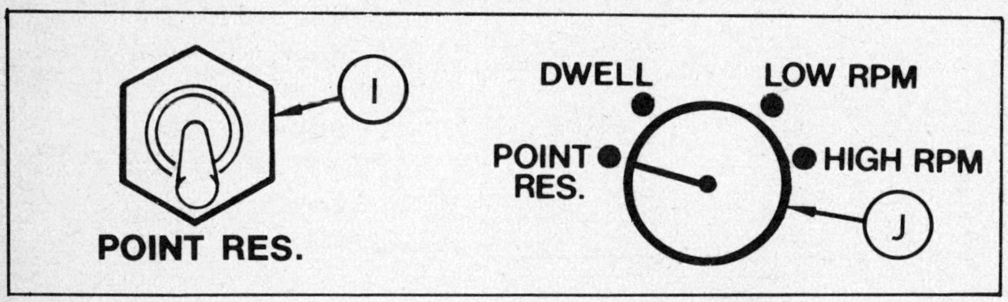

If a dwell tachometer is equipped for checking breaker point resistance, it will have either a separate switch marked POINT RES. (I) or include both the setting and designation as part of a multiple-function switch (J).

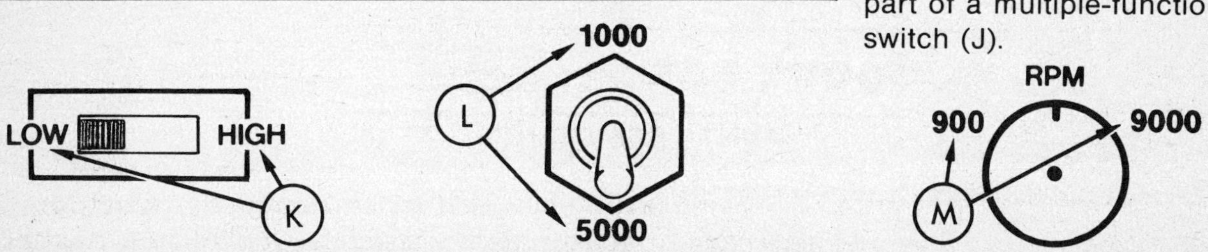

When using the unit as a tachometer, a LOW/HIGH RPM switch is provided if the meter has a dual range. This may be marked LOW/HIGH (K), 900/9000 RPM (M) or 1000/5000 (L). While the numbering on your meter may not be exactly like (L) and (M), the spread between the two numbers is sufficiently great so that you will be able to tell which is low and which is high. The LOW range is ideal for accurate setting of engine idle speeds and multi-carburetor adjustments, while the HIGH range reads cruising and high-speed rpm. One scale is adequate, but it is usually much more cluttered, more difficult to read and interpret and less accurate than a dual-range tachometer.

DWELL TACHOMETER SCALES

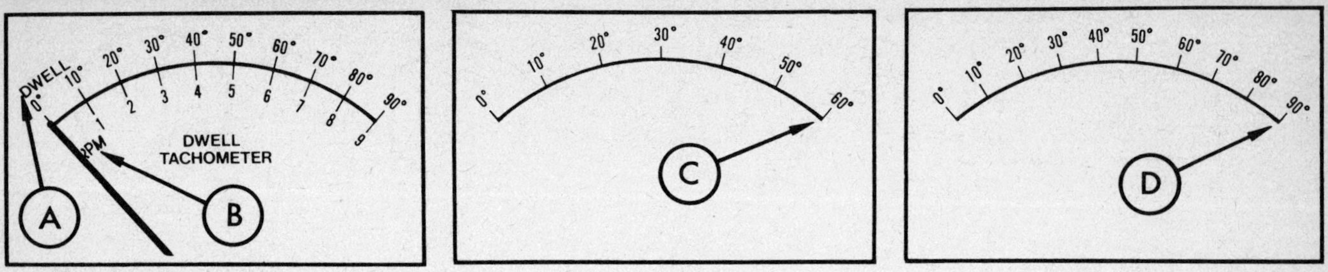

The dwell tachometer unit will have at least two scales, one marked DWELL (A) and the other marked RPM (B).

If a single DWELL scale is used, it will be marked from 0° to 60° for six- and eight-cylinder engines (C) or extend to 90° if designed for testing four-cylinder engines (D) as well.

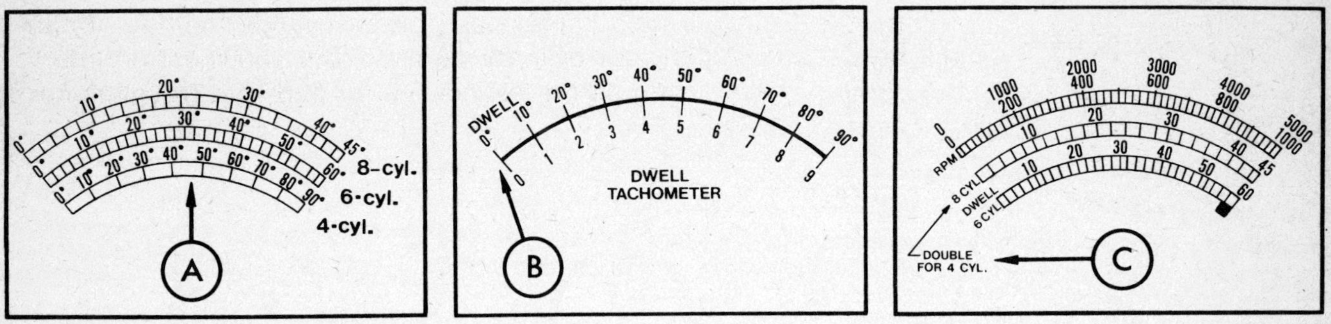

You may find three separate dwell scales, marked 8-CYL., 6-CYL. and 4-CYL. (A), but where a four-cylinder test function is provided, it is more common to find a single scale (B) or a double scale marked 6-CYL. and 8-CYL. with a notation on the latter to "DOUBLE FOR 4-CYL." (C).

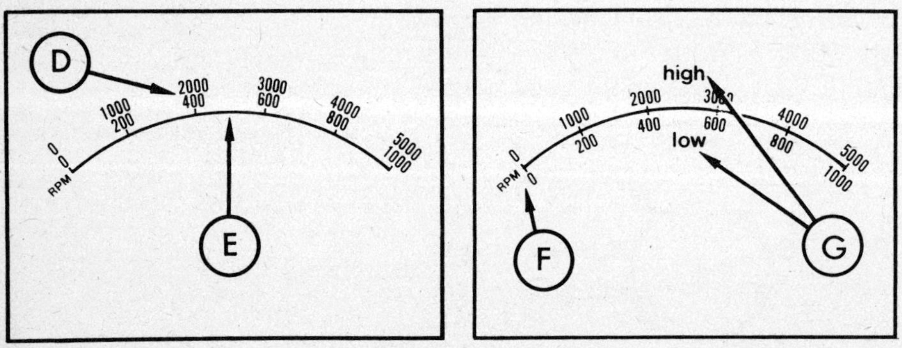

Engine rpm may be read on a single scale (E), but if the tachometer has a dual-range function, the single scale will be marked with two sets of numbers (D) or use upper/lower numbering (F). It may even be marked according to range (G).

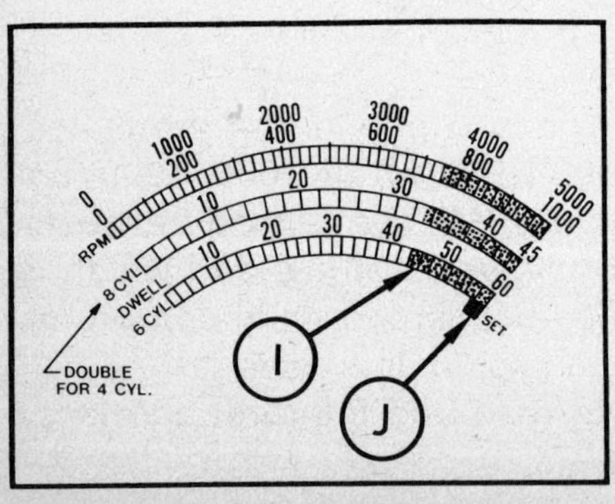

When a two-color GOOD/BAD range (H) is not provided for reading breaker point resistance, you'll find a color-coded area which serves as the GOOD indicator (I). This is usually located at the right side of the meter dial. Any reading that falls outside this colored area should be interpreted as BAD when reading point resistance.

Meters equipped with a SET switch also have a color-coded SET position located at the extreme right side of the dial face (J). This may be the *same* color-coded area used to read GOOD point resistance.

CONNECTING THE DWELL TACHOMETER FOR USE (CONVENTIONAL IGNITION SYSTEMS ONLY)

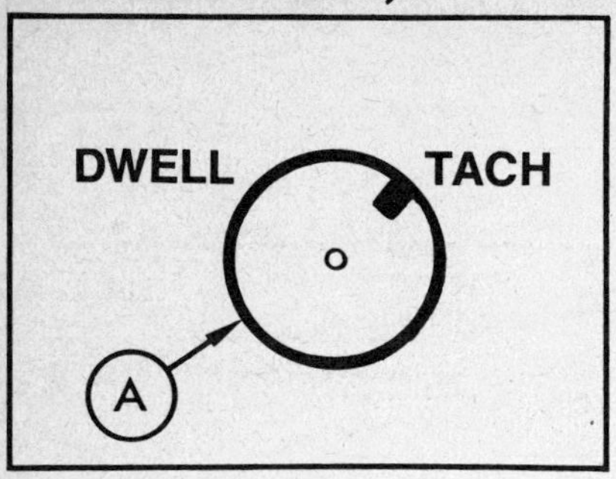

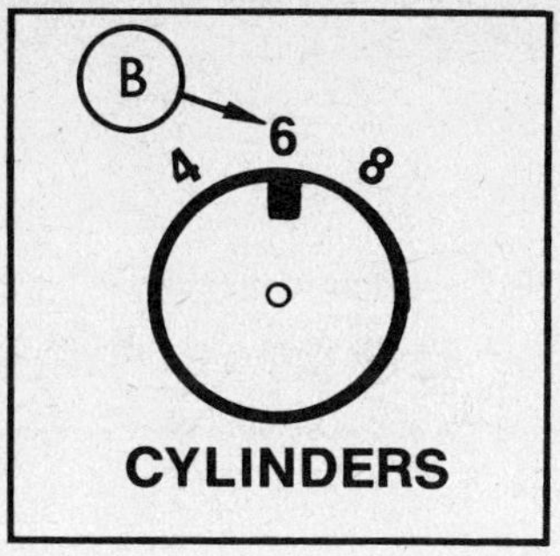

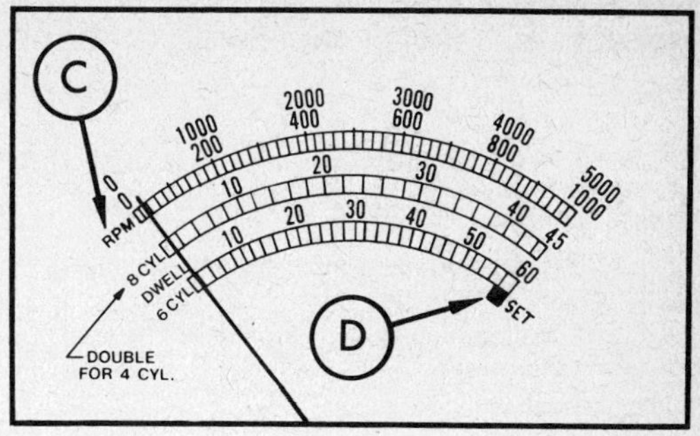

Before connecting the dwell tachometer to a conventional ignition system, set the control switch to the function you intend to use—DWELL or TACH (A). Set the cylinder or lobe switch according to the engine being adjusted (B).

On units without a SET switch, make sure the meter needle is right on the zero on the RPM scale (C). Units with a SET switch should be adjusted until the meter needle is on the SET line shown on the scale (D).

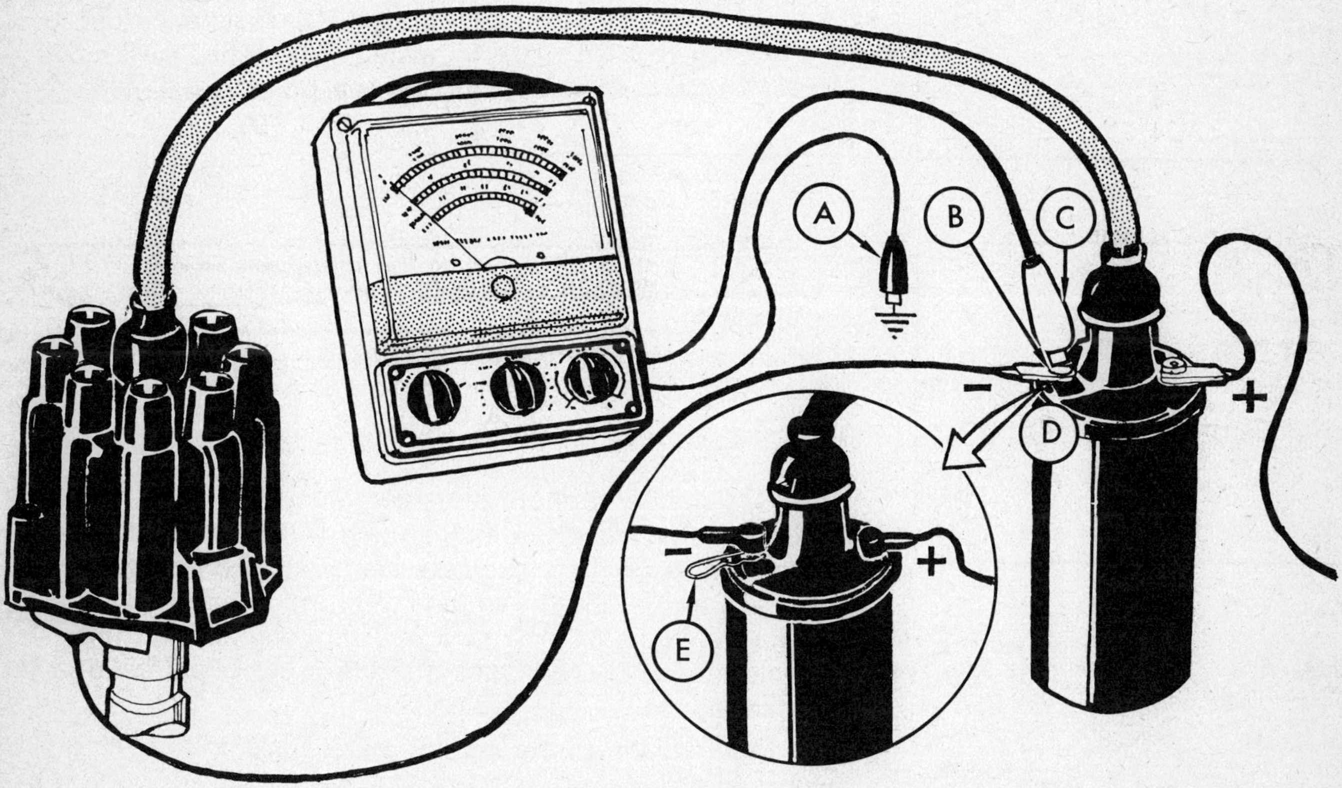

Connect the colored (red, yellow or white) alligator test clip (C) to the primary terminal (B) on the distributor side of the coil (D). This may be marked DIST. or simply (-). Because the push-on wire connector used by Ford completely covers the terminal, some dwell tachometers include a slide-on circuit adapter (E) to assure a good connection. Any small piece of metal with a hole large enough to fit over the terminal post can be used if the adapter is lost. Remove the connector, place the adapter over the terminal post and replace the connector. Attach the alligator test clip to the adapter.

Connect the black alligator test clip (A) to a good ground on the engine. With ignition systems that use a positive ground, the position of the clips must be reversed if the meter is to read upscale. Failure to connect the two clips correctly will result in reversed polarity; the meter needle will move instantly to the extreme right of the dial when the ignition switch is turned on and remain there until the clips are removed and reconnected correctly.

BREAKER POINT RESISTANCE TEST

Once the dwell tachometer is properly connected, you can perform a breaker point resistance test—if the dwell function provides for it.

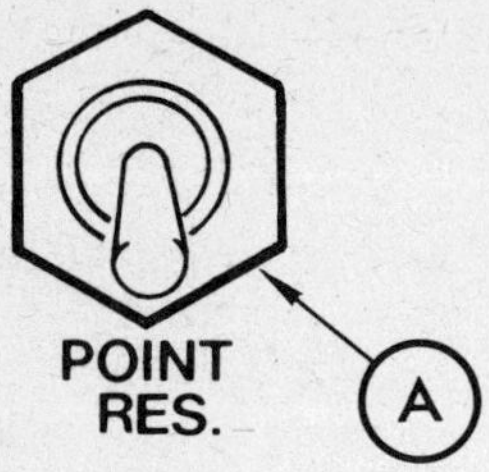

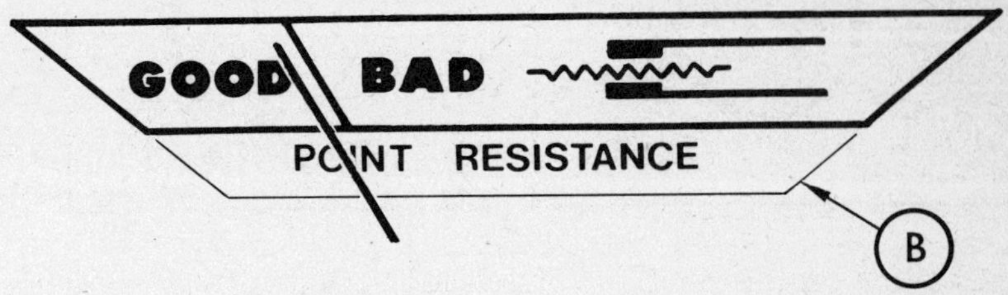

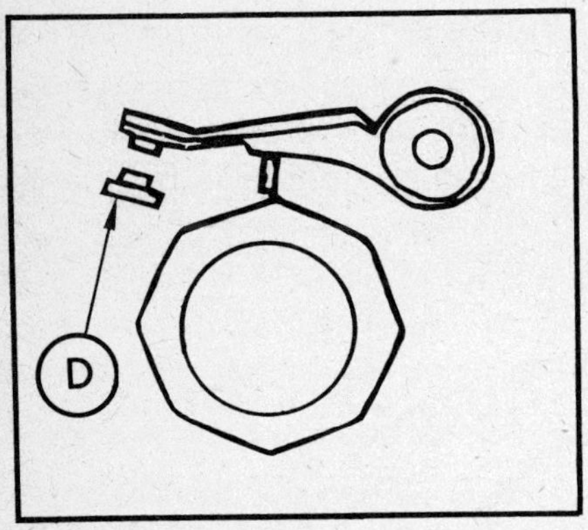

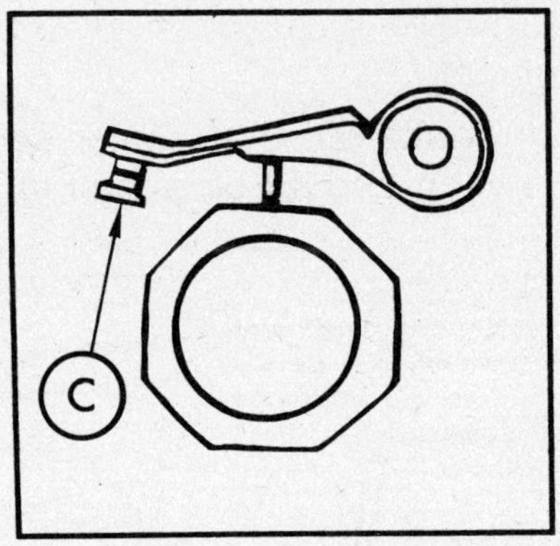

If the test unit has a setting marked POINT RES., move the switch (A) to that position. Turn on the ignition switch and watch the meter needle. If the points are closed (C), resistance can be tested. The needle will move into the point resistance scale limit (B).

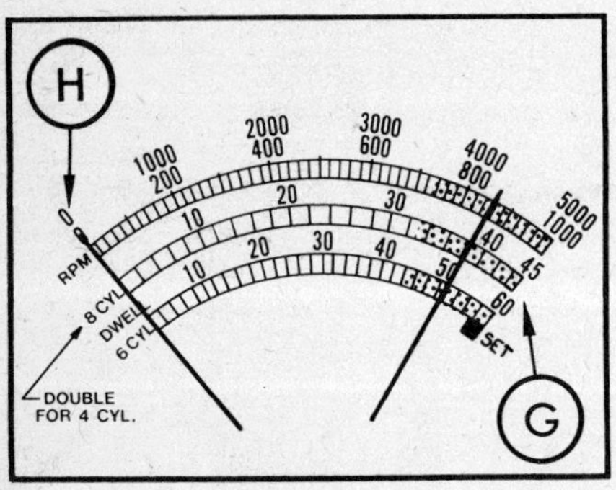

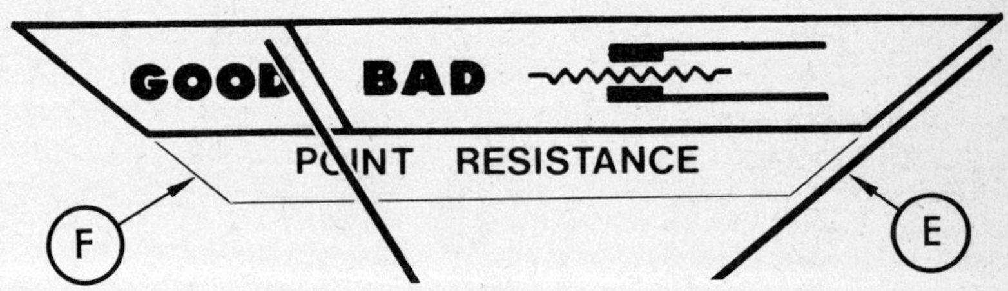

Resistance cannot be tested when the points are open (D); the needle will move rapidly to the far right of the scale (E). Use a remote start switch (see that chapter) and crank the engine until the needle moves within the limits of the point resistance scale (F). The needle movement indicates that the breaker points are closed, as well as their condition.

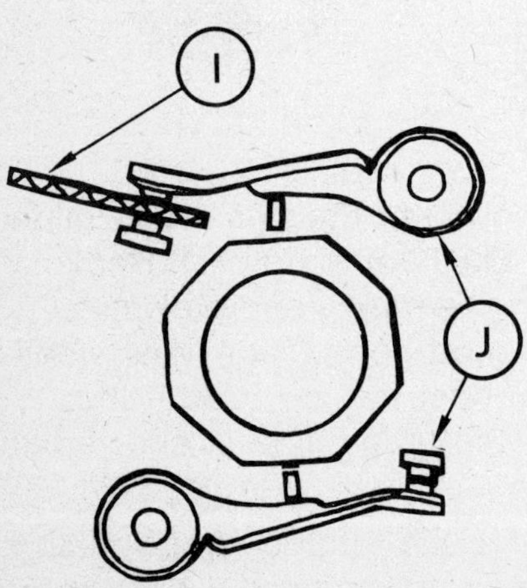

Meters without the marked point resistance setting will read at the zero setting (H) if the points are open. As resistance cannot be checked until they are closed, crank the engine slightly to close them. If breaker point resistance is good, the meter needle will move into the color-coded area at the extreme right of the dial (G).

When checking point resistance on a dual set of breaker points (J), block open one set with a piece of cardboard (I) while the other set is being tested.

If point resistance reads BAD, look for improperly installed, misaligned or burned points or loose lead/ground wires in the distributor. Don't forget to replace the secondary coil wire when testing is completed.

CHECKING DWELL

DISTRIBUTOR
INITIAL TIMING.................................8° BTDC
ROTATION ....Clockwise
DWELL ANGLE
Range ....28°-32°
Set to..30°
CAPACITOR 18-.23-MFD
FIRING ORDER..........................1,5,6,3,4,2,7,8
CENTRIFUGAL ADVANCE
Start.....................................0°-2° @ 500 RPM
Maximum......................11°-13° @ 2,000 RPM
VACUUM ADVANCE
Start ....8.0″-10.0″
Full ....13.0″
Maximum Advance................................12.60°

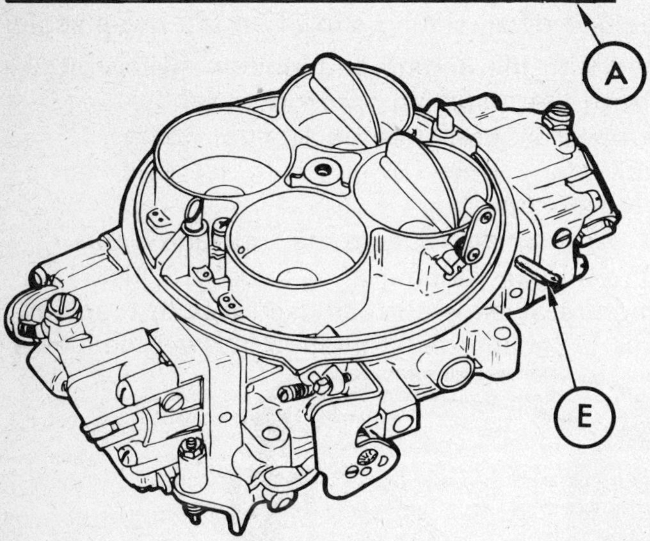

Before dwell can be checked, you must know the correct dwell setting as specified by the manufacturer. This information can be found either in your owner's handbook or on the Vehicle Emissions Control decal (A) located in the engine compartment. If the decal is missing and the handbook does not contain the dwell setting, consult the factory shop manual or a copy of Petersen's HOW TO TUNE YOUR CAR for the necessary information.

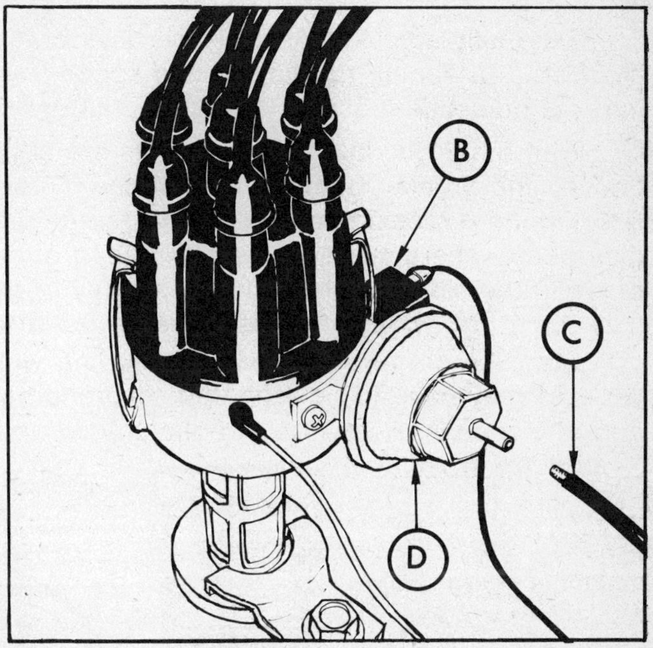

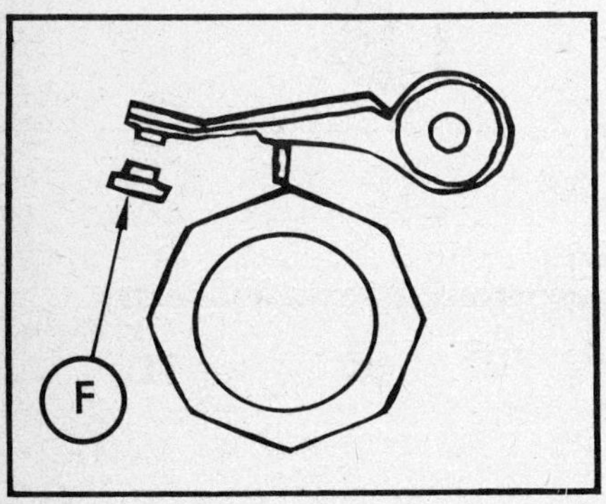

Unless otherwise stated in the specifications, disconnect and plug the hose (C) connected to the distributor vacuum advance unit (D), because dwell changes with the operation of the vacuum advance on some distributors. If equipped with an advance/retard solenoid (B—Chrysler Corporation cars), disconnect its wire at the carburetor (E).

With the dwell tachometer connected as before and the secondary coil wire disconnected, run the engine at idle and read the dwell setting on the scale that corresponds to the engine being tested.

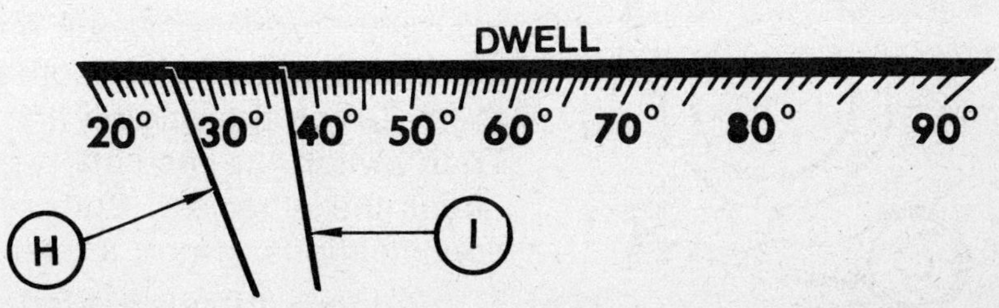

If the dwell reading is too low (H), the breaker point gap is too great (F); if the reading is too high (I), the gap is too narrow (G). Either condition requires that dwell adjustment be reset.

DWELL TACHOMETER

DWELL ADJUSTMENTS

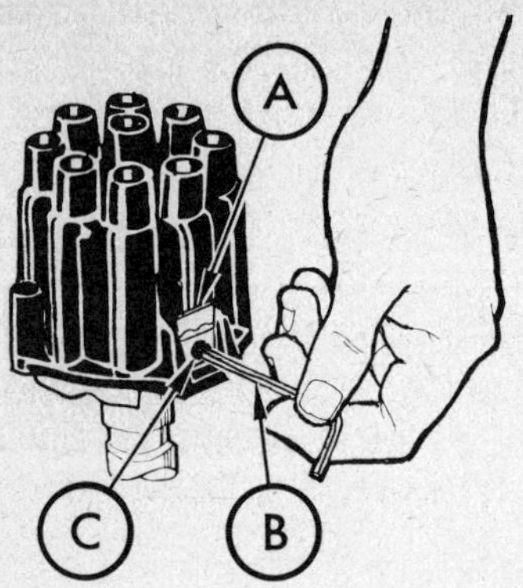

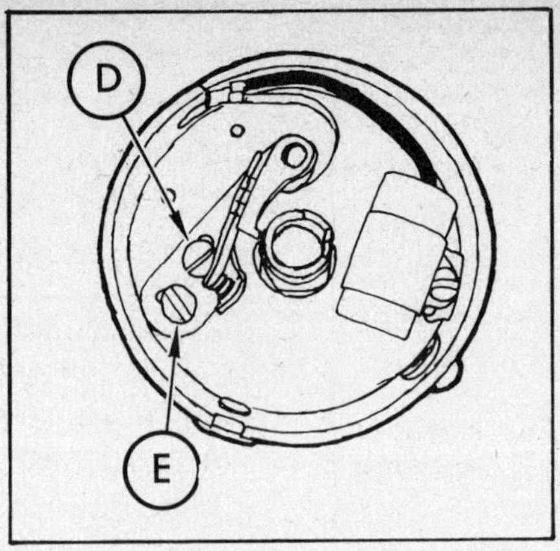

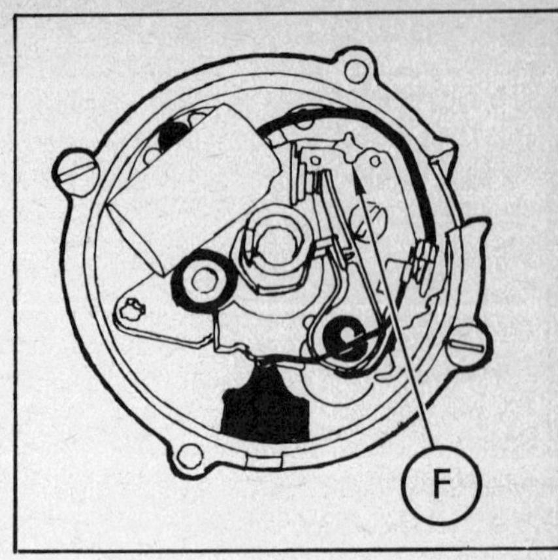

Externally-adjusted Delco-Remy distributors have a sliding window in the distributor cap (A). With the engine running at idle speed, raise the window and insert an Allen wrench (B) into the adjustment screw (C). Turn until the specified dwell setting shows on the dwell scale.

With other distributors, the points are adjusted with the engine and ignition switch off. Loosen the breaker point locking screw (D) and turn the adjusting screw (E) slightly. Retighten the locking screw (D) and recheck the dwell by cranking the engine. Some distributors use a slotted hole (F) instead of an adjusting screw (E). Insert the screwdriver blade in the slotted hole and turn slightly to the left or right.

Once the dwell setting is correctly set at cranking speed, replace the distributor cap.

Start the car and recheck the reading with the engine running at idle speed. This process may have to be repeated several times before dwell is correctly set at both cranking and idle speeds. Remember to check ignition timing, because a 1° change in dwell causes a 1° change in timing.

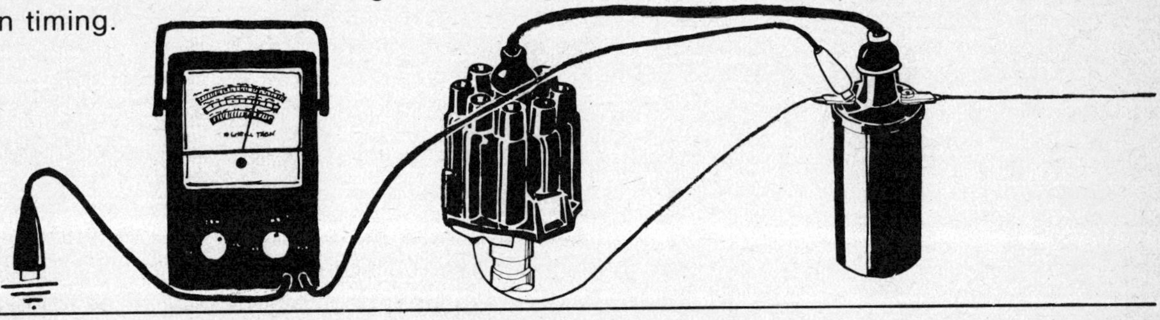

CHECKING DWELL—DUAL BREAKER POINT DISTRIBUTORS

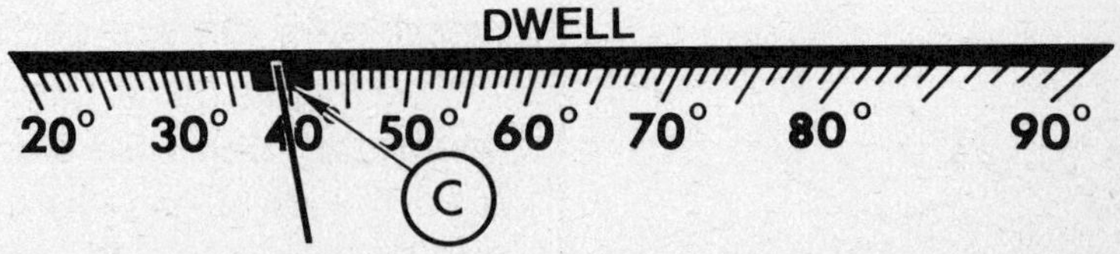

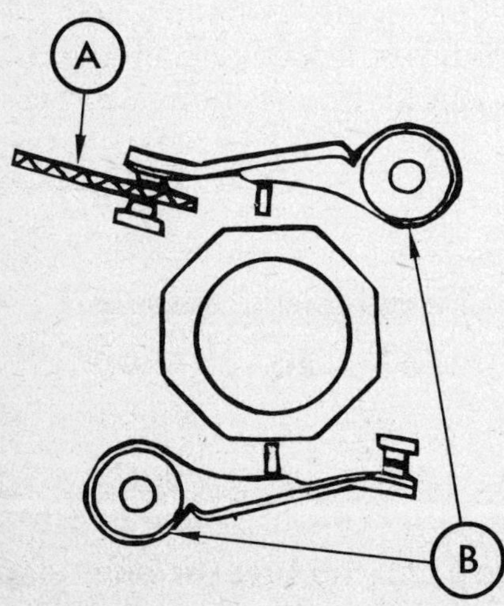

Some distributors use two sets of breaker points (B); one to make and the other to break the current to build up secondary voltage. Block out one set with a piece of cardboard (A), paper or other insulation that will not damage them. Crank the engine and take a reading of the first set. Then switch the insulation to the tested points while you check the other set. Both readings should fall within factory specifications. If not, adjust one or both sets until they do.

When both sets are adjusted correctly, take a reading of the two operating together and check the total dwell (C) on the meter scale against factory specifications. If the individual settings are within specifications, total dwell should also be correct.

DWELL VARIATION TEST

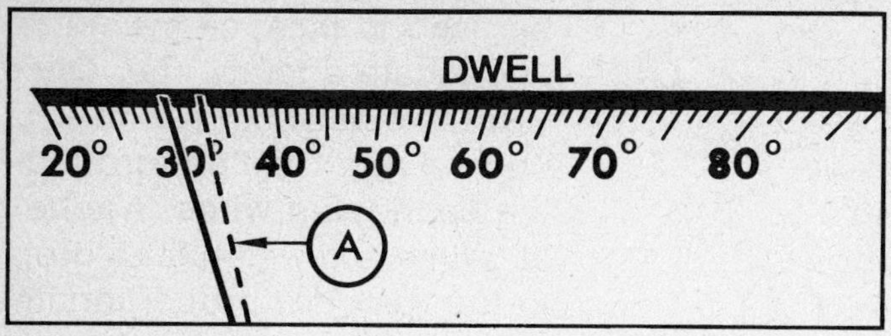

With the dwell tachometer connected, run the engine at idle and read the dwell setting on the scale that corresponds to the engine being tested. Continue watching the scale as you increase engine speed to about 1,200 rpm and then drop back to idle. The meter needle should not move more than 3° (A) for any engine, regardless of the number of cylinders.

Be sure to check factory specifications before running this particular test. While most engines must be tested with the vacuum advance hose disconnected, a few require that it be reconnected for the dwell variation test. If this is the case, the maximum allowed variation is 6° instead of 3°.

A variation in needle movement of more than 3° (6° if the test requires reconnecting the vacuum advance) indicates excessive wear in the distributor bushing, shaft or breaker plate, and repair is necessary.

TACHOMETER

The dwell tachometer connects to the ignition system in the same way, regardless of whether you are reading dwell or engine rpm. To use the tachometer function, move the control switch to the tachometer setting (A). The cylinder or lobe switch (B) is set according to the engine being adjusted. On meters with a dual range, the RPM switch (C) adjusts to read LOW or HIGH according to use.

CARBURETOR ADJUSTMENT

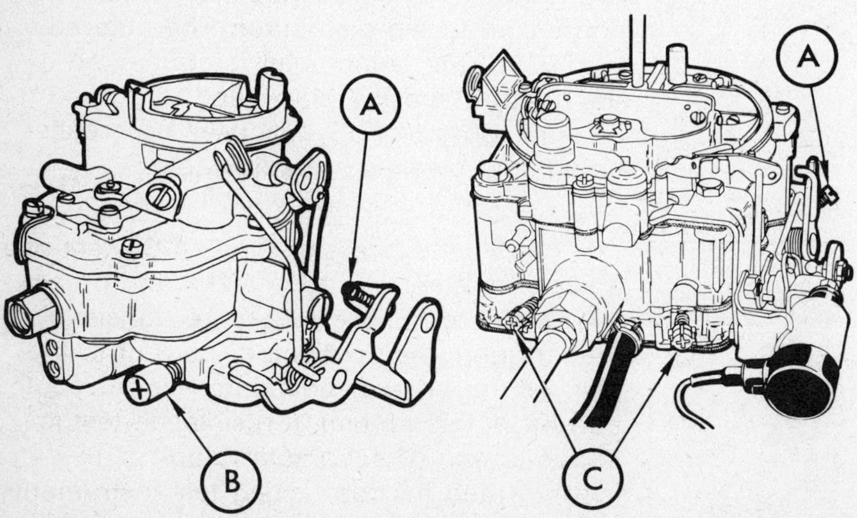

With the engine running at normal operating temperature and the tachometer set to read LOW range, adjust the idle speed screw (A) to obtain the specified idle speed. Then adjust the idle mixture screw (B) for the highest steady rpm reading.

If the carburetor uses two idle mixture screws (C), adjust each one in and out equally until the highest steady rpm is reached. The idle speed screw (A) may have to be readjusted if adjustment of the mixture screws increases the idle speed.

When adjusting 1968 and later models, follow the manufacturer's instructions and specifications exactly. As these may differ considerably from manufacturer to manufacturer and model year to model year, make certain that you know precisely what is required. Check the factory shop manual if necessary.

BALANCING CYLINDERS

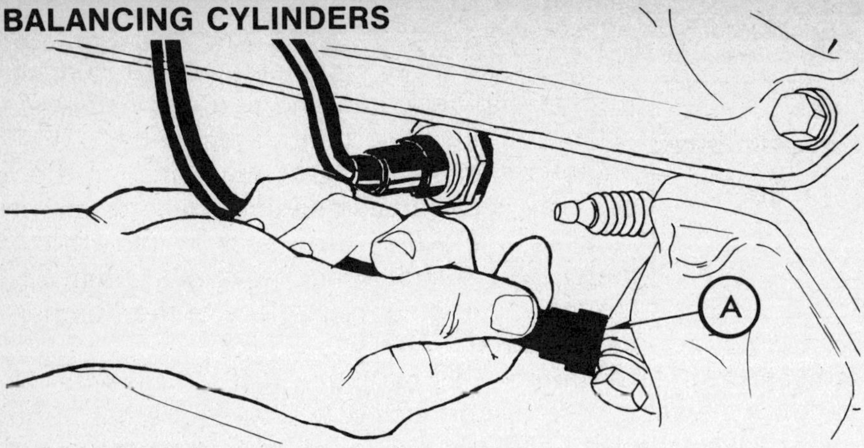

The tachometer can also be used to balance cylinders and locate a faulty one. Once you have established the correct rpm reading, ground the spark plug wires (A) one at a time and watch for a drop in the rpm. If a proportionate drop does not occur when you ground one particular wire, that cylinder is faulty.

NON-CONVENTIONAL IGNITIONS

Numerous transistorized and capacitive-discharge ignition systems have been made available over the years. With the exception of 1972 and later Chrysler Corporation cars and a small minority of Ford and GM autos, most of these have been an add-on or option. Because of the rarity of most such ignition systems, it is only practical to deal here with the three most likely to be encountered.

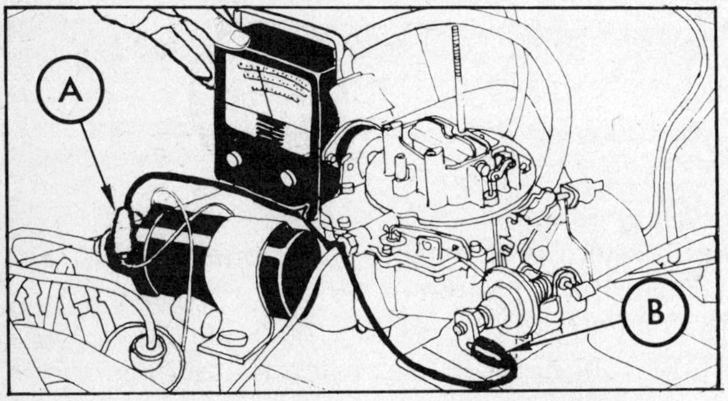

The Chrysler Electronic Ignition System (EIS) is found on some 1972 and all 1973 and later Chrysler Corporation cars. Since EIS dwell is electronically controlled, it is not adjustable and there is no need for a dwell meter with this system. Tachometer readings are obtained by connecting the colored lead of the unit to the negative (-) or battery (BAT) terminal of the ignition coil (A) and the black lead to a ground on the engine (B).

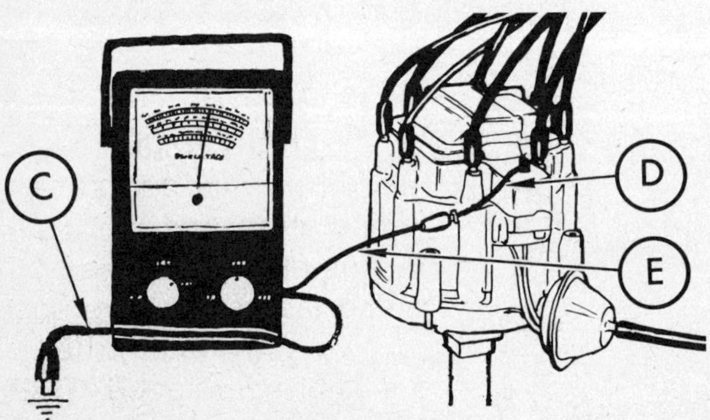

The Delco-Remy Unitized Ignition System incorporates the ignition coil directly on the distributor cap, but with the use of a Delco UIS adapter (D) the tachometer can be connected to the distributor. The colored lead (E) of the tachometer is connected to the adapter and the black lead to ground (C). Like the Chrysler EIS, UIS uses no breaker points and dwell is not adjustable.

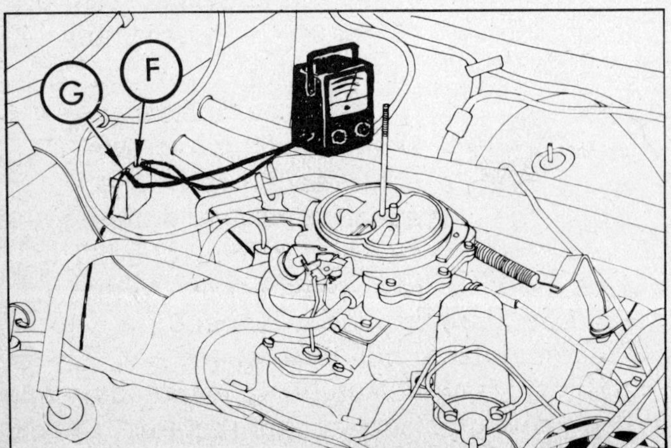

The Ford Transistor Ignition System uses a set of breaker points to make and break the "base" current through the transistor. This triggers the "collector" current to the coil. As the breaker points do not carry full current in this system, a resistance test is unnecessary. A tachometer block is incorporated for connecting test instruments; the colored clip goes to the red-coded terminal (F) and the black clip to the black-coded terminal (G). Connections made in any other way are likely to ruin the system's transistors.

TIMING LIGHTS

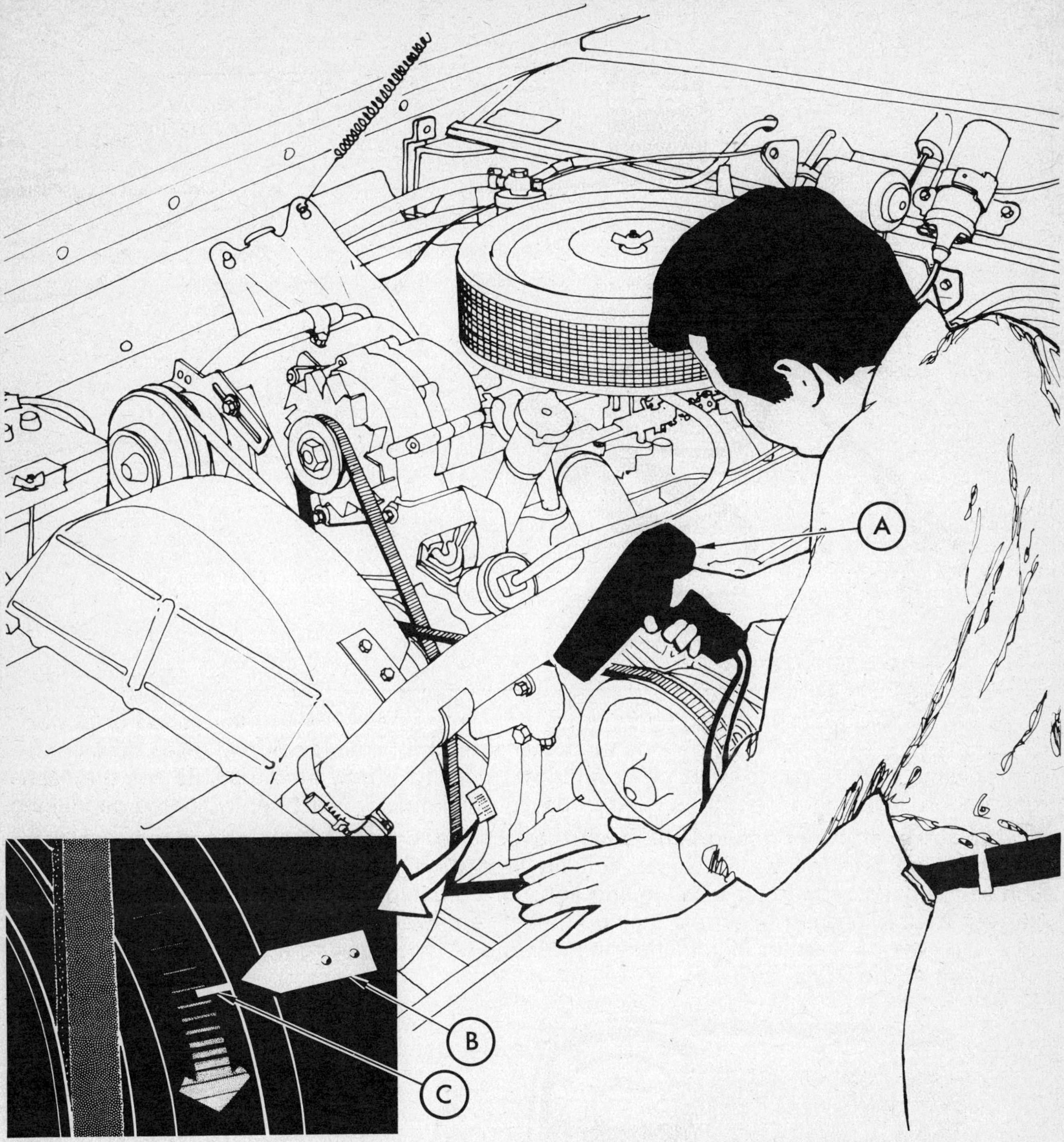

A properly timed automotive engine is one in which the spark created by the spark plug ignites the fuel/air mixture at the precise instant required by the compression cycle. This is known as ignition timing, and a correct setting is essential. Ignition timing is the final step in tuning an engine and is done to make certain that all adjustments made to the ignition system contribute to correct spark and fuel/air ignition. Improper ignition timing can lead to loss of power, overheating, poor acceleration and performance, excessive fuel consumption and generally shorten the service life of your engine.

To check initial ignition timing, the use of a timing light that flashes a stroboscopic (rapidly repeating) light (A) is required. Aiming this strobe light at the front of the engine illuminates a timing mark on the revolving crankshaft pulley, harmonic balancer or damper flywheel (C) and a stationary (unmoving) reference pointer on the engine block (B).

TIMING LIGHTS

TYPES OF TIMING LIGHTS

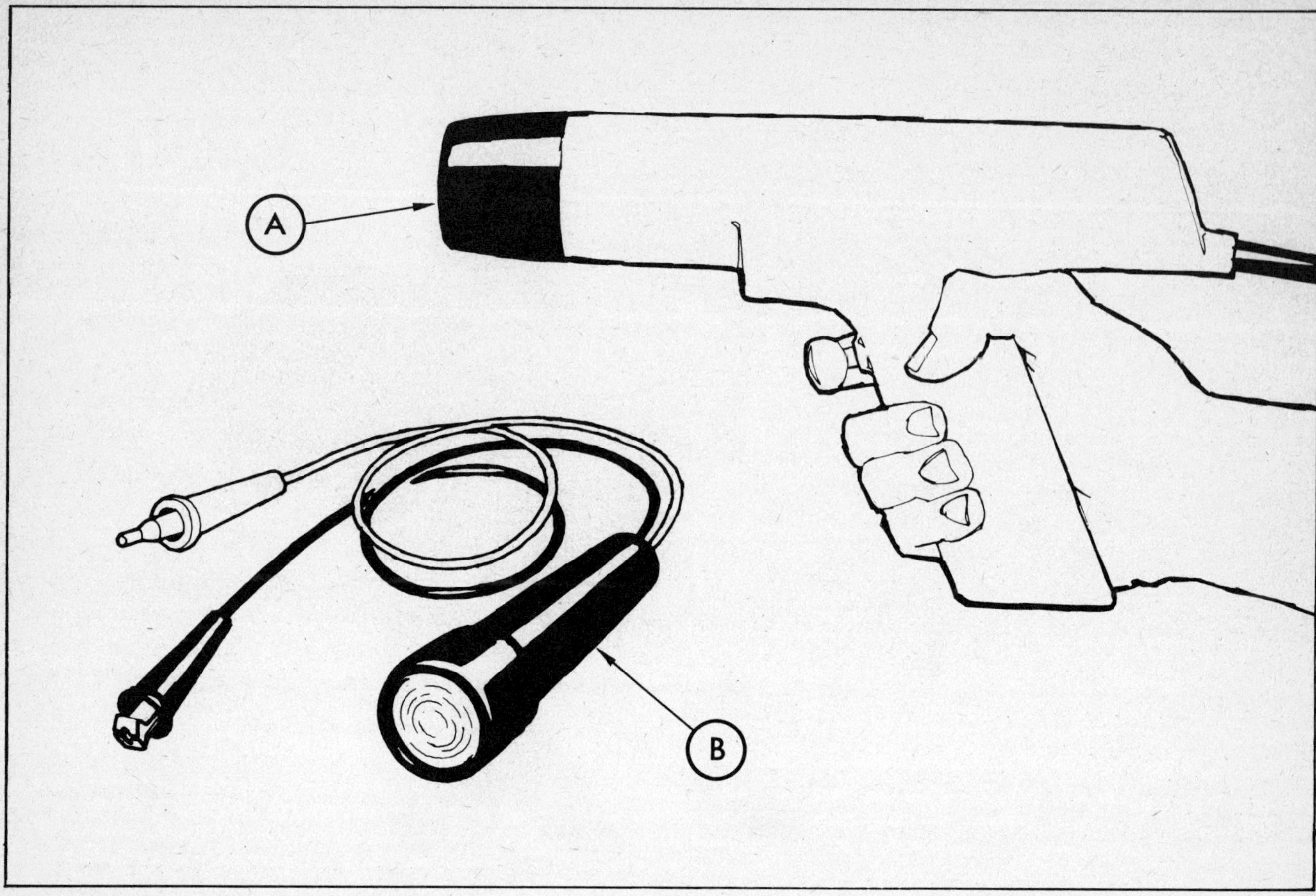

There are two types of timing lights in use today: the neon (B) and the xenon or power timing light (A). The neon light obtains its power from the ignition coil and gives off a reddish-white light. Because of its color and low intensity, it may be difficult to see the flash from a neon timing light when it is used in daylight or under other brightly-lighted conditions.

The xenon or power timing light is more expensive, but uses the high voltage from the coil only to trigger the flash. Power to actually flash the light is drawn from another source, such as the car's battery (for a DC timing light) or an external 115-120 volt AC power supply like your house voltage (for an AC timing light). The power timing light gives off a bright white flash with far greater illumination than the neon type, making it much easier and more convenient to use in daylight.

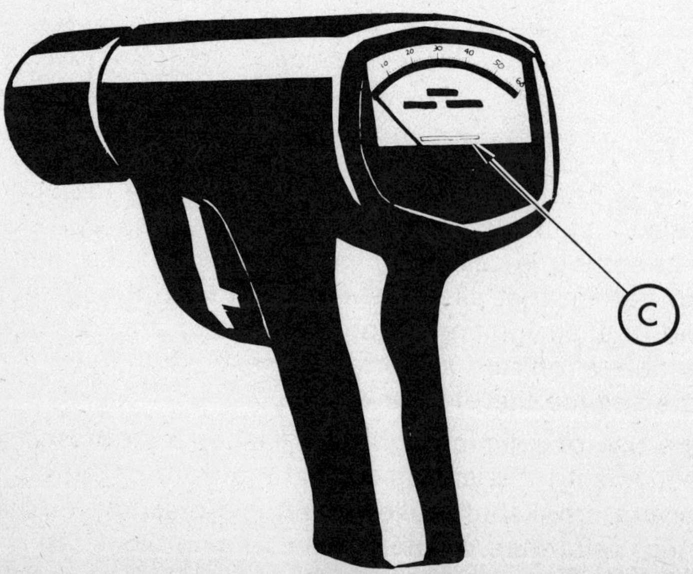

While any timing light can also be used to check the approximate accuracy of distributor centrifugal and/or vacuum advance (called dynamic tuning) in addition to establishing initial ignition timing, a precise measurement of these functions requires a power timing light with a built-in timing advance meter (C) calibrated in degrees, usually from 0° to 60°. Such timing lights are generally AC powered and are quite expensive.

HOW THE TIMING LIGHT WORKS

Because of the very brief duration of each flash, the timing light synchronizes *light* and *motion*. This causes the moving mark to appear as if it were standing still. Timing lights are designed to flash at exactly the same instant as the No. 1 spark plug fires, and timing marks are positioned to show the proper location of the No. 1 piston at this same instant.

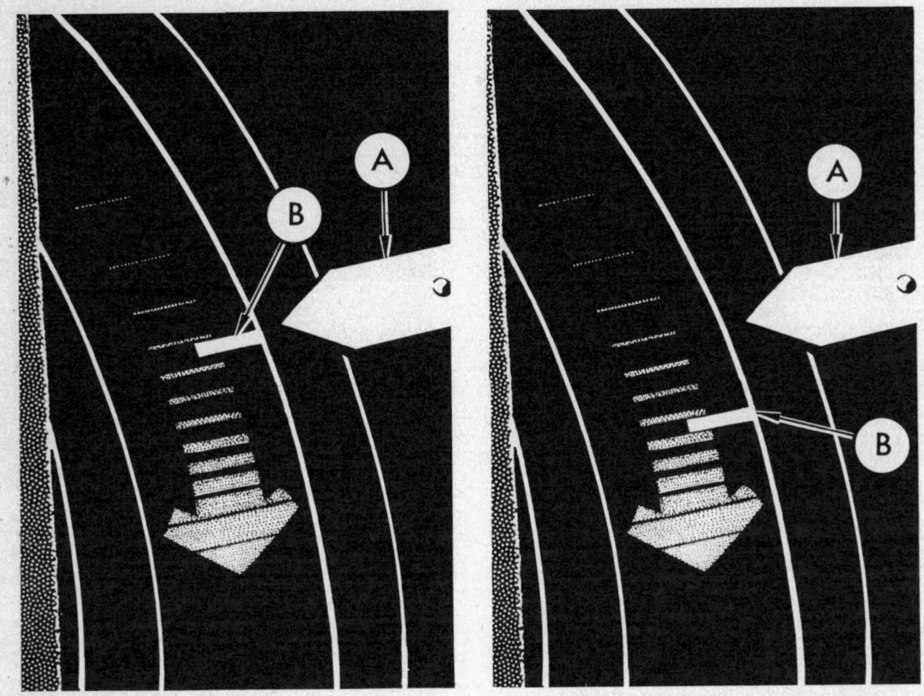

If the moving mark (B) is positioned exactly opposite the stationary reference pointer (A) when the engine is running at an rpm specified by the manufacturer, the engine is said to be "in time." This means that the No. 1 spark plug fires each time the distributor breaker points begin to open.

But if the moving mark (B) stands still at one side or the other of the stationary reference pointer (A), the engine is "out of time" and the instant of ignition must be changed by rotating the distributor.

TIMING MARKS

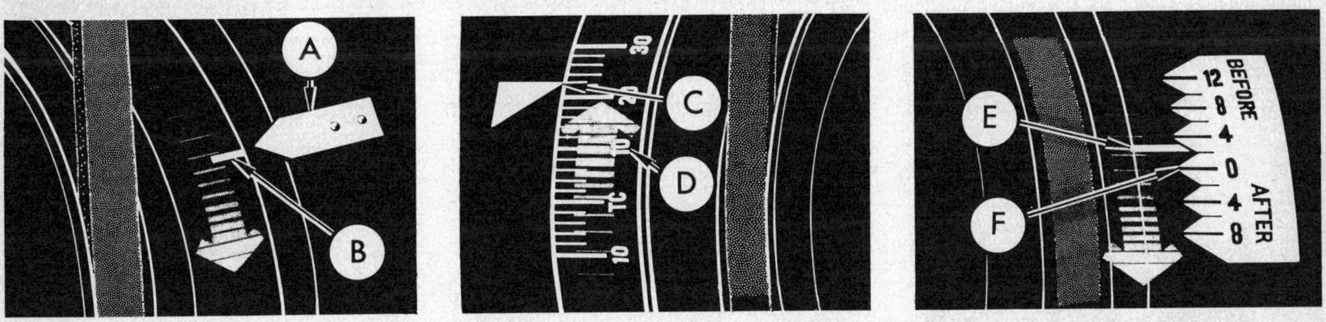

Timing marks differ in appearance from engine to engine. Some engines (usually four-cylinders) use a simple fixed reference pointer (A) and a moving mark (B). But because of the complexity of timing larger engines, their timing marks are often more complicated. The Ford 351 CID (cubic inch displacement) engine uses a stationary arrowhead reference pointer (C) and a revolving scale marked in degrees (D). Still other engines, like the Chevrolet 454 CID, reverse this order. The stationary reference pointer (F) contains a scale marked in degrees, while the moving timing mark (E) is simply a groove in the crankshaft pulley.

Where there is just a single fixed reference pointer (A) and a moving mark (B), the two need only be aligned horizontally (side-by-side) to "time" the engine correctly. But with the more complicated timing mark arrangements, it is necessary to know the exact number of degrees before top dead center (BTDC) or after top dead center (ATDC) as specified by the engine manufacturer in order to properly retard (move back) or advance (move forward) the spark for correct timing. In either case, you can tell if the engine requires such adjustment by using a timing light. You can also correct an out-of-time condition by rotating the distributor while watching the timing mark and reference pointer until they line up as specified.

IGNITION SPECIFICATIONS

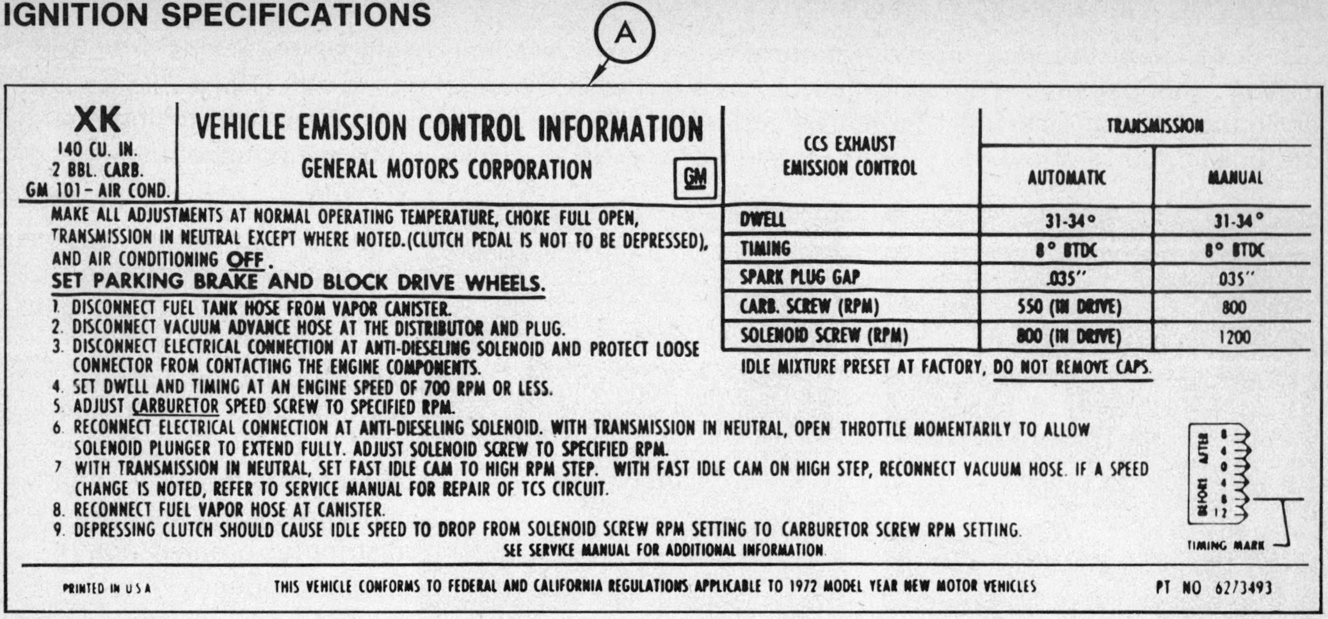

XK
140 CU. IN.
2 BBL. CARB.
GM 101 – AIR COND.

VEHICLE EMISSION CONTROL INFORMATION
GENERAL MOTORS CORPORATION

MAKE ALL ADJUSTMENTS AT NORMAL OPERATING TEMPERATURE, CHOKE FULL OPEN, TRANSMISSION IN NEUTRAL EXCEPT WHERE NOTED. (CLUTCH PEDAL IS NOT TO BE DEPRESSED), AND AIR CONDITIONING OFF.

SET PARKING BRAKE AND BLOCK DRIVE WHEELS.

1. DISCONNECT FUEL TANK HOSE FROM VAPOR CANISTER.
2. DISCONNECT VACUUM ADVANCE HOSE AT THE DISTRIBUTOR AND PLUG.
3. DISCONNECT ELECTRICAL CONNECTION AT ANTI-DIESELING SOLENOID AND PROTECT LOOSE CONNECTOR FROM CONTACTING THE ENGINE COMPONENTS.
4. SET DWELL AND TIMING AT AN ENGINE SPEED OF 700 RPM OR LESS.
5. ADJUST CARBURETOR SPEED SCREW TO SPECIFIED RPM.
6. RECONNECT ELECTRICAL CONNECTION AT ANTI-DIESELING SOLENOID. WITH TRANSMISSION IN NEUTRAL, OPEN THROTTLE MOMENTARILY TO ALLOW SOLENOID PLUNGER TO EXTEND FULLY. ADJUST SOLENOID SCREW TO SPECIFIED RPM.
7. WITH TRANSMISSION IN NEUTRAL, SET FAST IDLE CAM TO HIGH RPM STEP. WITH FAST IDLE CAM ON HIGH STEP, RECONNECT VACUUM HOSE. IF A SPEED CHANGE IS NOTED, REFER TO SERVICE MANUAL FOR REPAIR OF TCS CIRCUIT.
8. RECONNECT FUEL VAPOR HOSE AT CANISTER.
9. DEPRESSING CLUTCH SHOULD CAUSE IDLE SPEED TO DROP FROM SOLENOID SCREW RPM SETTING TO CARBURETOR SCREW RPM SETTING.

SEE SERVICE MANUAL FOR ADDITIONAL INFORMATION.

CCS EXHAUST EMISSION CONTROL	TRANSMISSION	
	AUTOMATIC	MANUAL
DWELL	31-34°	31-34°
TIMING	8° BTDC	8° BTDC
SPARK PLUG GAP	.035"	.035"
CARB. SCREW (RPM)	550 (IN DRIVE)	800
SOLENOID SCREW (RPM)	800 (IN DRIVE)	1200

IDLE MIXTURE PRESET AT FACTORY, DO NOT REMOVE CAPS.

BEFORE 12 8 4 0 4 8 AFTER
TIMING MARK

PRINTED IN U.S.A. THIS VEHICLE CONFORMS TO FEDERAL AND CALIFORNIA REGULATIONS APPLICABLE TO 1972 MODEL YEAR NEW MOTOR VEHICLES PT NO 6273493

Manufacturers publish ignition timing specifications for each engine type they make. These may remain the same from year to year or they may vary considerably between model years. Don't rely on your memory; check the specifications table for the engine in question. If this information is not included in your owner's handbook, you should find it on the Vehicle Emissions Control decal (A) located somewhere in the engine compartment. If not, check the factory shop manual at a garage for your make, model and year of car, or refer to a copy of Petersen's HOW TO TUNE YOUR CAR—this will also show you exactly what the timing marks look like and where to find them.

Determine the exact setting for the make, displacement and model year of engine before proceeding to work with the timing light. An engine may *seem* to sound and run better when it is slightly out-of-time, especially those late models equipped with emission control systems. However, this defeats the entire reasoning behind emission controls and causes the engine to give off illegal quantities of pollutants, so time the engine to specifications.

THE NO. 1 CYLINDER/SPARK PLUG

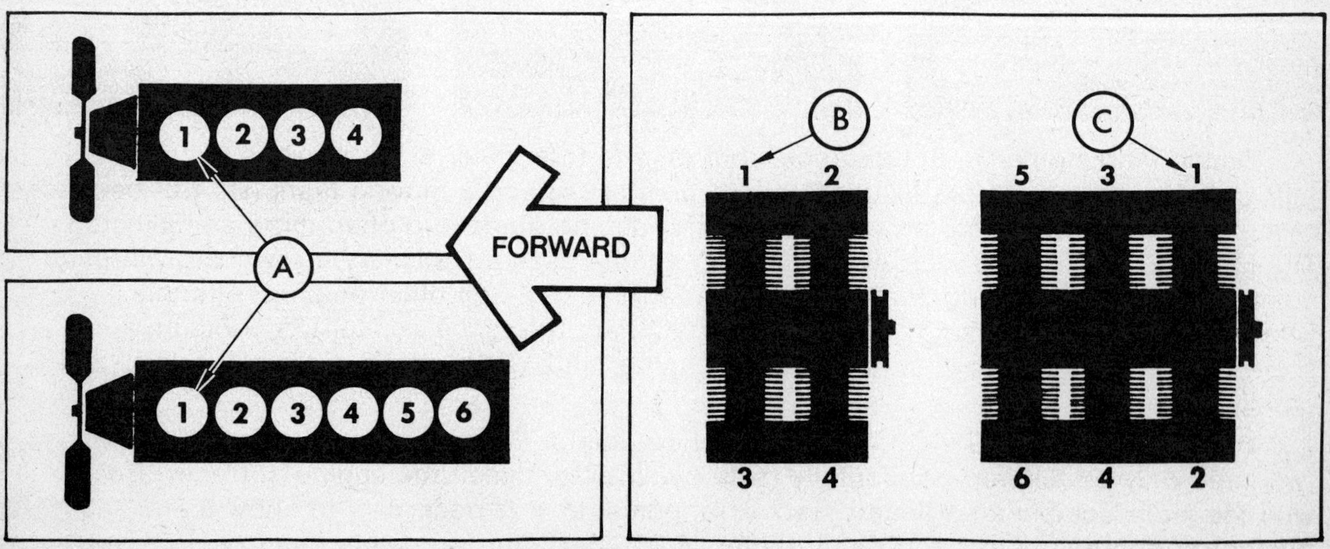

The No. 1 cylinder is the key to timing an engine, so you must also find the location of the No. 1 spark plug. This is always the spark plug directly behind the radiator on inline four- and six-cylinder engines (A). These cylinders are always numbered from front to rear in sequence. Chevrolet Corvair (C) and Volkswagen (B) engines use horizontally-opposed cylinders, and No. 1 is positioned differently from any other automotive engine.

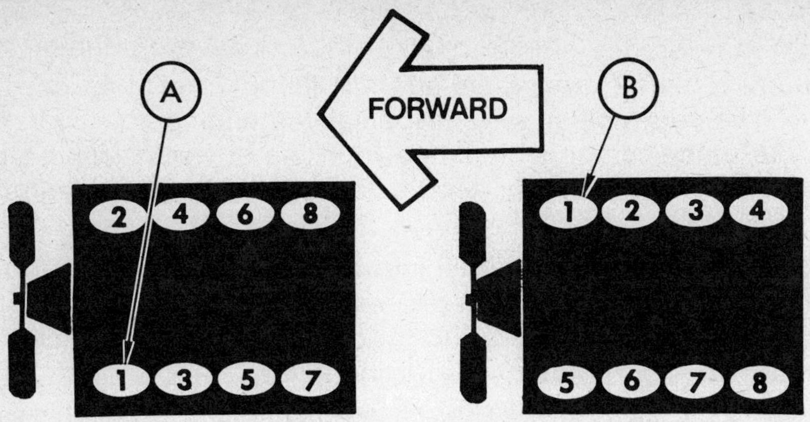

The No. 1 spark plug on V-8 engines is also directly behind the radiator, but on the left-hand side of all GMC (except Cadillac), AMC and Chrysler Corporation eight-cylinder engines (A) and on the right-hand side of all Ford, Lincoln, Mercury (and Cadillac) eight-cylinder engines (B).

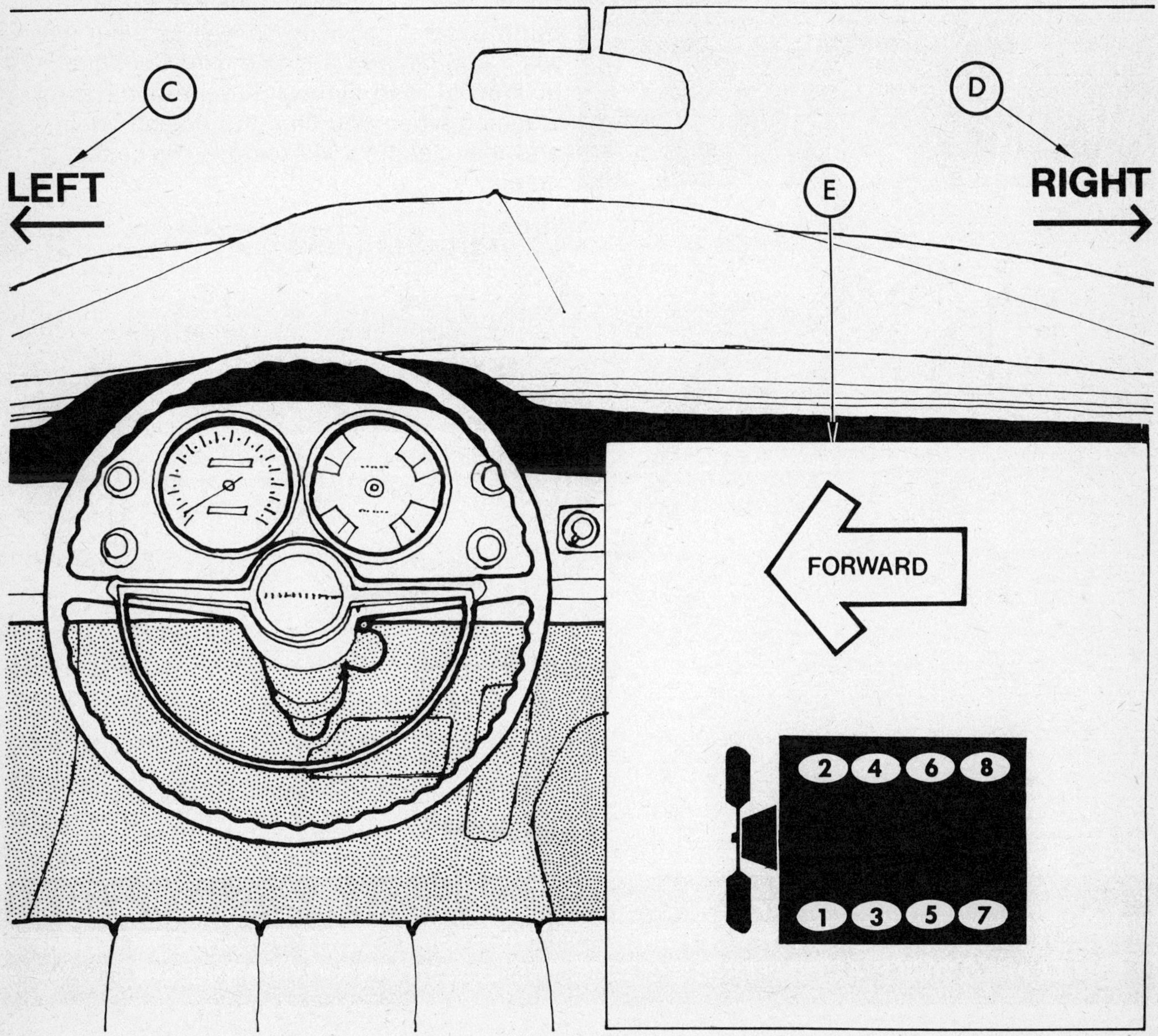

Determination of right- and left-hand sides of a V-8 engine is made from the *driver's seat*. Thus the side of the engine nearer to the driver's side is the left-hand (C) side, while the side nearer the passenger's side of the car is the right-hand side (D). Many engines have their cylinder numbers marked on the intake manifold above the cylinders, on the distributor cap or in some other prominent place. If in doubt, check a cylinder numbering diagram (E) before working on a V-8 engine.

PREPARATIONS FOR TIMING

When you find the timing mark and reference pointer, they will probably be too dirty to work with. Because of their location, a film of grease, oil and dirt tends to build up on them. This must be cleaned away to give you a clear view, especially in engine compartments where the placement of other components limits your line of vision. Wiping with a clean cloth will usually be sufficient, but you may need a stiff bristle brush to clean away heavy deposits.

If the timing mark or reference pointer still cannot be seen clearly after cleaning, paint a ⅛-inch-wide line on the reference pointer (A) and the timing mark (B) with a quick-drying white enamel to make them show up brighter and clearer. A piece of white chalk can be used instead of paint, but it will not show up as well as paint nor will it last beyond a single use. Continued buildup of road film will require another cleaning when you time the engine again, and this cleaning will remove the chalk marks.

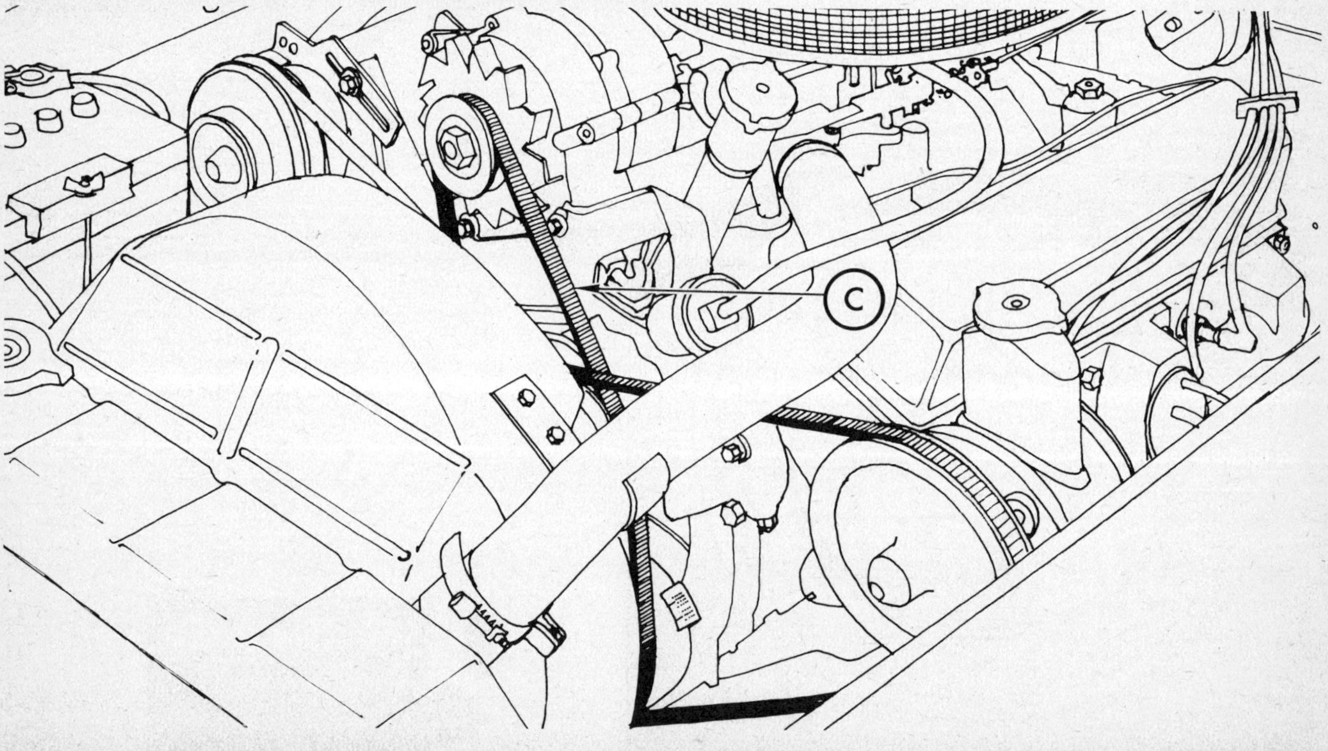

WARNING: THE TIMING MARKS ON MOST ENGINES ARE CLOSE TO THE FAN BLADES. BE CAREFUL NOT TO GET YOUR HANDS OR THE TIMING LIGHT TOO CLOSE TO THE FAN BLADES WHILE ATTEMPTING TO TIME THE ENGINE. INJURY CAN RESULT.

It's a good idea to loosen the fan belt (C) before proceeding. This will prevent the fan from revolving and possibly injuring your hands or damaging the timing light if you become careless or distracted while working. On late-model cars equipped with air conditioning, air pump emission control systems, etc., though, loosening the fan belt may prove to be more difficult and time-consuming than it is worth because of the number of belts and the need for a belt tensioning device to reset them correctly. If this is the case, exercise caution whenever working around the fan blades, even when they are surrounded by a shroud.

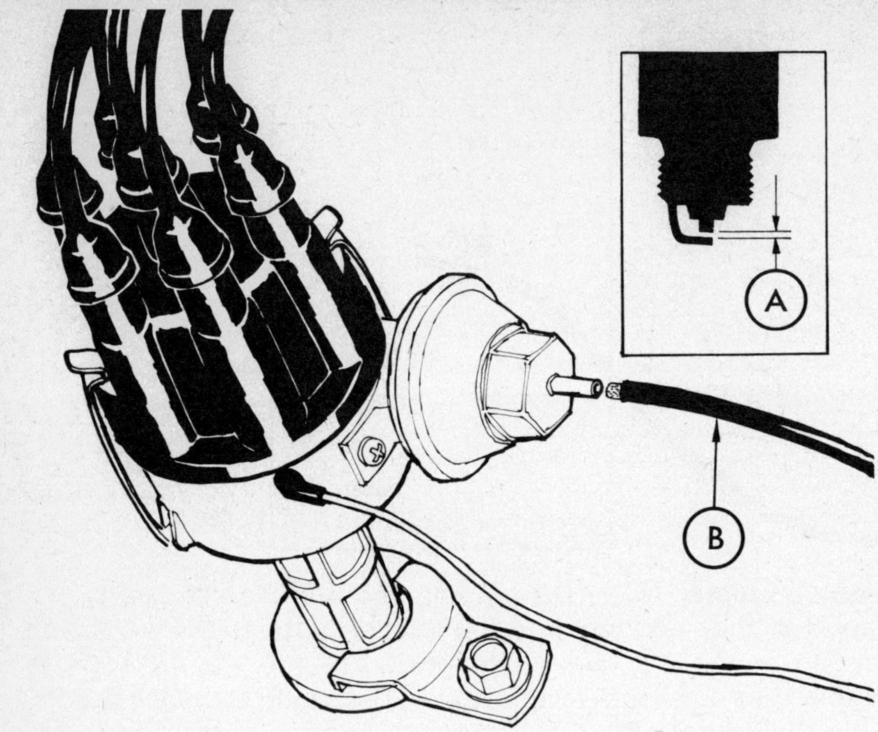

Because initial ignition timing is directly related to the opening and closing of the distributor breaker points, the points should be in good condition and properly adjusted to specifications prior to timing the engine. Disconnect and plug (a wooden pencil works well) the distributor vacuum advance line (B) to prevent possible vacuum advance or actual vacuum advance on full vacuum systems.

Spark plugs must also be clean and their gap (A) correctly set. Connect a remote start switch (see that chapter). It lets you crank the engine from the engine compartment without help from someone in the car.

CONNECTING THE TIMING LIGHT

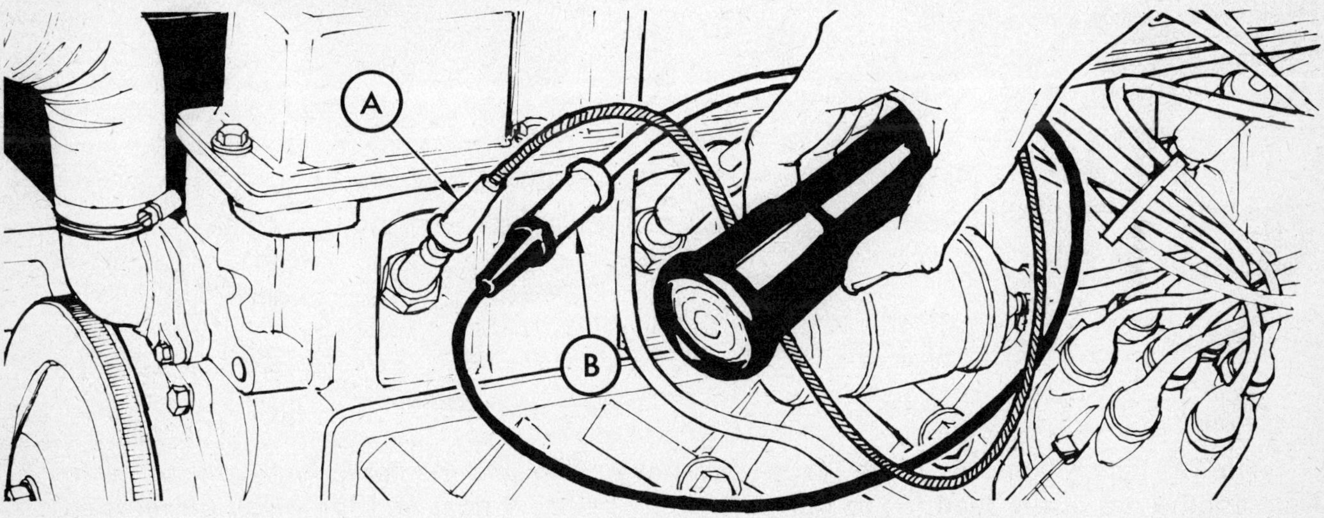

The neon timing light uses two leads. To connect it into the ignition system, remove the spark plug wire from the No. 1 spark plug. Attach one lead (B) to the spark plug wire. Make sure that it does not ground (touch any solid metal surface) when you set it aside. Now connect the other lead directly to the No. 1 spark plug (A). Start the engine and let it warm up to normal operating temperature. Use a tachometer (see that chapter) if necessary to adjust the engine speed to the manufacturer's specified idle rpm. Be sure to keep the engine idle at or below that speed to prevent the distributor's centrifugal advance from operating, because you want to check the *initial* timing setting and not the *advance* curve. Some manufacturers even recommend resetting the idle speed to lower than their normal specifications just to make certain that the centrifugal advance is not active.

Neon timing lights usually start flashing when the engine is started, but some neon units are equipped with either a pushbutton or trigger switch. If so, push the button or pull the trigger as required. Then aim the flashing light at the timing mark and reference pointer with the engine running at the specified idle speed. You can now check initial ignition timing.

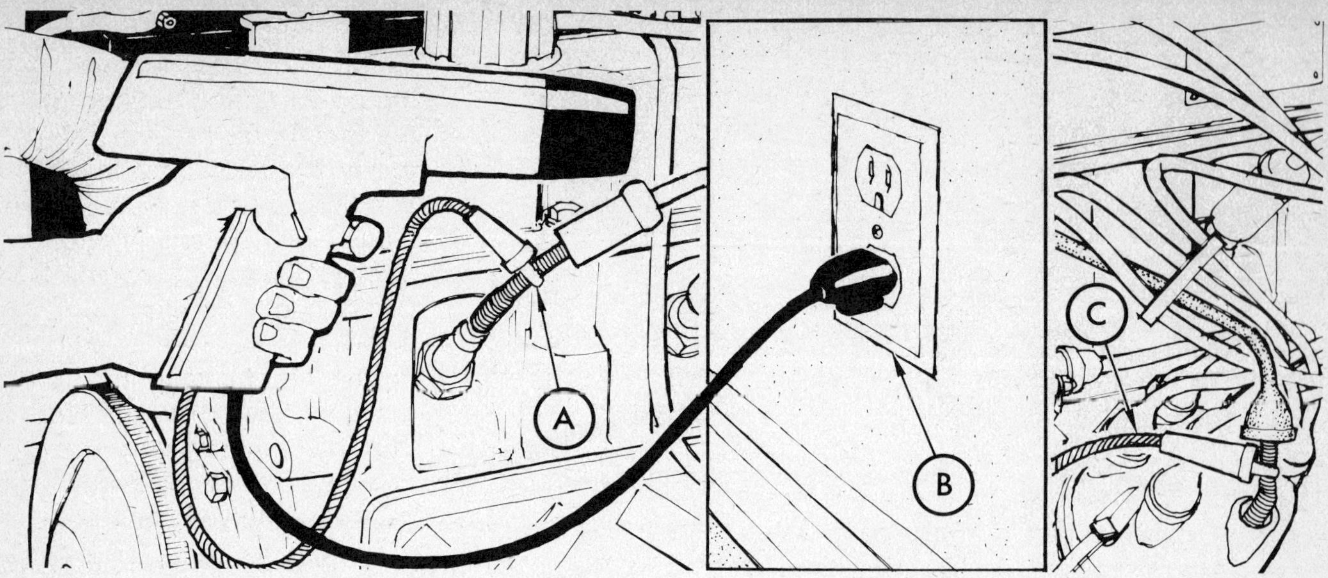

Power timing lights are available in either AC or DC models and are used in exactly the same way as the neon light except that they connect into the ignition circuit. The AC timing light has two leads—one two-pronged power lead that plugs into a 115-120 volt AC outlet (B) and one insulated high-voltage cable that connects to the No. 1 spark plug terminal (A) or to the No. 1 plug connection at the distributor cap (C).

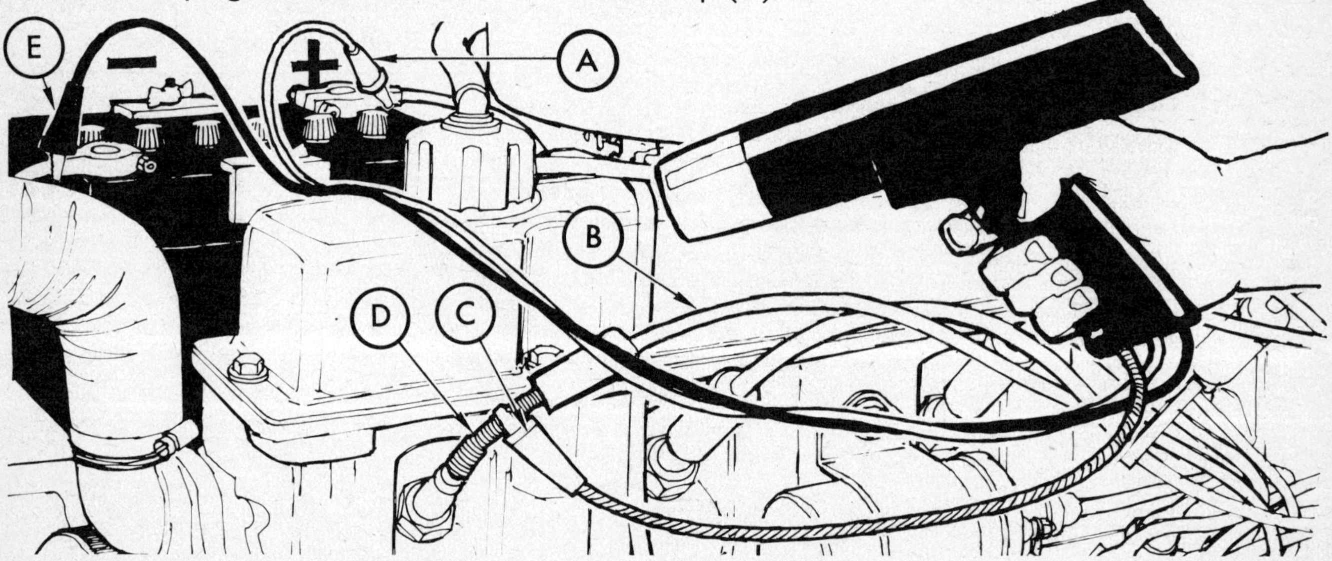

The DC timing light has three leads—two power and one insulated high-voltage cable. Connect the red power lead (A) to the positive (+) battery post and the black power lead (E) to the negative (−) battery post. If the car in question has a positive instead of negative ground, the red lead connects to the negative (−) post and the black to the positive (+) post. If the battery terminal post is not marked with negative and positive symbols on its top, or if the symbols do not appear on the battery case near each post, trace the battery cables to see where they connect. Generally speaking, the ground (−) cable is the shorter and is secured to the frame or sheet metal at a point quite close to the battery.

Remove the No. 1 spark plug wire (B) and attach the plug adapter (D) that comes with the timing light. Then place the plug wire over the adapter. Attach the insulated high-voltage cable clamp (C) to the adapter.

CAUTION: ON SOME CARS, CONNECTING THE ADAPTER AND INSULATED CABLE AT THE DISTRIBUTOR CAP INSTEAD OF THE SPARK PLUG WILL KEEP IT AWAY FROM A HOT MANIFOLD AND PREVENT MELTING OF THE CABLE'S INSULATION.

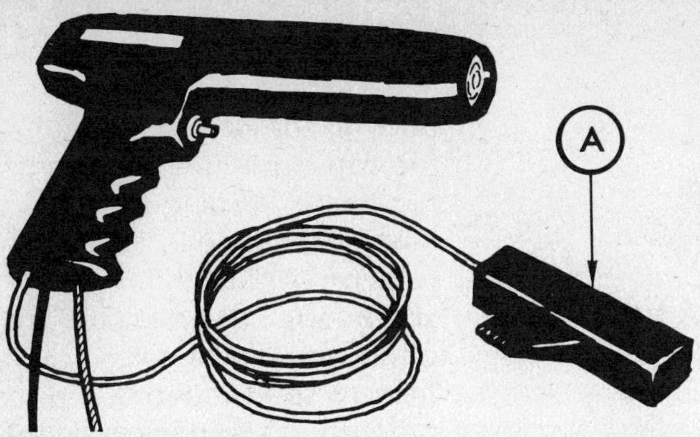

Other DC timing lights use a current pickup coil to trigger the flash. Instead of a high-voltage cable with an alligator clamp, this inductive pickup (A) snaps around the No. 1 spark plug wire at any point between the plug and the distributor cap. There's no need to get at the plug or use an adapter, and the timing light can be connected while the engine is running. Power timing lights with inductive pickups will not interfere with the operation of other tune-up equipment that may already be connected to the engine.

IF TIMING LIGHT OPERATION IS ERRATIC

If the timing light is properly connected and does not flash when triggered, the No. 1 spark plug is not firing and must be replaced. A less likely but also possible cause is that the xenon bulb in the timing light may be burned out. If installing a new spark plug does not cause the light to flash, the problem is a burned-out bulb.

Intermittent flashing at long intervals can be caused by a defective timing light or a poor connection—internally, at the power lead attached to the battery post or where the insulated cable attaches to the spark plug. A poorly grounded distributor, cracked distributor cap, spark plug gap that is too narrow and a faulty or badly adjusted set of breaker points can also cause intermittent flashing of the light.

The light will operate when the spark plug gap is set too wide, but the plug may misfire. If so, you will hear an arcing of the safety gap inside the timing light. Extremely wide gaps in the other plugs, as well as defective plug wire insulation, can cause the light to double flash. Such multiple flashing can also be caused by the plug wire resting on or close to the other plug wires. Whenever the timing light does not operate correctly, the odds are great that tune-up adjustments made prior to ignition timing were done incorrectly. Retrace your steps and correct the problem before attempting to time the engine.

INITIAL IGNITION TIMING

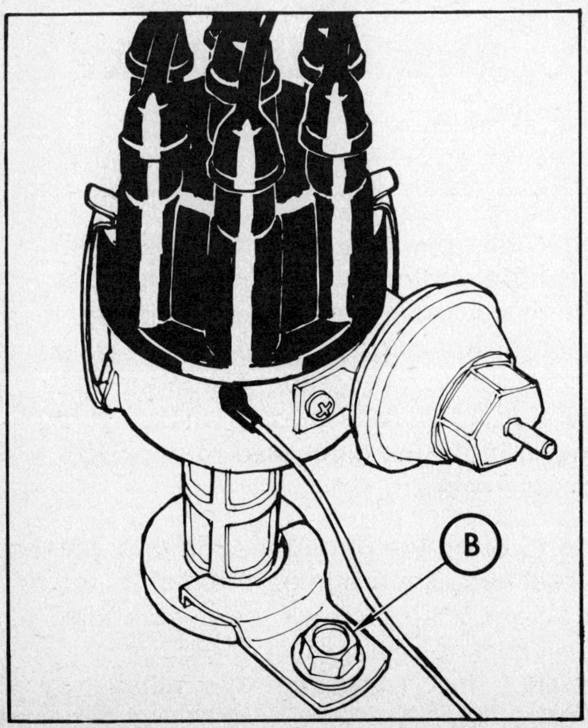

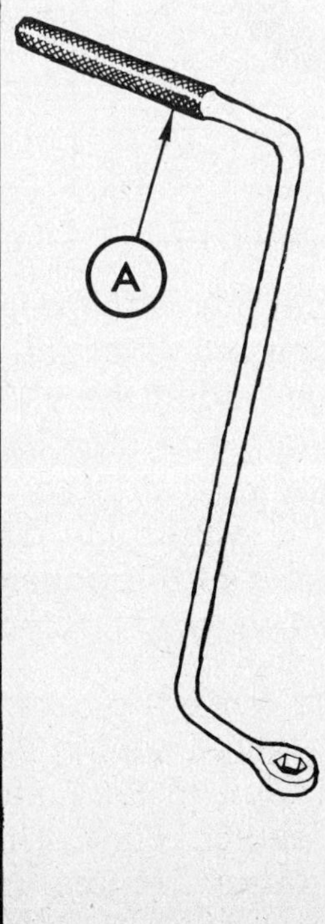

Regardless of the type of timing light used, the light is aimed at the timing mark and reference pointer. If the mark and the pointer line up at the setting specified by the engine manufacturer, initial ignition timing is correct. If the mark and reference pointer do not line up as specified, the distributor body must be rotated until they do. To adjust timing, loosen the hold-down bolt (B) located at the base of the distributor. Distributors on many late-model V-8 engines are positioned in such a manner that a special offset wrench (A) is required to reach the hold-down bolt.

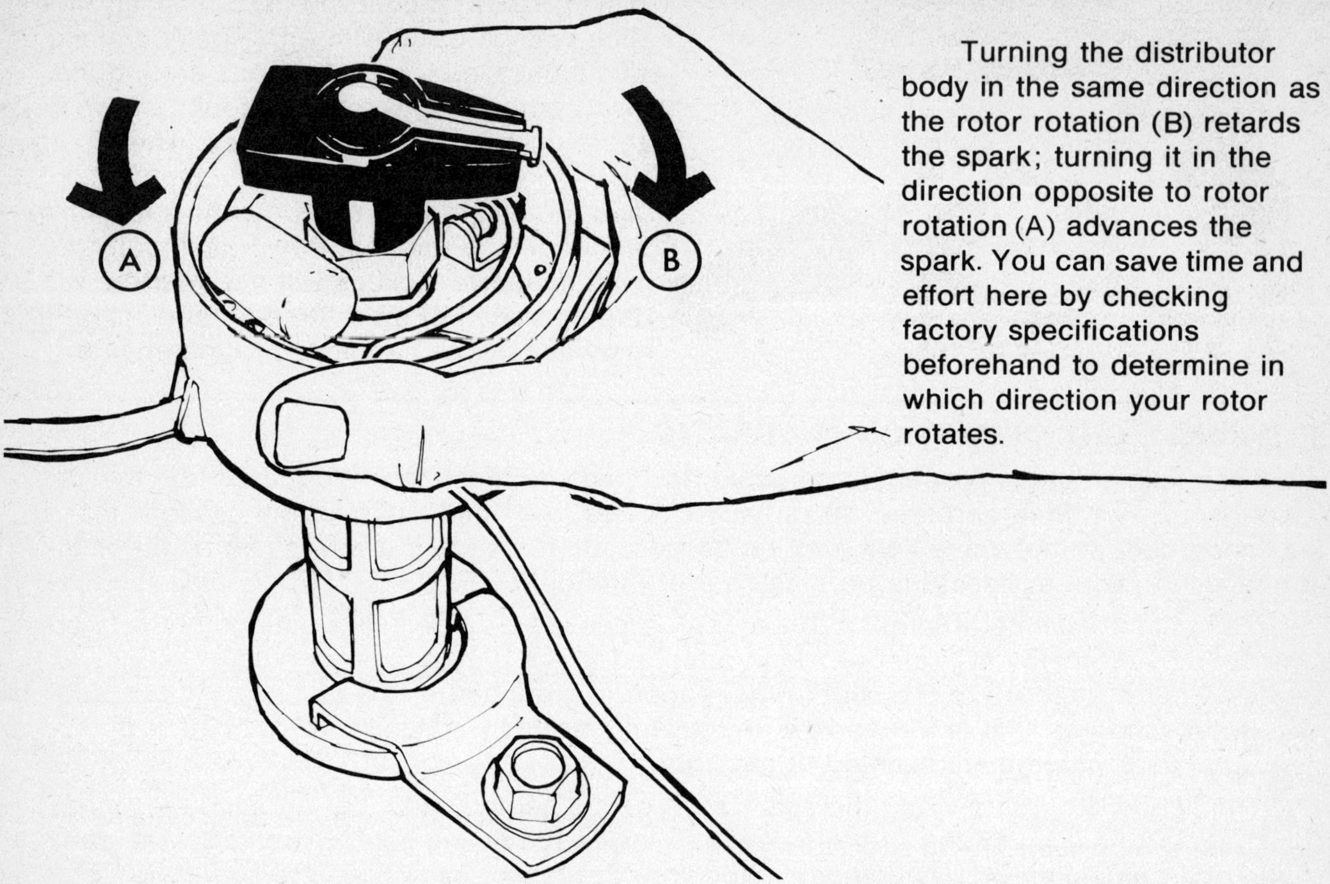

Turning the distributor body in the same direction as the rotor rotation (B) retards the spark; turning it in the direction opposite to rotor rotation (A) advances the spark. You can save time and effort here by checking factory specifications beforehand to determine in which direction your rotor rotates.

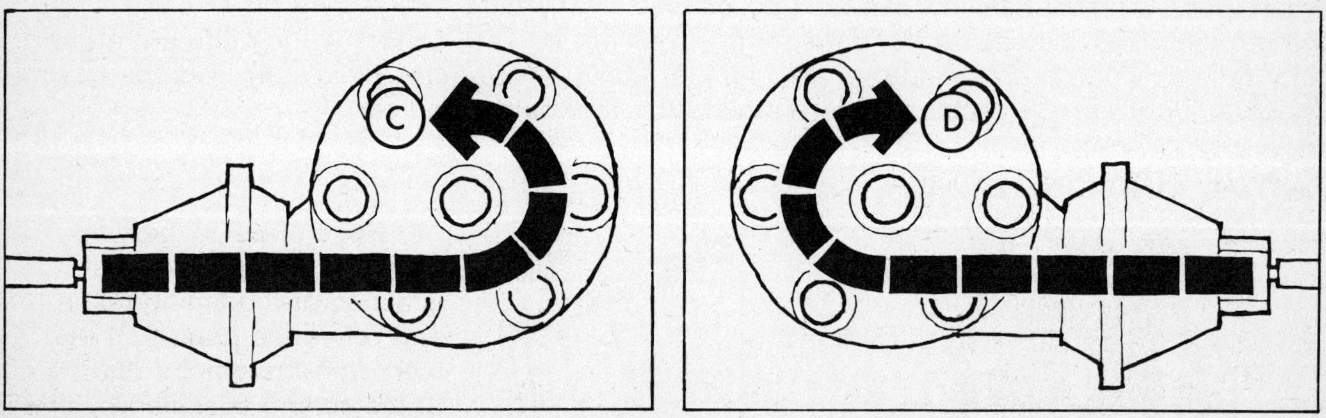

If you cannot locate the specifications when needed, there is a rule of thumb you can follow without leaving the engine compartment. You can determine the rotation of the distributor by pointing with your finger along the vacuum advance unit so that the curve of your finger corresponds with the curve of the distributor housing. The direction in which your finger points will be the direction of distributor shaft and cam rotation. Since the rotor is located on the distributor shaft, this will be the direction in which the rotor rotates. If it is your left finger that follows the curve of the housing, it indicates clockwise rotation (D). If it is your right-hand finger, that shows counterclockwise rotation (C).

Turn the distributor slowly in the correct direction to bring the timing mark into alignment with the reference pointer as the specified setting is reached. Then tighten the distributor hold-down bolt and recheck the timing with the timing light. Also recheck the engine idle speed and reset it if it varies from specifications. Turn off the engine, disconnect the timing light and reconnect the distributor vacuum advance line. This completes initial ignition timing.

DYNAMIC TIMING

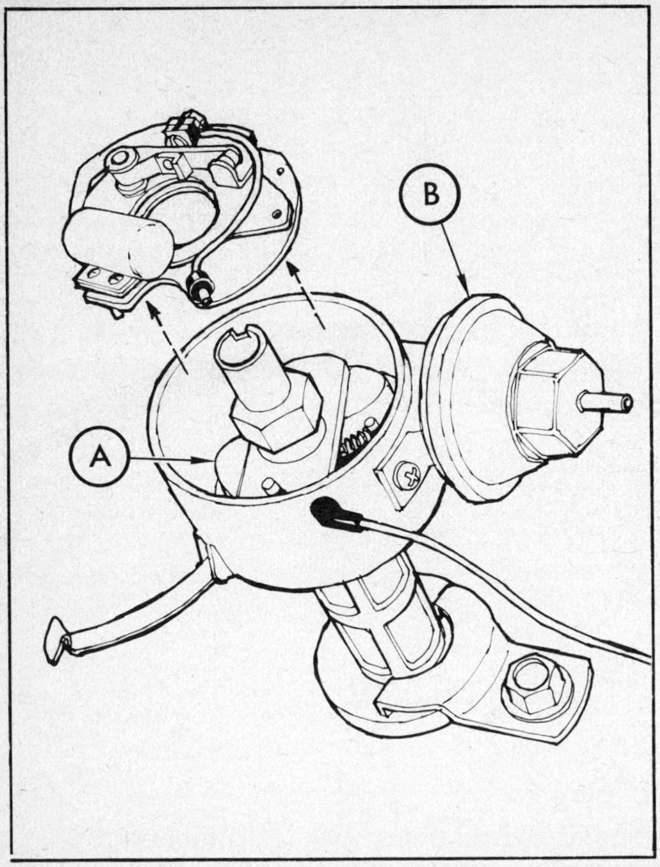

The timing of the spark relative to the stroke of the pistons must be varied according to speed and load conditions. This timing change is called an advance curve and is automatically adjusted by the distributor's internal spark advance system. This spark advance system consists of a centrifugal mechanism (A) that advances or retards the distributor cam relative to the distributor driveshaft and a vacuum diaphragm (B) which advances or retards the position of the breaker points relative to the distributor cam.

Dynamic timing means checking the operation of the centrifugal and vacuum advance mechanisms to see that they are functioning properly. The distributor must be removed from the car and placed on a distributor tester for accurate calibrations, but you can determine whether or not the advance mechanisms are working by combining the use of a tachometer (see that chapter) with a timing light. Determining just how *accurately* they are working, however, is beyond the capacity of this equipment.

CAUTION: CHECKING ADVANCE MECHANISMS WITH JUST A TACHOMETER AND TIMING LIGHT GIVES ONLY AN APPROXIMATION, NOT ACCURATE RESULTS. DO NOT TRY TO MODIFY ADVANCE CURVES AS A RESULT OF YOUR TESTING. SUCH TAMPERING CAN PERMANENTLY DAMAGE THE ENGINE DUE TO INCORRECT SPARK ADVANCE.

If the centrifugal and/or vacuum advance mechanisms do not seem to be working properly, pay a visit to your local dealer, where trained mechanics working with the proper equipment can test their action.

CENTRIFUGAL ADVANCE CHECK

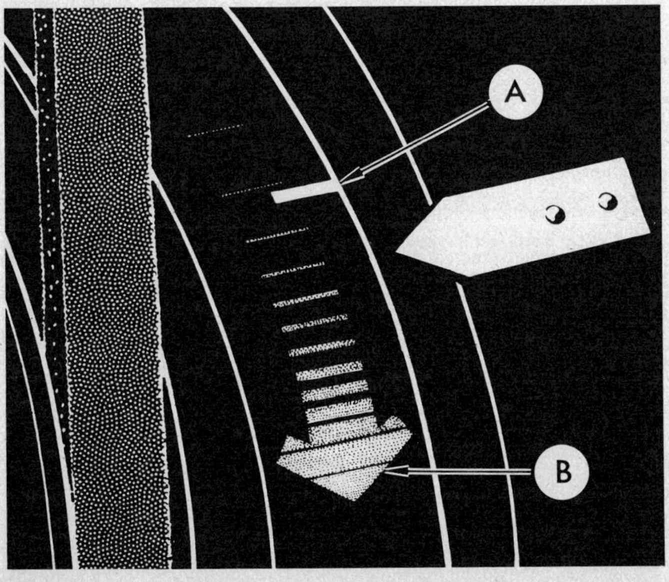

To check operation of the centrifugal advance once initial ignition timing has been set to specifications, increase engine speed gradually with the vacuum advance line disconnected. When advance begins, the timing mark (A) will gradually move opposite to the direction of engine rotation (B) as you watch the timing light flash on it. Decreasing engine speed should also gradually return the timing mark to its original position.

If the timing mark refuses to remain fixed in position when engine speed is held constant, the problem could range from worn camshaft lobes to loose, broken or disconnected advance weights. Further diagnosis by a trained mechanic with the proper test equipment is needed.

VACUUM ADVANCE CHECK

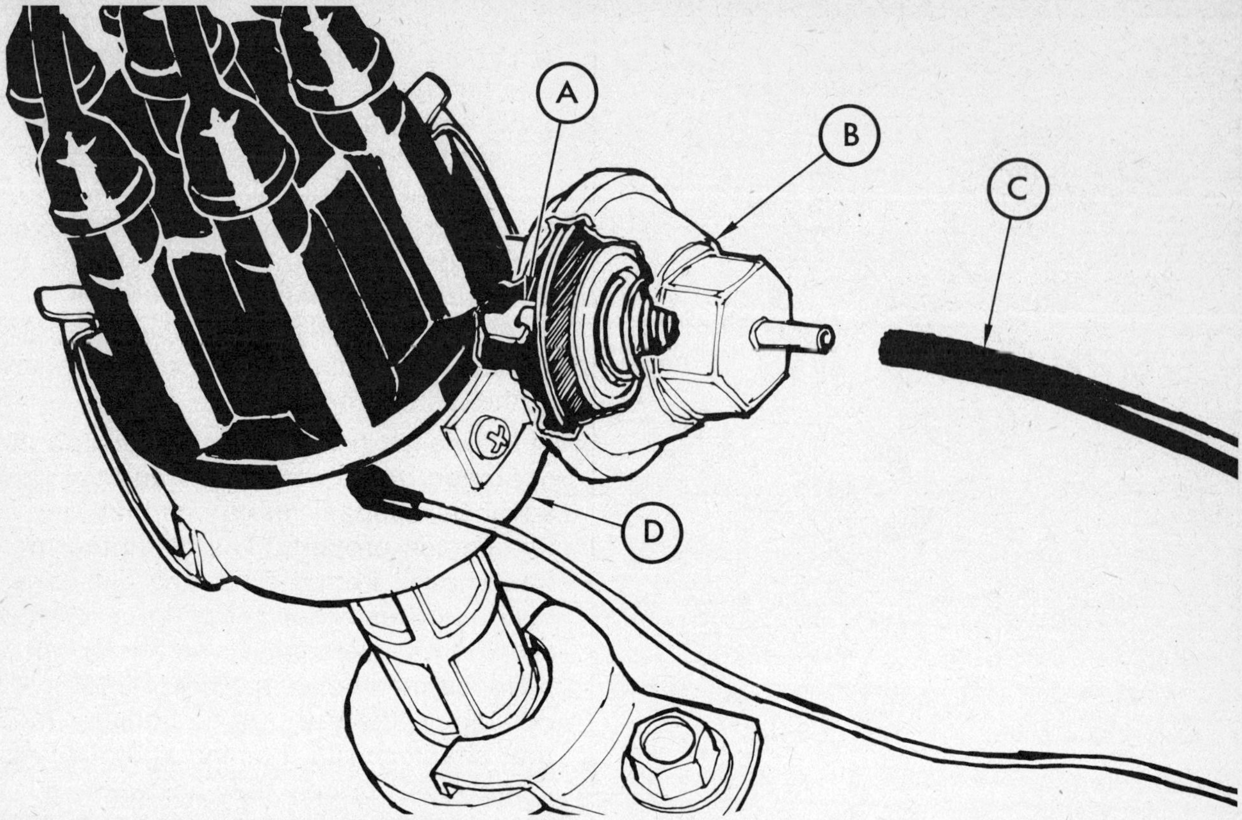

Vacuum spark control operates by a spring-loaded vacuum chamber (B) attached to the side of the distributor (D) and connected to the breaker plate by linkage (A). Vacuum is supplied to the line (C) from the carburetor .

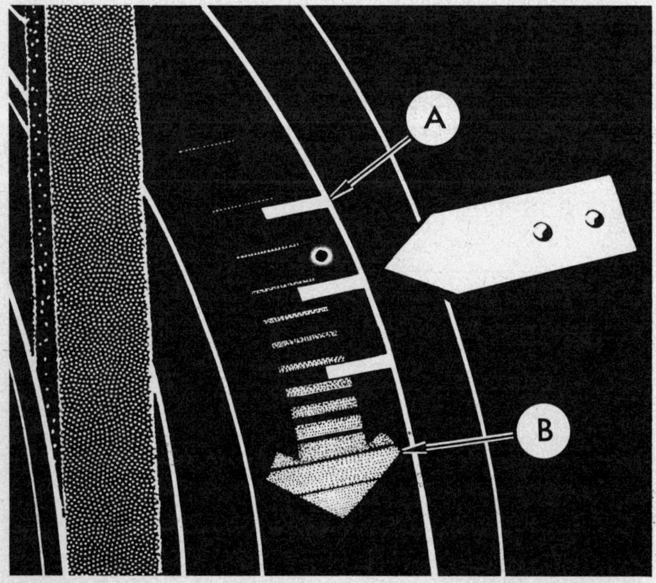

Operate the engine at normal idle and gradually increase its speed to approximately 1,500 rpm. The timing mark (A) should move opposite to the direction of engine rotation (B) as centrifugal advance is at work. Reconnect the vacuum line to the distributor vacuum advance unit and hold the engine speed constant. The timing mark (A) should move farther along the scale. Now remove the vacuum line and watch to see if the timing mark moves back. If it does not, the vacuum advance device is incorrectly adjusted or defective. This is also true if the timing mark does not move when the vacuum line is reconnected. We recommend the services of a trained mechanic to repair a bad vacuum advance unit.

Depending upon the design of the timing mark and reference pointer, the timing mark may move beyond the range of the pointer scale while advance operation is being checked. If so, extend the degree scale with white chalk or paint to watch the amount of timing mark movement.

Vacuum shutoff at idle can be checked by reducing engine speed from 1,500 rpm back to idle speed. Aim the timing light at the timing mark and recheck its position. If initial timing is still the same as it was when the engine was timed with the vacuum line disconnected, the shutoff is operating correctly. If not, you still need that mechanic. Keep in mind, though, that this test does not work with engines using full vacuum advance at idle.

OTHER TEST FUNCTIONS

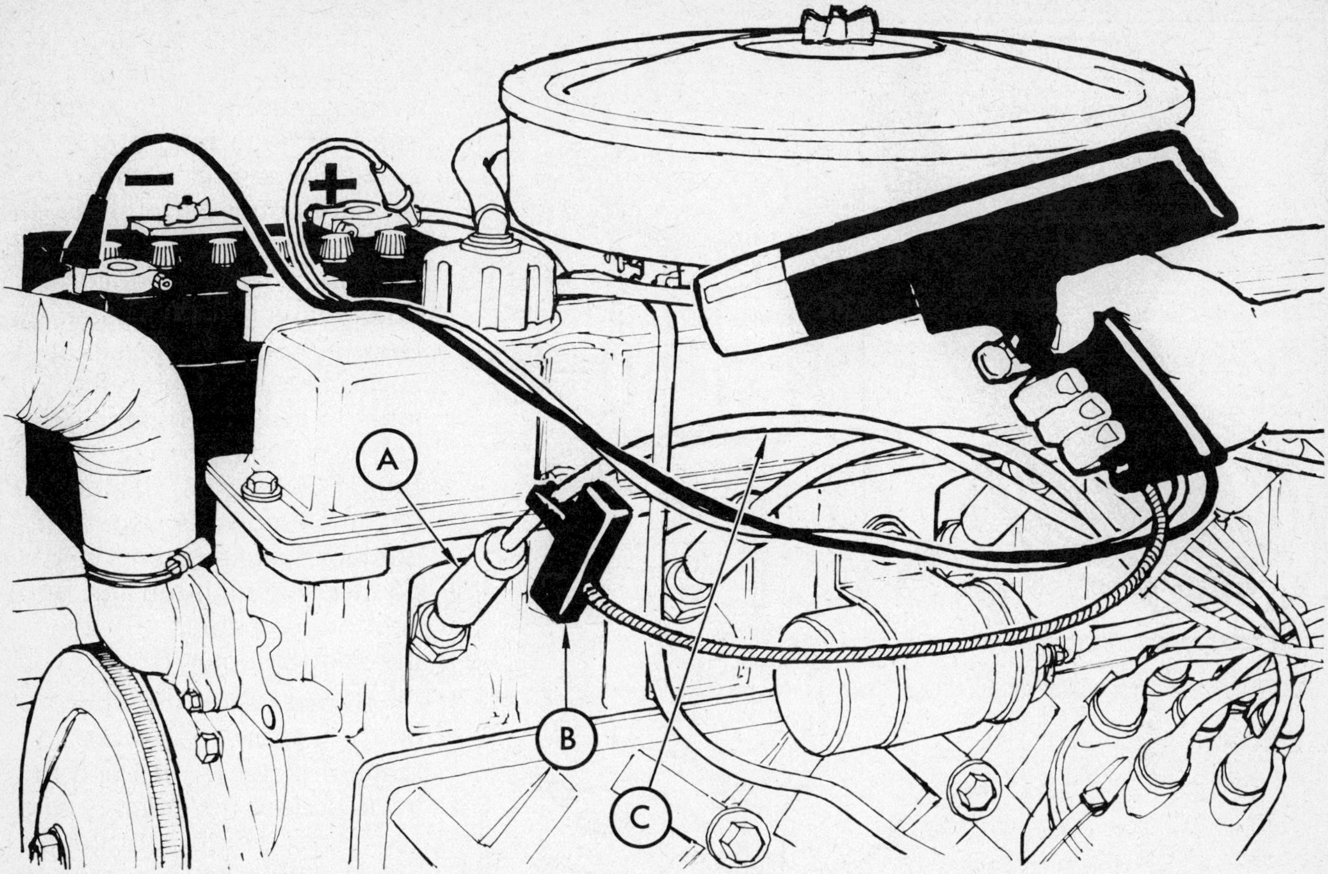

Power timing lights can be used for test functions other than initial ignition timing. Testing spark plugs, especially those with insulating plug boots (A), can be simplified considerably with the help of a power timing light equipped with an inductive pickup (B) or a capacitive clip. Because these pickup units are so sensitive that they can simply be clipped to the spark plug wire instead of requiring the use of an adapter, you can test any or all plugs you suspect are not working properly just by putting the clip on the plug wire (C) and watching the light. If the spark is very weak, the plug is shorted out. If it is so fouled that no spark is present, the timing light will not flash. Remove and replace the faulty plug, then retest to make certain that the new plug operates correctly.

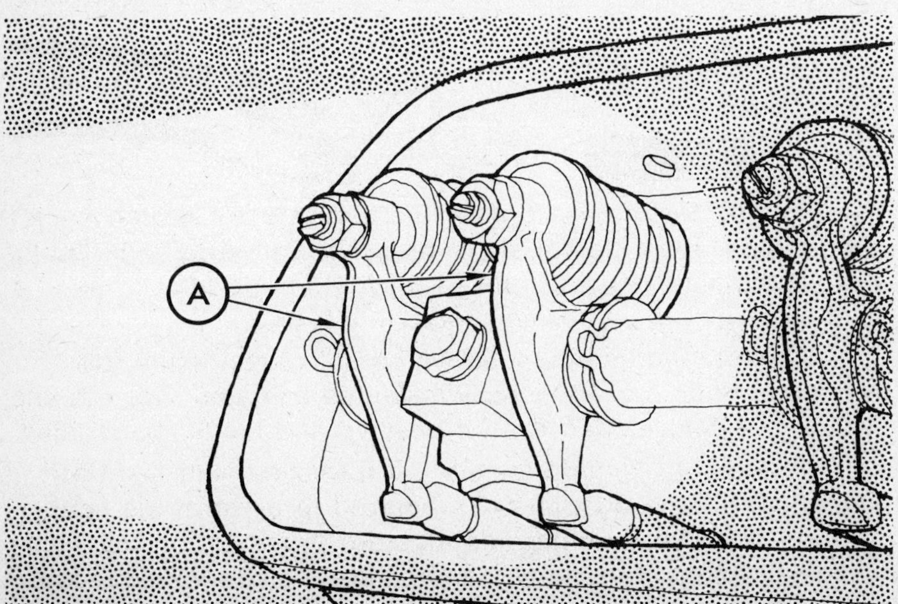

A power timing light can also be used to check valves suspected of malfunctioning. To do so, you must remove the rocker arm cover so that valve action is visible. Connect the high-voltage cable to the spark plug of the cylinder in question. With the engine idling, the timing light flashes will make both intake and exhaust valves appear to be frozen in their closed position (A) if they are working properly; if one is out of visual line with the other, you have a bad valve.

IGNITION TESTER

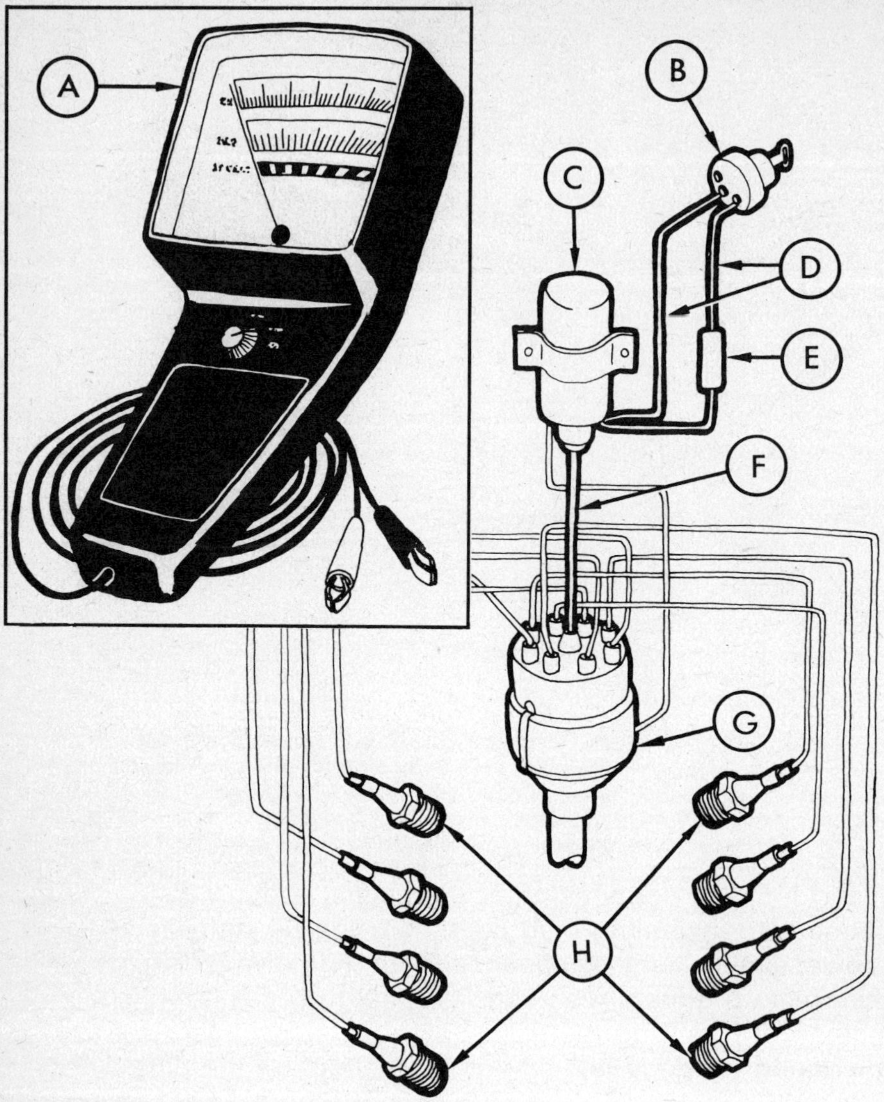

Designed to fire the spark plugs, the ignition system must provide a hot spark at each plug at the correct time and in the proper sequence to ignite the air/fuel mixture in the cylinders. The conventional automotive ignition system (six- or 12-volt) consists of seven essential parts: the ignition switch (B); ignition coil (C); ballast resistor (E); primary wiring (D); distributor (G), including the cap, rotor, contact breaker points and condenser; high-tension or secondary cables (F) and the spark plugs (H).

Often called the "poor man's oscilloscope," the ignition tester (A) is used to troubleshoot the primary and secondary portions of the automotive ignition system. In addition to checking the ignition coil for opens (broken wires), shorted windings and internal insulation breakdown, the ignition tester locates excessive resistance and/or external insulation leakage.

IGNITION TESTER CONTROLS

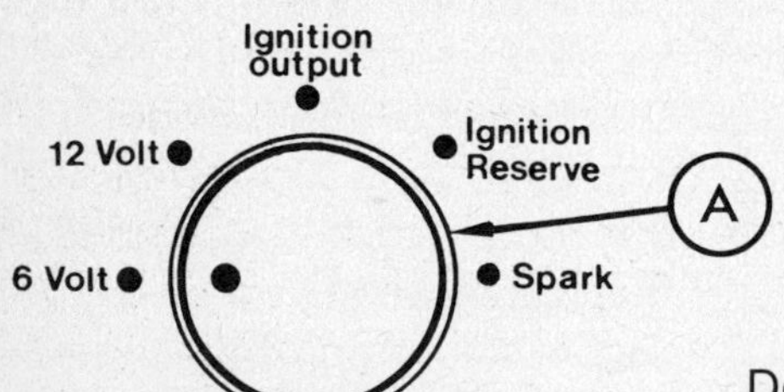

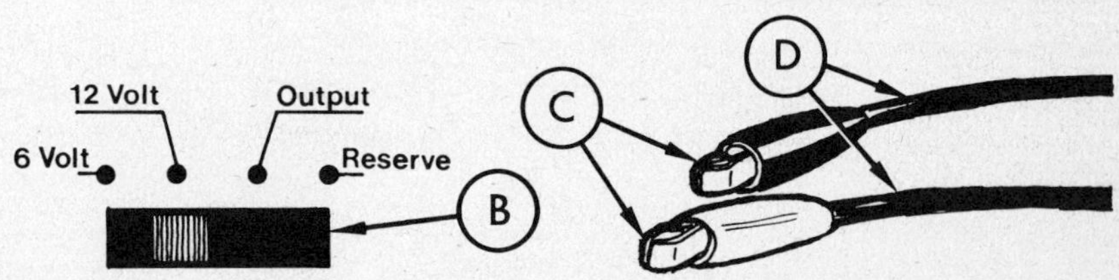

Ignition testers use only a function selector switch. Depending upon the variety of tests performed by a particular tester, this can be as simple as a sliding switch (B) or as elaborate as a multiposition rotary knob (A).

The test unit is equipped with a set of test leads (D), whose connectors are usually insulated alligator clips (C) and which are color-coded red for positive and black for negative.

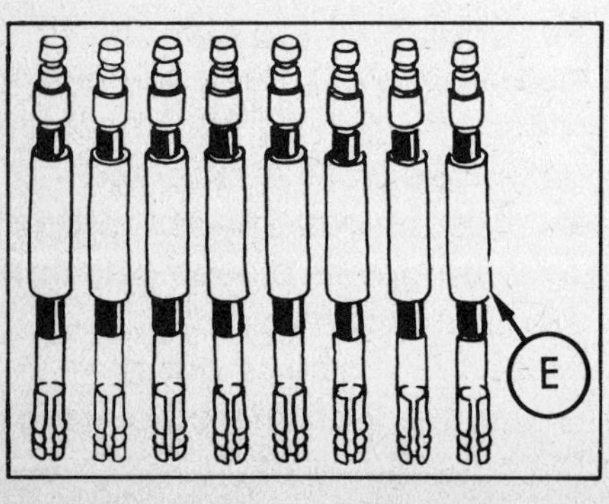

In addition, ignition testers are furnished with a set of metal spark plug adapters (E) required to perform the spark output test. Usually eight adapters are provided for use with domestic plugs, but many manufacturers also include at least four adapters for use with foreign plugs.

IGNITION TESTER METER BANDS

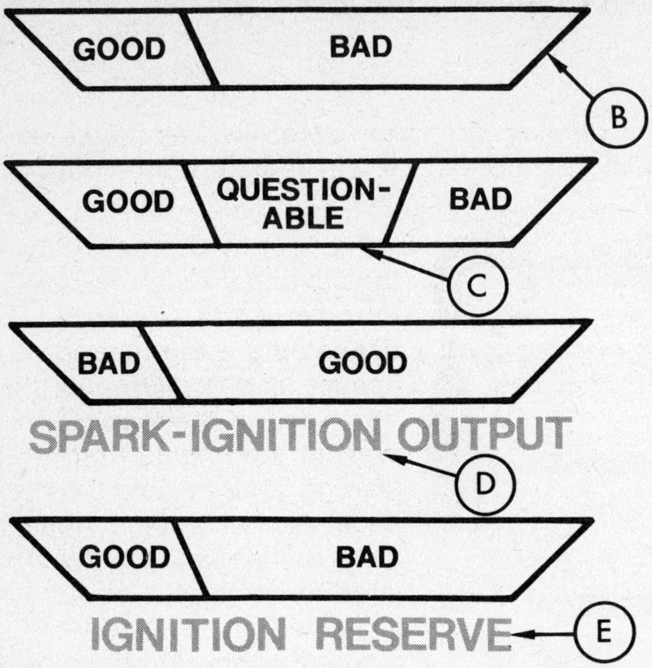

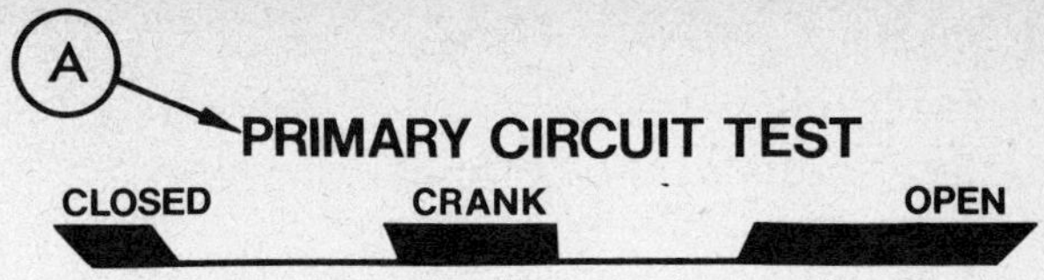

Ignition testers report their findings directly on color-coded bands marked GOOD/BAD (B), although some manufacturers may also include a QUESTIONABLE zone (C) between the two.

Meters should include at least two such bands, one marked SPARK & IGNITION OUTPUT (D) and the other designated IGNITION RESERVE (E).

The more expensive testers also provide a voltmeter function for testing the primary circuit condition. This direct-reading band is usually labeled PRIMARY CIRCUIT TEST (A) and divided into three test segments marked CLOSED, CRANK AND OPEN.

CONNECTING THE IGNITION TESTER

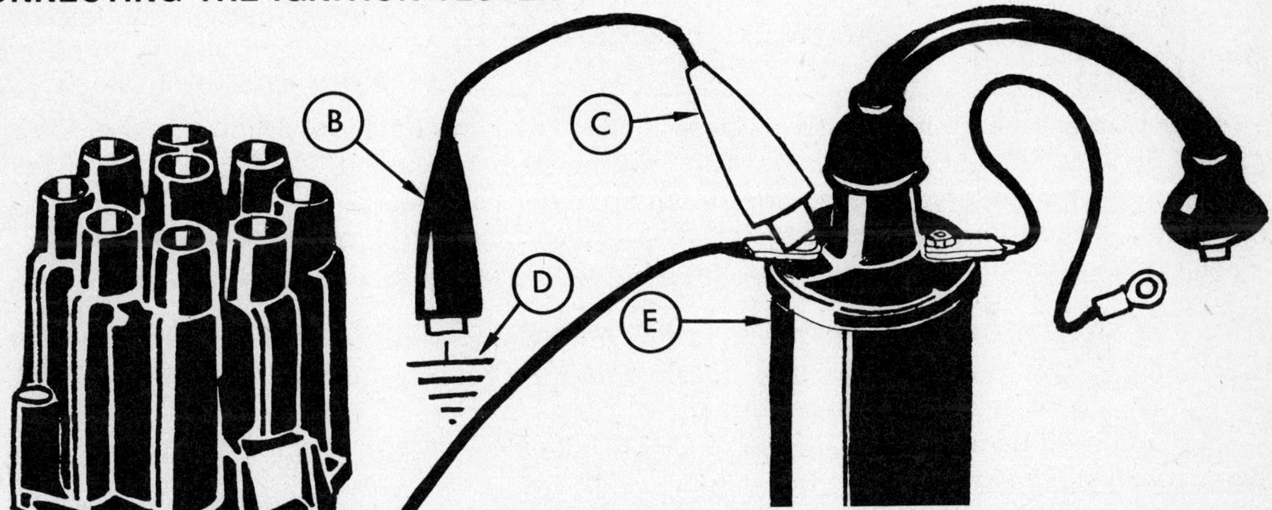

With the exception of the spark output test, a single connection pattern is used for all ignition tester functions with negative ground ignition systems. The red tester lead (C) is connected to the distributor primary terminal, either at the coil (E) or at the distributor (A). The black tester lead (B) is connected to a good ground (D) on the engine block or body of the car. If the car being tested uses positive ground, reverse the leads from the pattern specified.

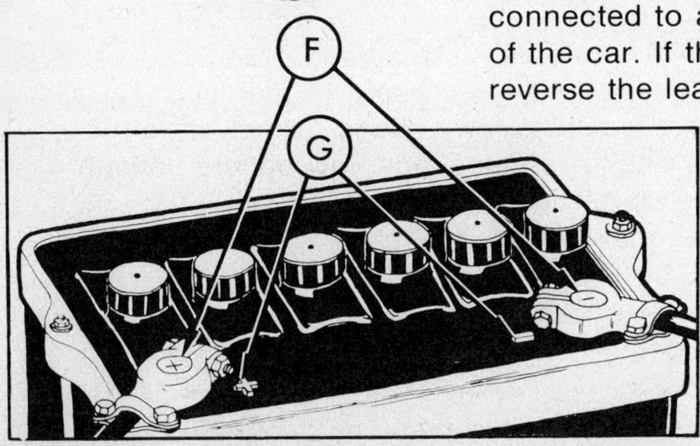

Ground can be determined by locating the battery cable which connects to the engine or frame. The battery terminal post to which the other end of this cable connects should be marked with a + (positive) or — (negative) sign. The marking may appear on top of the battery post (F) or on the battery case beside the post (G). Generally speaking, all American cars are negative ground.

PRIMARY CIRCUIT TEST

If the ignition tester is equipped with a PRIMARY CIRCUIT TEST band, you can test:

- Battery-to-distributor circuit condition.
- Voltage available for starting.
- Distributor resistance.

These three tests will check the entire ignition primary circuit for excessive resistance and should be performed before other tests to simplify test procedure and reduce the amount of time involved.

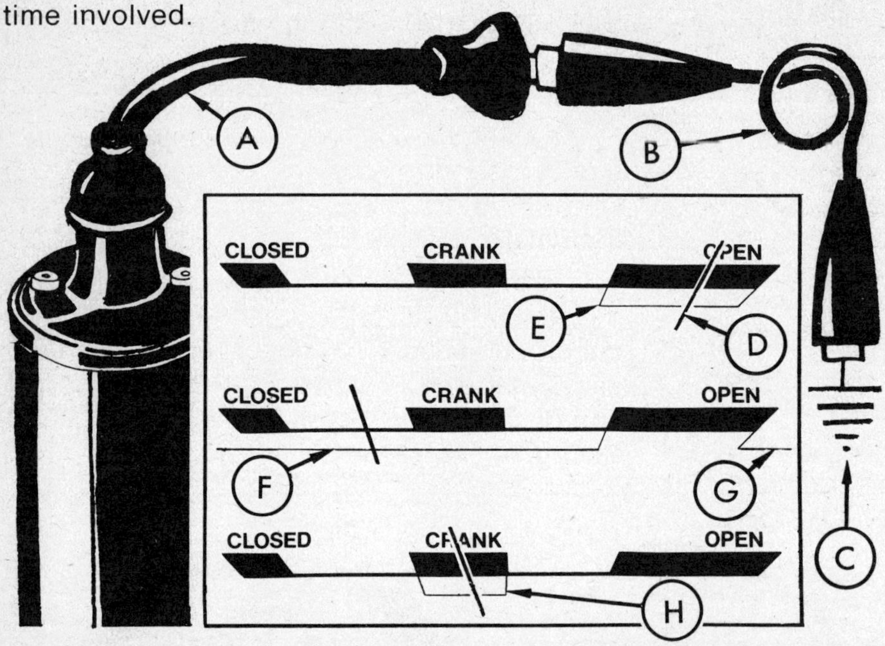

Begin by disconnecting the secondary coil wire (A) from the distributor and connecting it to one end of a jumper lead (B). Ground the other end of the lead (C) to prevent the engine from starting. This is not necessary with some ignition tester units, because their internal circuitry is designed to prevent the engine from starting without grounding the secondary coil wire. Check your ignition tester instruction manual to be sure which kind you have.

Now connect the ignition tester as described and set the function switch to either the 6-VOLT or 12-VOLT test position, according to the car being tested.

To check the battery-to-distributor circuit condition, connect a remote start switch (see that chapter) and turn the ignition switch ON. "Bump" the starter to open the distributor contact breaker points. This will cause the test needle (D) to read in the band segment marked OPEN (E).

If the needle reads to the left (F) of OPEN, it indicates excessive resistance or an open in the primary circuit. But since it is also possible that the breaker points are not really open, you should repeat the test to make certain.

A reading off-band to the right (G) of OPEN indicates either a shorted coil primary or shorted ballast resistor.

Now check the voltage available for starting by cranking the engine momentarily with the ignition switch ON. Watch the tester needle, which should read in the band segment marked CRANK (H).

If the needle reads to the left (A) of CRANK, it indicates low voltage. Check for an incorrect breaker point gap, faulty battery cables and/or connections, low battery voltage, excessive starter circuit draw or a malfunctioning starter ignition bypass circuit.

A reading to the right (B) of CRANK indicates that the points are either gapped too wide or are not closing at all.

The final primary circuit test is for distributor resistance. Turn the ignition switch ON and "bump" the starter to close the contact breaker points. This will cause the test needle (C) to read in the band segment marked CLOSED (D).

If the needle reads to the right (E) of CLOSED, the breaker points are burned, have high resistance or are improperly aligned. It is also possible that the distributor body or breaker plate is not properly grounded. If the breaker plate has a broken ground wire, it could cause a high resistance reading, but this problem is more likely to cause trouble when the engine is running and the plate is moving in response to the vacuum advance.

IGNITION OUTPUT TEST

WARNING: BE CAREFUL WHEN WORKING ON A HOT ENGINE. DO NOT TOUCH THE METAL PARTS OF THE ENGINE.

Having completed the primary circuit sequence, set the ignition tester function switch to the IGNITION OUTPUT position and connect a tachometer (see that chapter). Start the engine with the remote start switch and run it at approximately 1,500 rpm. The meter needle should give *no* reading at this point.

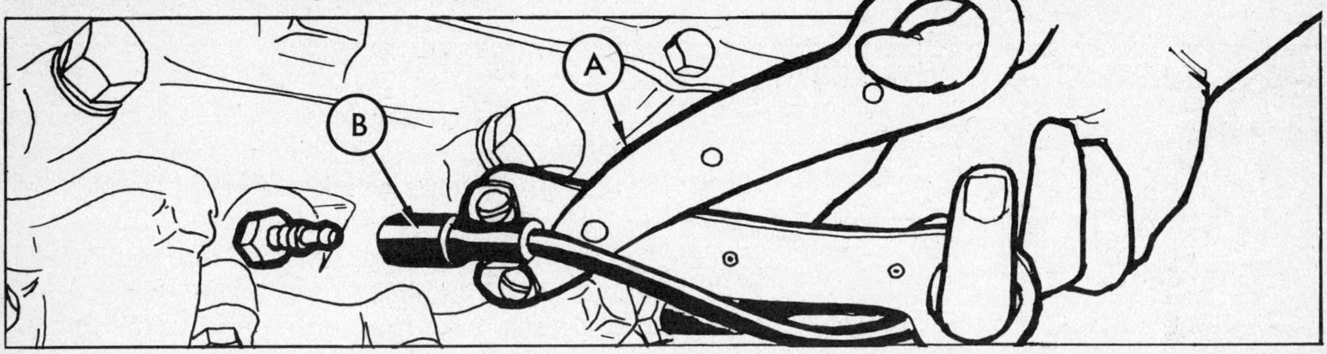

Use a pair of insulated pliers (A—these are furnished with some tester units) to remove the No. 1 spark plug cable (B—see chart in timing lights chapter) from its plug. Hold it in the air so that no spark is given off. This causes the coil to put out maximum secondary voltage to that particular wire, with a corresponding rise in the induced voltage in the coil primary.

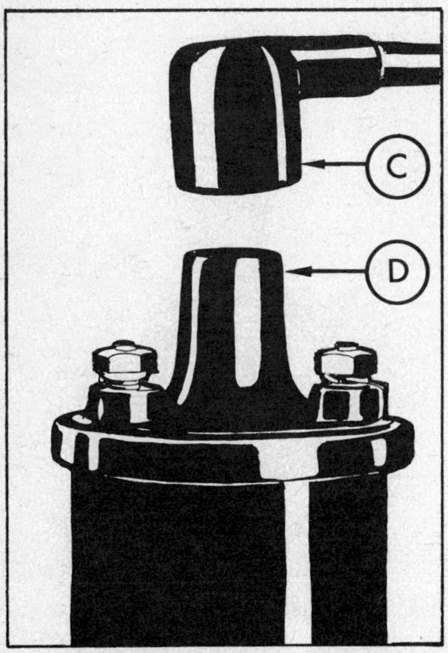

Read the SPARK & IGNITION OUTPUT band on the meter; the needle should rest in the GOOD segment (F). Replace the spark plug cable and repeat the test with the next cylinder. Each time a plug cable is removed, the meter needle should give a reading; when the cable is replaced, the needle should drop off the band completely.

A reading in the segment marked BAD (E) on one or more cylinders indicates leakage to the ground. This can be caused by a defective distributor cap, a condenser lead that is too near the rotor or insulation leaks in the spark plug cables.

If the meter reads BAD or gives an unsteady reading on *all* cylinders, look for a defective coil, distributor cap, rotor, condenser or excessive primary circuit resistance.

After completing this test, you can check the condition of the coil without changing the tester setting. Pull the high-tension wire (C) from the coil tower (D) and crank the engine (ignition switch ON) while watching the meter.

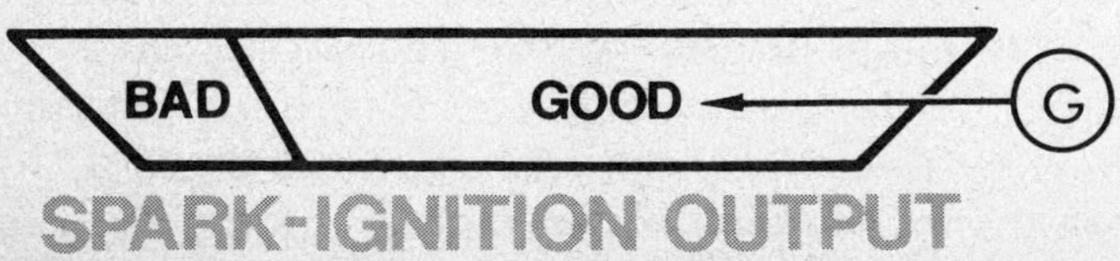

If the coil is good, the needle should read in the SPARK & IGNITION OUTPUT band segment marked GOOD (G). If it does not, the coil (or a primary circuit leading to it) is defective or the condenser is shorted.

IGNITION RESERVE LOAD TEST

The ignition output test must be performed and all cylinders must register a GOOD reading before this test can be done. Set the function switch to the IGNITION RESERVE position and start the engine. Run it at normal idle speed and check each plug wire to make certain that it is securely fastened to the plug.

Accelerate the engine by depressing the accelerator pedal quickly and releasing it—the pedal should not be held down. Read the IGNITION RESERVE band on the meter.

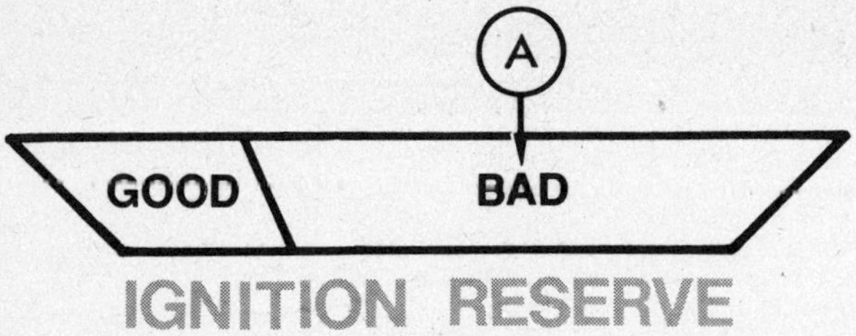

If the needle reads in the area of the scale marked BAD (A), you have a definite misfire. This could be caused by excessive resistance or opens in the circuit or spark plug gaps that are set too wide. The spark current or output test should be completed to locate the exact problem area.

SPARK CURRENT (OUTPUT) TEST

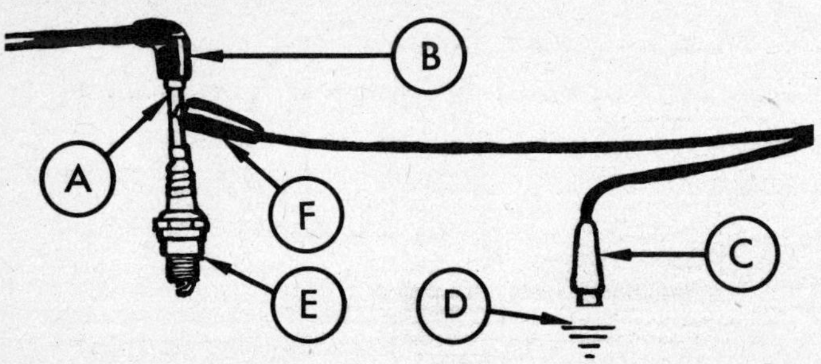

To run a spark current test, set the tester function switch to the SPARK position. One at a time, install a spark plug adapter (A), which is furnished with the tester unit, between each plug (E) and its connecting cable (B). Connect the red tester lead (C) to ground (D), start the engine and touch the black lead (F) to each plug adapter in turn.

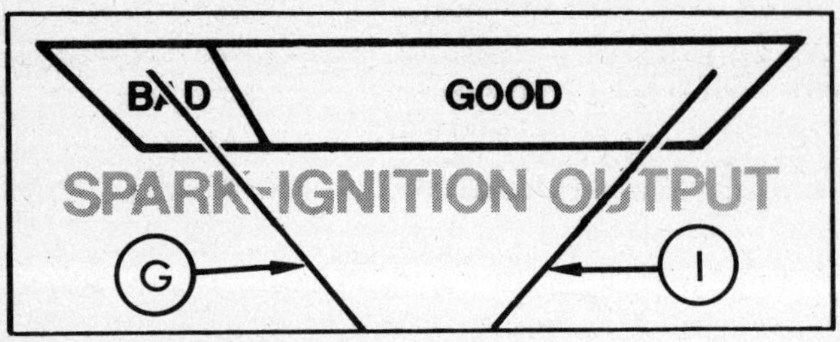

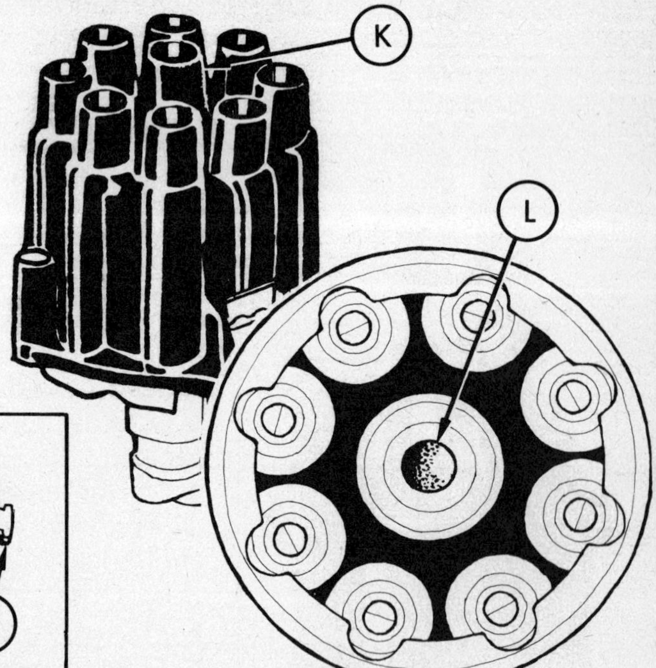

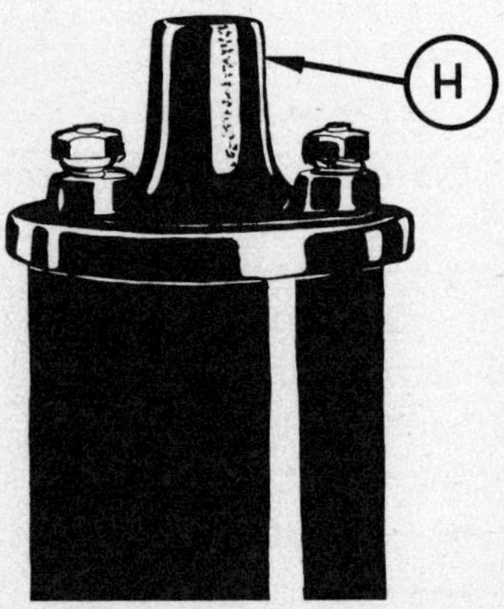

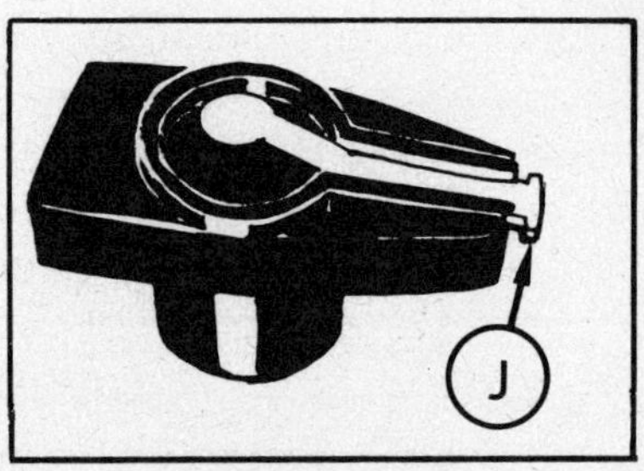

The SPARK & IGNITION OUTPUT band on the meter will indicate the amount of current reaching the plug as a GOOD or BAD reading. If all readings are excessively high in the GOOD band (I), you may have installed the wrong coil or a ballast resistor may be incorrectly wired in the coil circuit.

A BAD reading (G) on one or more cylinders indicates excessive resistance, an open in the spark plug cable(s) or corroded distributor cap terminal inserts. A BAD reading on *all* cylinders is caused by excessive resistance, an open in the coil or coil lead wire or corrosion in the coil tower (H) or distributor cap center tower (K). Less likely but also a possibility is a burned rotor tip (J) or carbon button (L) in the center of the distributor cap.

COMPRESSION TESTER

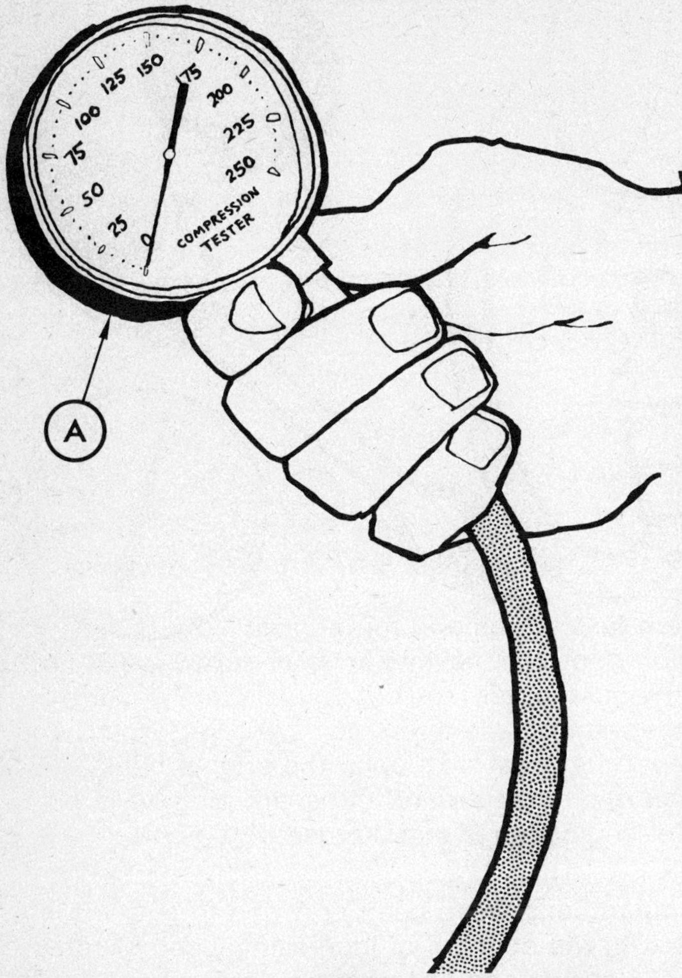

The compression tester (A) is a very simple instrument designed to tell you one important thing—whether or not the engine needs a piston ring and/or valve job. Lost compression means an overhaul is just around the corner, and spending time and money on a tune-up is not likely to do much that will help the car run better. For that reason, the compression test is usually the first step in any tune-up. Here's why.

The air/fuel mixture in an automotive engine has to be greatly compressed within each cylinder before ignition. This is done by upward movement of the piston, which squeezes the mixture into a much smaller volume. The exact reduction in volume is expressed as the engine's compression ratio. Thus an 8:1 ratio means the air/fuel mixture is compressed into ⅛ of its original volume; a 12:1 ratio means the mixture is reduced to 1/12 of its original volume.

Such compression of the air/fuel mixture increases the pressure in the combustion chamber from 15 psi (pounds per square inch) to 120 psi in the case of an 8:1 ratio, or 180 psi in the case of a 12:1 ratio. (Example: 15 psi x compression ratio = combustion chamber pressure.) But when the air/fuel mixture is ignited, this pressure is multiplied again by four or five times. Full power and performance from the engine are dependent upon effective sealing of both compression and combustion pressure by the piston rings (E), intake and exhaust valves (B, C) and cylinder head gasket (D).

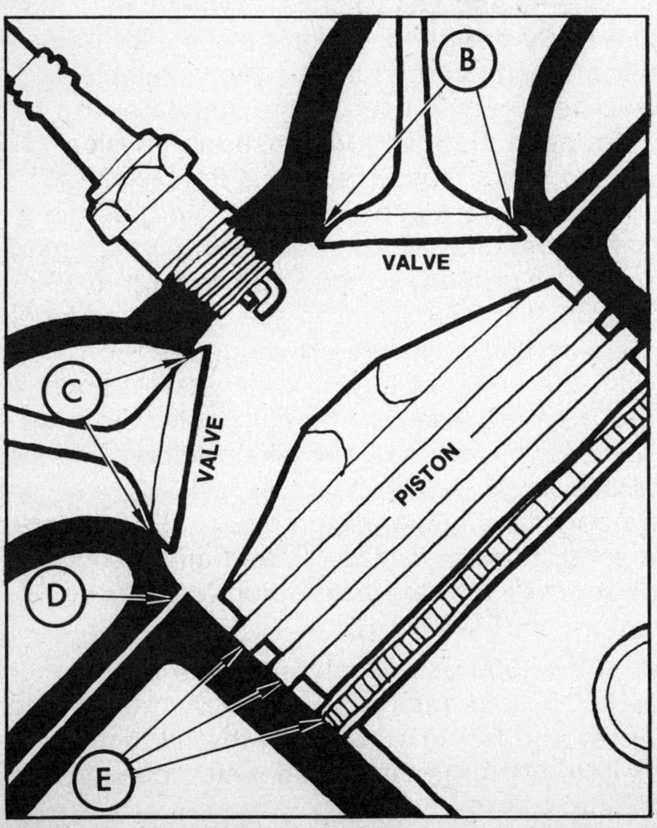

The compression test tells us when there is a loss of sealing in these areas and helps locate such a loss exactly. Because such mechanical defects cannot be remedied by adjustments of the engine or its accessory systems, we use the compression test to determine the state of the engine and whether it will respond to a tune-up. To do so, we connect a gauge in such a way as to measure the amount of compression in the cylinders.

COMPRESSION TESTER

TESTER DESIGN

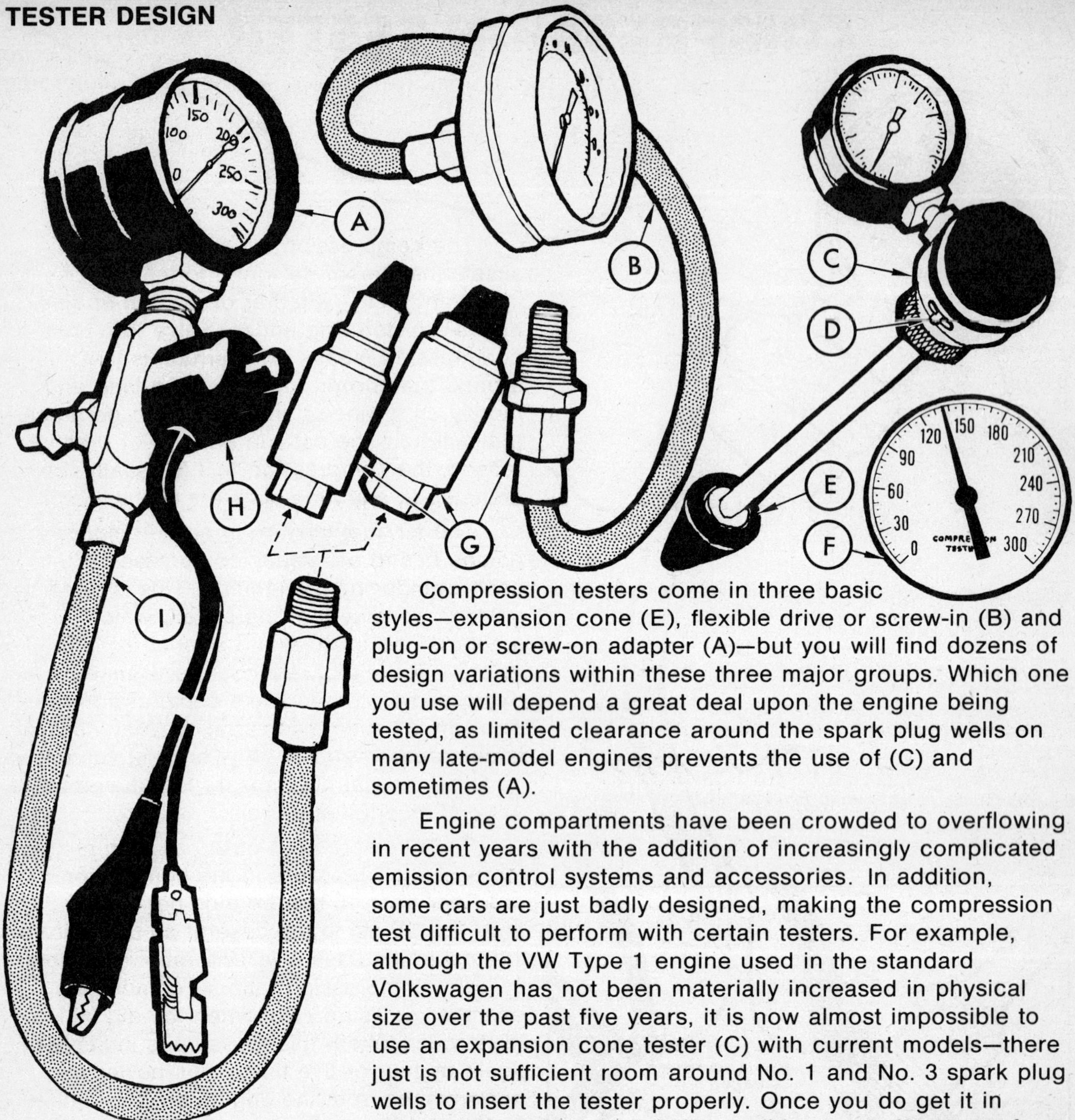

Compression testers come in three basic styles—expansion cone (E), flexible drive or screw-in (B) and plug-on or screw-on adapter (A)—but you will find dozens of design variations within these three major groups. Which one you use will depend a great deal upon the engine being tested, as limited clearance around the spark plug wells on many late-model engines prevents the use of (C) and sometimes (A).

Engine compartments have been crowded to overflowing in recent years with the addition of increasingly complicated emission control systems and accessories. In addition, some cars are just badly designed, making the compression test difficult to perform with certain testers. For example, although the VW Type 1 engine used in the standard Volkswagen has not been materially increased in physical size over the past five years, it is now almost impossible to use an expansion cone tester (C) with current models—there just is not sufficient room around No. 1 and No. 3 spark plug wells to insert the tester properly. Once you do get it in place, with many units it is impossible to see the gauge face.

Those who own or work on more than one car will find a flexible drive compression tester (B) the most convenient, as it does away with the need for buying a new tester each time cars are traded. It also eliminates the need to hold the tester in the plug hole with a lot of force. With a flexible drive or other screw-in testers, you can actually take compression tests without using a remote start switch if necessary.

All compression testers consist of three basic parts: the compression gauge (F), which is fitted with a check valve or release button (D); a shaft, hose or tubing (I) and an expansion cone, plug fitting or adapter (G,E) that is fitted or screwed into the spark plug hole for an airtight connection. Some even have a built-in remote start switch (H).

The compression gauge face (F) reads up to 250 or 300 psi, depending upon the manufacturer, and is marked in 5- or 10-psi divisions for accurate readings of any automotive engine you will need to test. The more expensive testers use an adapter tip that automatically seals itself in the plug hole by suction when the engine is turned over and which may be provided with a finger-grip handle for ease in holding.

PREPARATIONS FOR USE

Check the engine oil for grade, amount and quality before making a compression test. Oil that is too thick may mean that an attempt has been made to compensate for piston ring or cylinder bore wear by adding overly thick oil. If the oil is extremely dirty, worn out or if the level is very low, the oil will not do its lubrication job properly and should be replaced.

You should also make certain that the battery is up to at least 75% of full strength; if not, the starter motor may not be able to turn the engine over fast enough to give an accurate compression reading on the gauge.

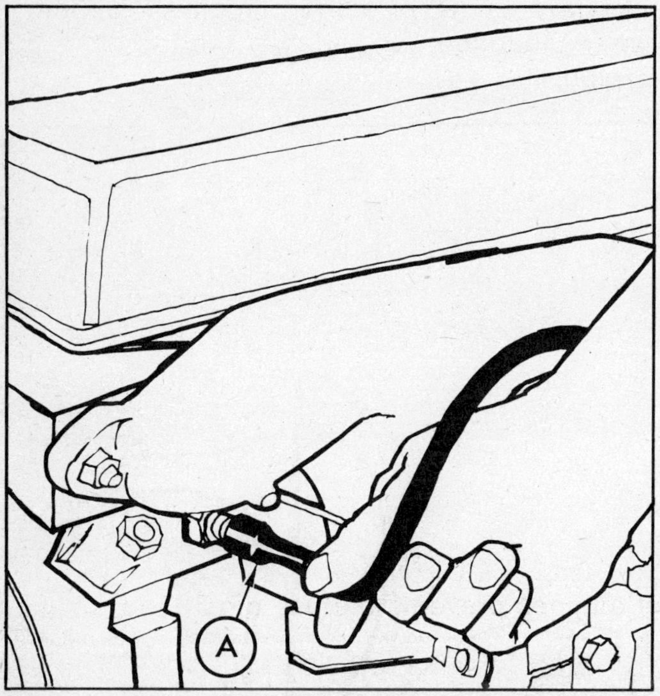

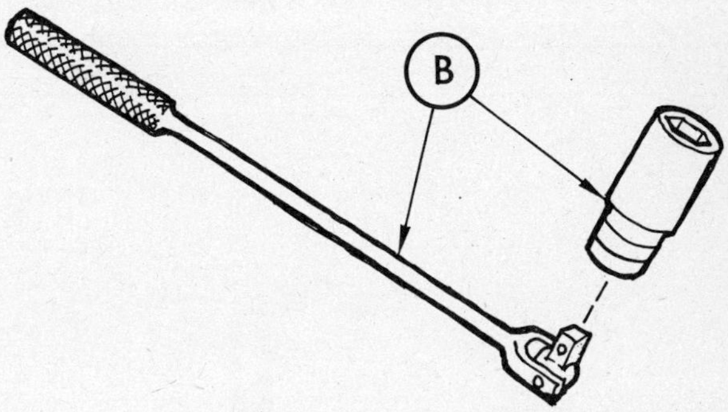

Run the engine until it reaches normal operating temperature, then shut it off. Mark and identify all spark plug wires (A) one at a time before disconnecting them, so that they can be replaced on the correct plug, then move them out of the way. Loosen each spark plug about one full turn with a spark plug socket and breaker bar (B)

CAUTION: BLOW THE SPARK PLUG WELLS CLEAN WITH COMPRESSED AIR. THIS PREVENTS ANY ACCUMULATED DIRT FROM FALLING INTO THE HOLES WHEN THE SPARK PLUGS ARE REMOVED. IF COMPRESSED AIR IS NOT AVAILABLE, USE A HAND TIRE PUMP OR WIPE WITH A CLEAN CLOTH. THIS IS ESPECIALLY IMPORTANT WHEN WORKING ON OLDER ENGINES THAT RECEIVED TUNE-UPS INFREQUENTLY AND WHERE DIRT AND GREASE BUILDUP MAY BE CONSIDERABLE.

Some auto manufacturers recommend loosening the plugs one turn, then starting the engine to blow out any carbon inside the cylinder that may have been dislodged by loosening the plug. Such carbon particles may lodge under a valve during the compression test, because there is no air flow through the combustion chamber to blow them into the exhaust as there is when the engine is running.

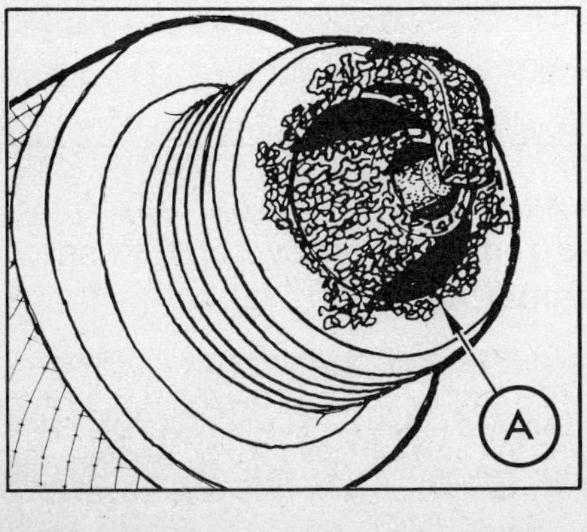

In order to relieve engine compression, this test should be made with all spark plugs removed. You *may* find a small gasket on the end of each plug as it is removed. *Don't lose them;* reuse is necessary for proper sealing. Many plugs do not use gaskets, however, so if they are not present, you are still okay. Check the electrodes and insulator at the tip of each plug for signs of excessive oil fouling (A). While oil-fouled plugs can be caused by a clogged PCV system, a leaking oil bath air cleaner element or just a simple case of too much oil in the crankcase, the odds are excellent that piston ring or valve problems are responsible.

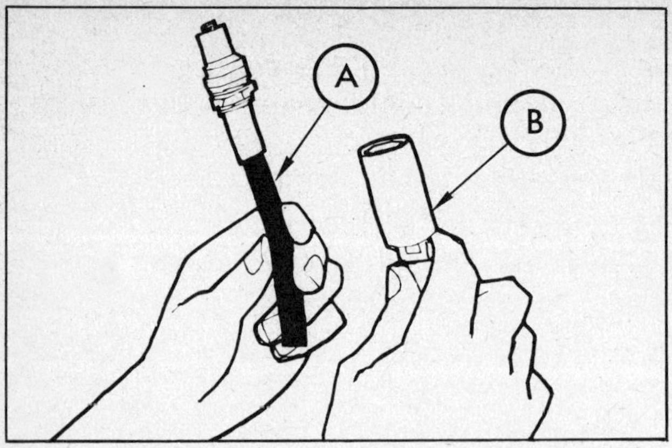

WARNING: THE ENGINE WILL REMAIN CLOSE TO OPERATING TEMPERATURE WHILE YOU ARE WORKING ON IT. BE CAREFUL NOT TO TOUCH THE HOT EXHAUST MANIFOLD, ENGINE HEAD OR SPARK PLUGS.

Spark plugs should be removed with a socket wrench (B) or a short length of rubber hose (A) that you can slip over the top of the plug.

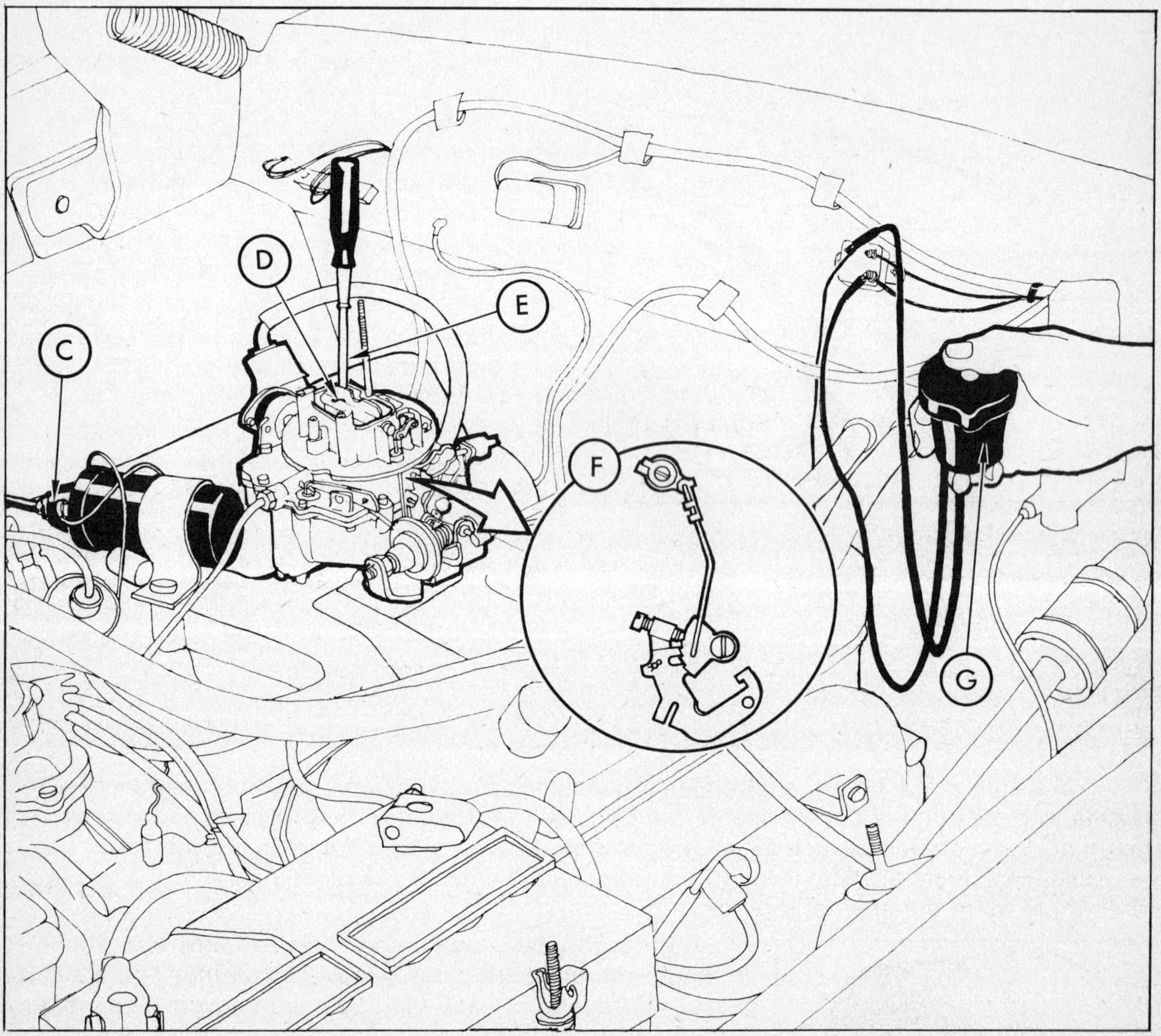

To prevent the automatic choke (D) from closing, insert a screwdriver in the carburetor air horn (E), then set the throttle valve (F) wide open. Connect a remote start switch (G—see that chapter) to let you turn the engine over from the engine compartment.

CAUTION: DISCONNECT AND GROUND THE IGNITION COIL HIGH-TENSION (SECONDARY) WIRE (C) TO PREVENT THE ENGINE FROM STARTING WHEN IT IS CRANKED.

ATTACHING THE COMPRESSION TESTER

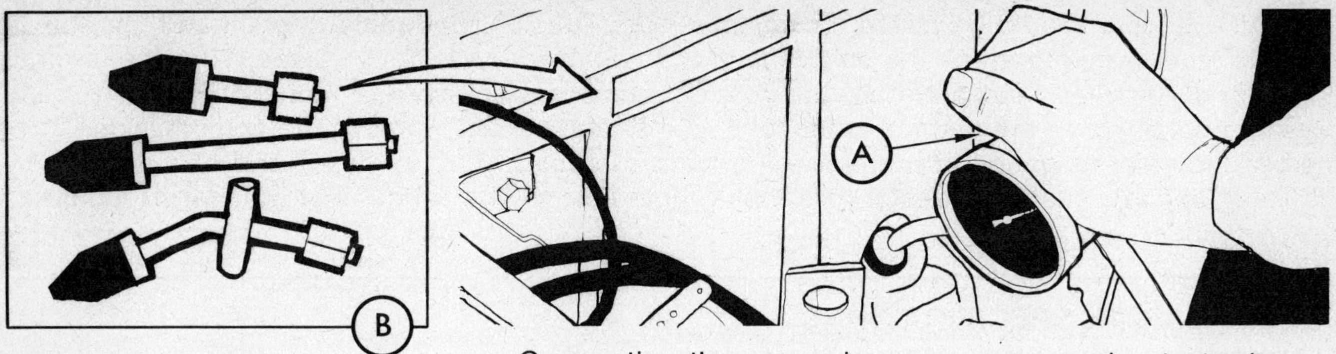

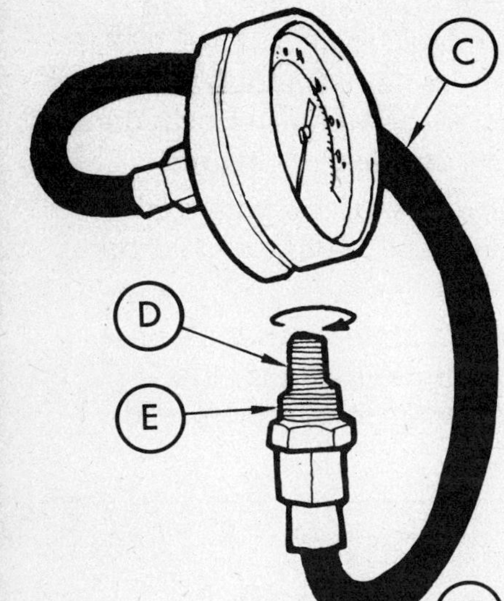

Connecting the expansion cone compression tester is easy. Just insert the cone fitting into the plug hole and hold the tester in place with firm hand pressure (A) while you crank the engine. The cone is tapered, so it fits all spark plug holes with equal ease and requires no adapters for various size plugs.

Some expansion cone testers use interchangeable cone shafts. These are available in different lengths and angles (B) for ease in testing when access to the spark plug well is limited. To change them, unscrew them from the tester gauge.

The flexible drive tester is equipped with a two-stage, screw-in plug fitting for use with 14mm (D) or 18mm (E) spark plug thread diameters. This will fit the majority of current engines. Simply screw it into the plug hole to finger tightness by turning the flexible hose (C).

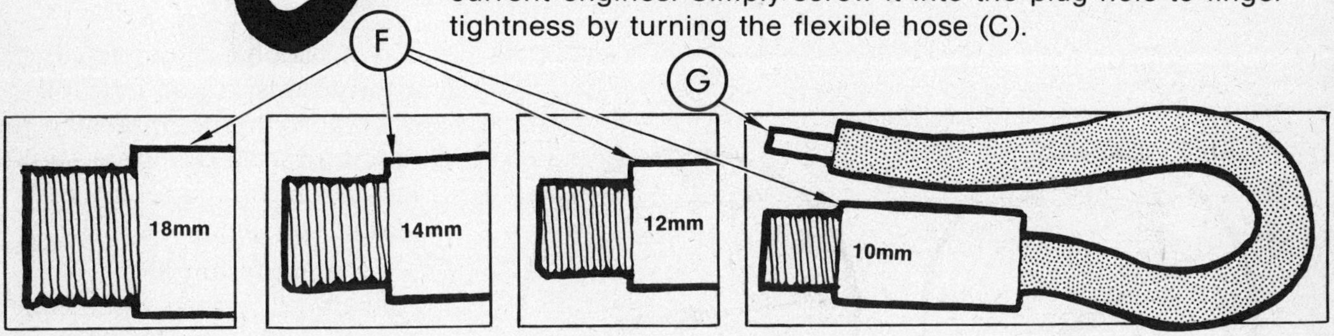

The plug-on adapter tester is the most versatile, because adapters (F) can be obtained for 18mm, 14mm, 12mm and 10mm thread diameters, allowing you to test *any* engine, old or new. Attached to the plug adapter is a length of hose with a fitting (G) that couples with the gauge hose.

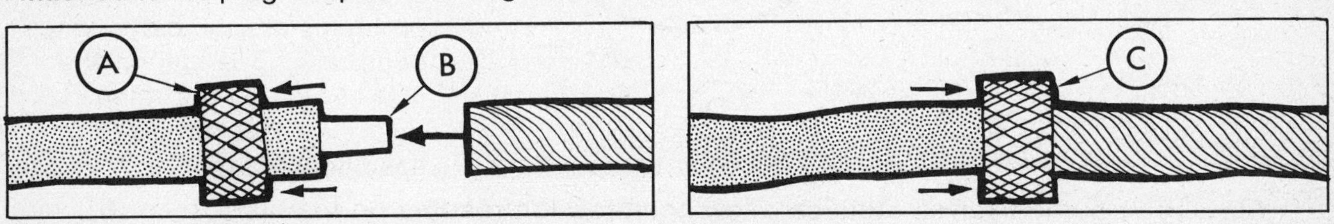

To use, screw the correct adapter size into the spark plug hole finger-tight. Pull back the sleeve (A) on the adapter hose end and push the gauge hose all the way onto the fitting (B). Then slide the sleeve (C) forward to make the connection airtight.

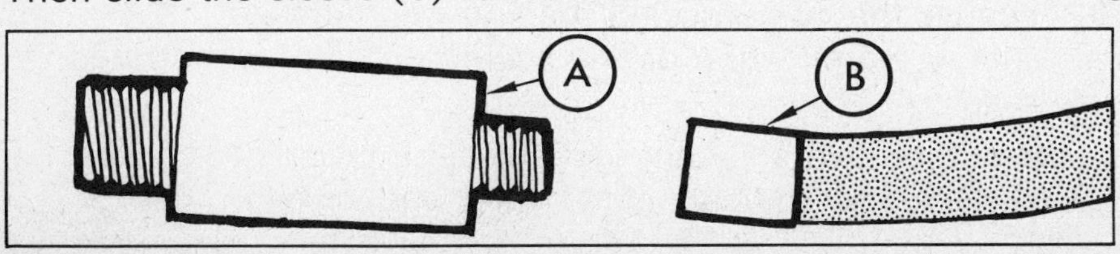

The screw-on adapter (A) is threaded on both ends. One end screws into the plug hole. The other end screws into a metal fitting on the end of the gauge hose (B).

COMPRESSION TESTER

COMPRESSION TESTING

All cylinders should be tested for compression. Turn on the ignition switch and use the remote start switch to crank the engine with the compression tester in place. Turn the engine over for at least four and preferably six to eight compression strokes. You should record the reading obtained on the first stroke as well as the final stroke. Remove the compression tester and clear its gauge of the remaining air by depressing the release valve or button (if it is so equipped). Attach the tester to the plug hole next in line to test that cylinder. Repeat for each cylinder.

COMPRESSION SPECIFICATIONS

Each manufacturer publishes compression specifications for his engines. These are usually found only in a factory shop manual or a book like Petersen's HOW TO TUNE YOUR CAR, although foreign car manufacturers often include them in the owner's handbook. As the value of a compression test is to determine variations in compression between the cylinders, these specifications are commonly expressed as a range instead of an absolute figure.

In many cases, the specifications require the lowest-reading cylinder to be within a percentage (usually 75%) of the highest reading. Example: If the highest reading is 120 psi, the lowest should be at least 90 psi.

Alternatively, the specifications may give a minimum figure with a stated variance or range between the cylinders. If the minimum is 110 psi and the variance is 40 psi, the compression in all cylinders is correct as long as the difference between the highest and lowest is 40 psi or less and none are below 110 psi.

TEST INDICATIONS

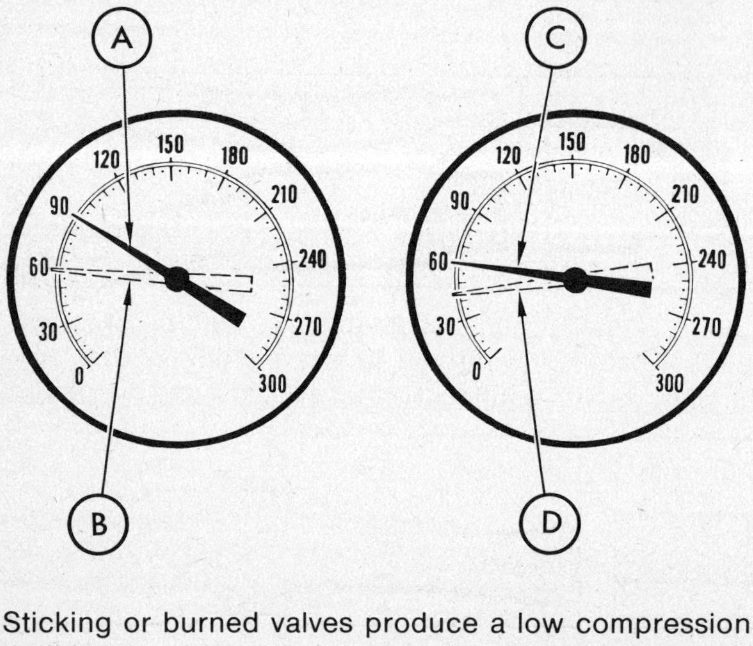

If pistons, rings, valves and gaskets are all in good condition, the compression test reading for all cylinders should fall within the manufacturer's stated specifications.

Compression is considered normal when the gauge reading builds up uniformly with each compression stroke to the specified figure. Piston ring problems are indicated when compression is low on the first stroke (B) and then builds up, but not up to specifications (A).

Sticking or burned valves produce a low compression reading on the first stroke (D), with little buildup on successive strokes (C). To accurately differentiate between ring and valve problems, pour a teaspoon of SAE 30 oil (engine off) into the spark plug hole of the low-reading cylinder to seal the rings and then retest. Little or no increase in the reading indicates that the valve is at fault, because the compression reading will increase considerably (10 psi or more) after adding the oil when poorly seated or worn piston rings are responsible for the original low reading.

If identical low readings are obtained on two adjacent (side-by-side) cylinders, gasket leakage between the cylinders is likely. Look for indications of water and/or oil in the cylinders.

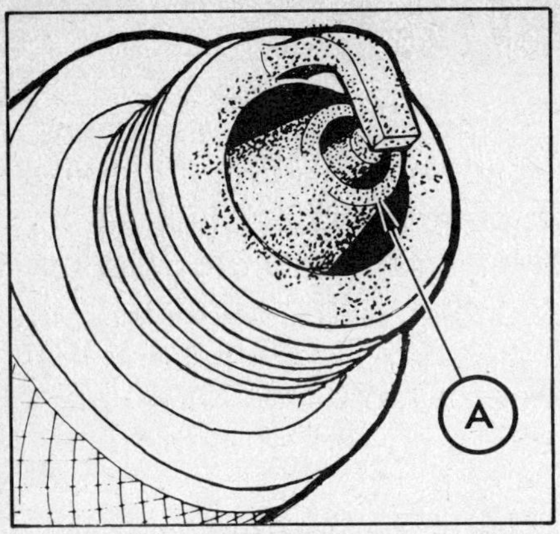

The accumulation of carbon deposits between the top of the piston and the cylinder head causes compression readings considerably higher than what the manufacturer specifies. One tip-off to this problem is a "pinging" sound when the engine accelerates under load (goes uphill or passes another car). Another is the presence of sooty, dry carbon covering the entire tip area of one or more spark plugs (A). It may be possible to correct this fault by using a plug with a higher heat range, but an excessively rich fuel mixture or a choke that is not opening fully might also be the cause. Check both possibilities before changing to a "hotter" spark plug.

The use of high-compression heads on a performance option engine will also cause compression readings higher than specified.

COMPRESSION TESTER MAINTENANCE

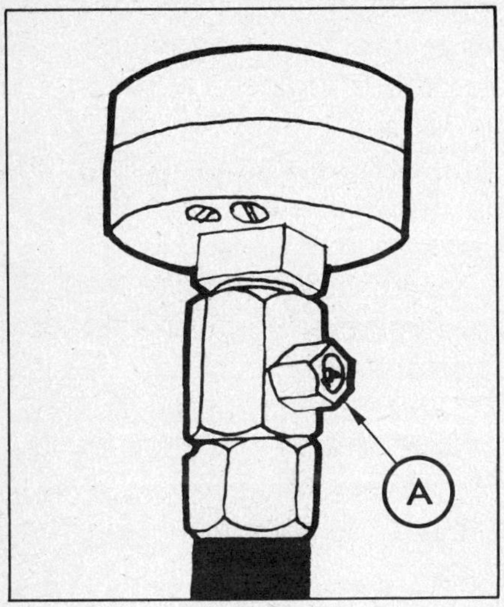

The major problem encountered with this simple test instrument is that of a leaking relief or check valve core (A). If the gauge does not seem to be working properly, remove the valve with a valve stem tool and look for dirt or contamination that may be interfering with its operation.

Cleaning the valve with solvent and compressed air and replacing it usually restores the compression gauge to proper operation, but if this does not stop the leak, it will have to be replaced. While any ordinary tire valve core will fit nicely in its place, avoid such a substitute. A tire valve core spring is too stiff and will keep gauge readings from 20 to 35 psi below what they should be. Replace the faulty core with the correct one as specified by the gauge manufacturer—this can usually be obtained from an auto supply store that handles the manufacturer's line of test equipment.

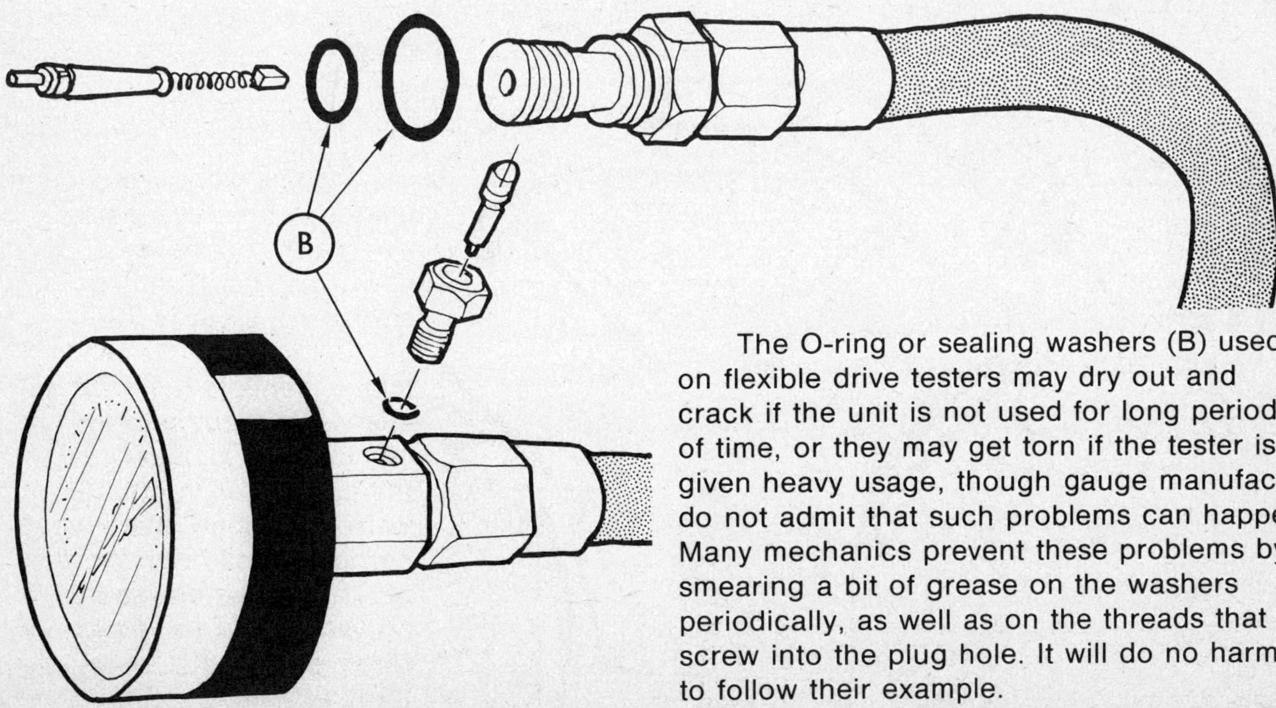

The O-ring or sealing washers (B) used on flexible drive testers may dry out and crack if the unit is not used for long periods of time, or they may get torn if the tester is given heavy usage, though gauge manufacturers do not admit that such problems can happen. Many mechanics prevent these problems by smearing a bit of grease on the washers periodically, as well as on the threads that screw into the plug hole. It will do no harm to follow their example.

VACUUM TESTER GAUGE

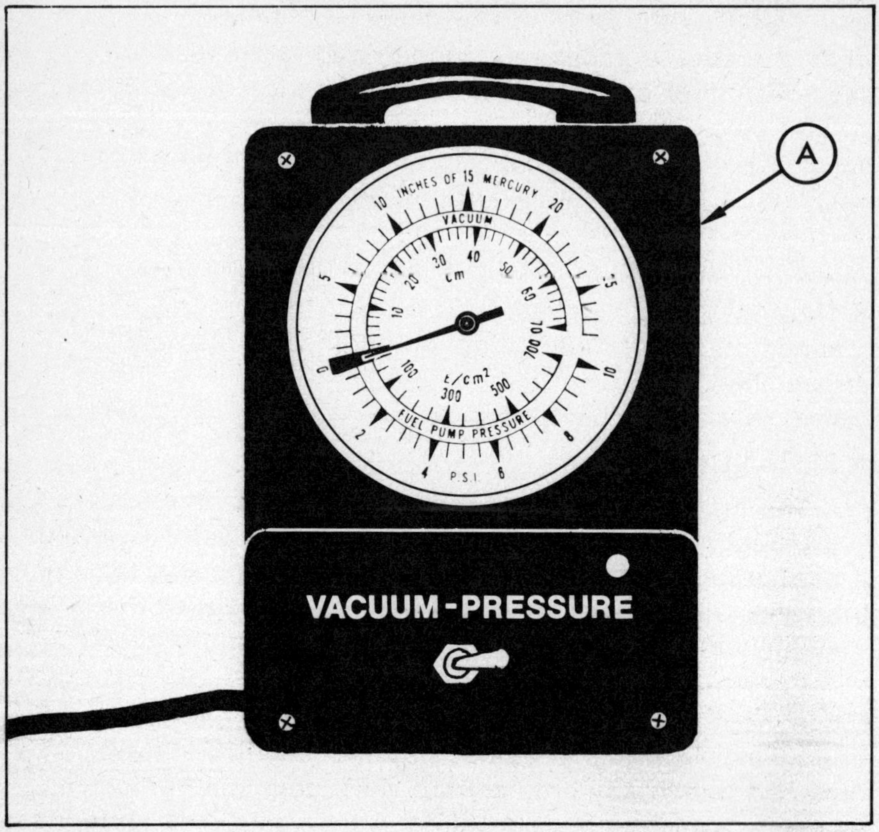

As mentioned elsewhere in this book, a properly tuned engine is one in which all of the components are working together with maximum efficiency. If you suspect that one or more are not working efficiently, for whatever reason, the quickest and most accurate way to pinpoint the malfunction is by the use of a vacuum gauge (A).

Few people realize just what engine vacuum is or its importance in diagnosing engine troubles. Yet vacuum is always present in any operating automotive engine and any change from the amount that should be present has a definite and predictable effect on engine performance.

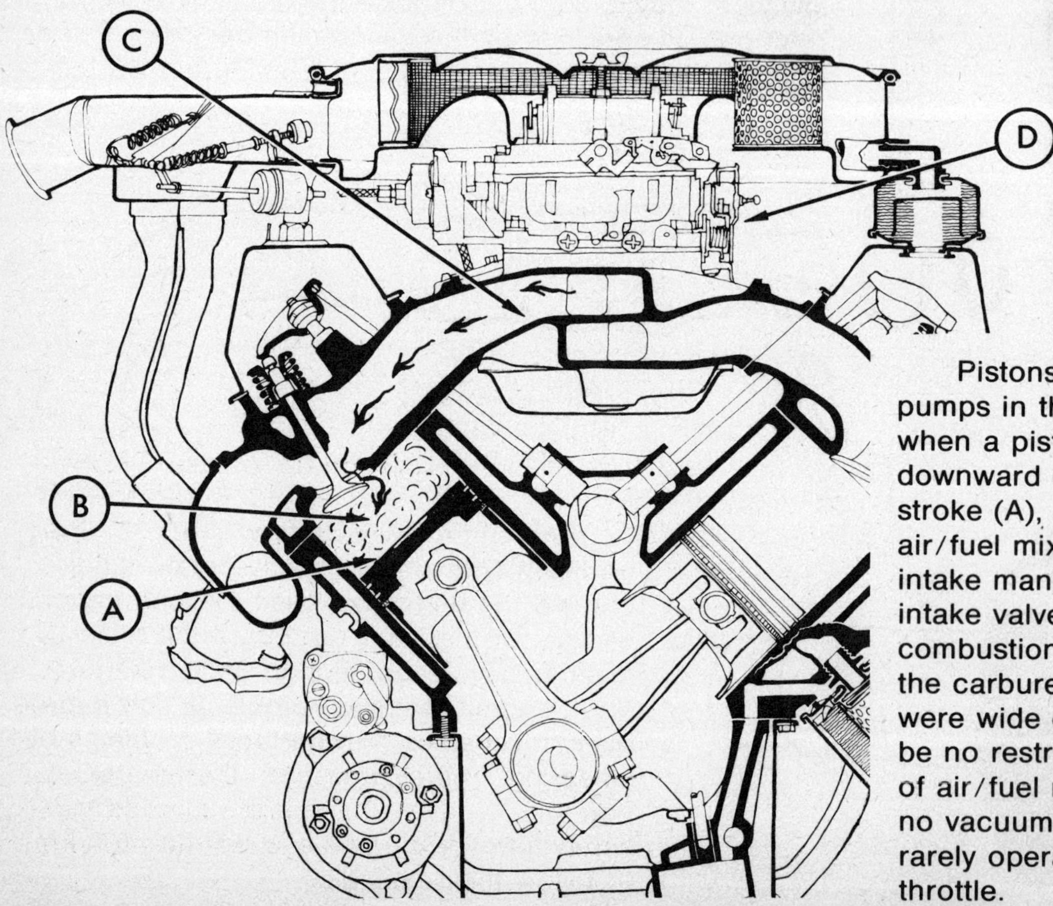

Pistons work like suction pumps in their cylinders; when a piston moves downward on its intake stroke (A), it sucks the air/fuel mixture into the intake manifold (C), past the intake valve and into the combustion chamber (B). If the carburetor throttle (D) were wide open, there would be no restriction on the flow of air/fuel mixture and thus no vacuum, but engines are rarely operated at wide-open throttle.

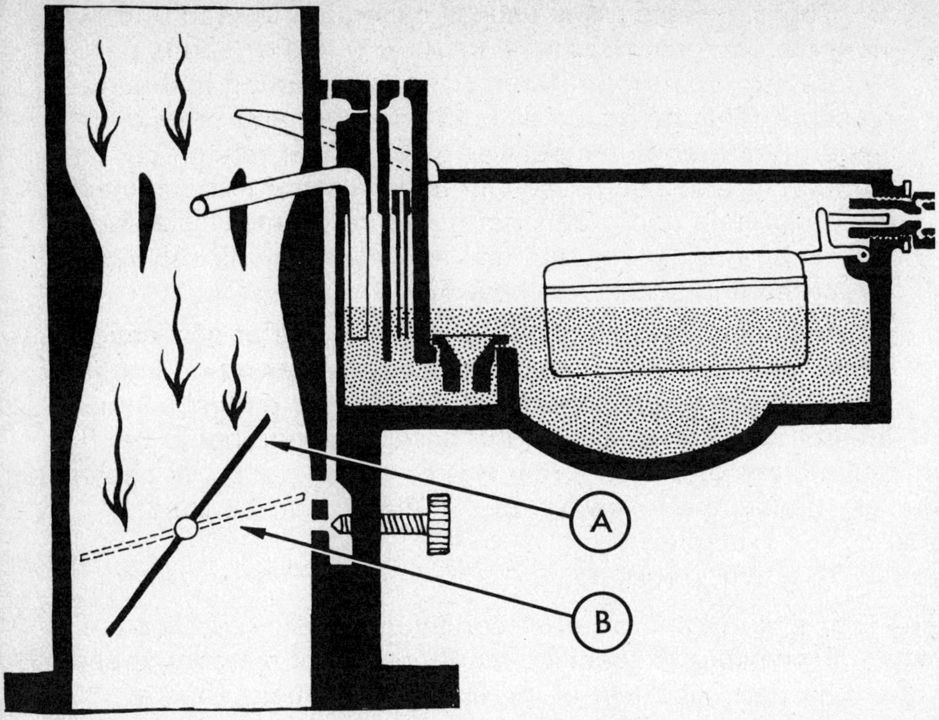

Instead, the throttle opening varies from almost closed under idle or deceleration conditions (B) to approximately half open at cruising speeds (A). As this partially closed throttle restricts the sucking action of the pistons, a vacuum results in the intake manifold and combustion chamber. The total amount of vacuum thus created is affected by how well such engine parts as piston rings, valves, ignition, carburetion and exhaust are doing their job. As each of these has a known and predictable effect on engine vacuum, any variations from the basic specifications furnished by the manufacturer can be used to diagnose engine condition and performance.

ENGINE DESIGN AND VACUUM

Engine design factors such as the number of cylinders, the length of stroke, displacement, valve overlap and lift and compression ratio all contribute to the amount of vacuum produced. As these factors vary from one engine design to another, there is also a variation in the normal vacuum produced by these different types of engines. Generally speaking, a V-8 engine will give a lower vacuum reading than six-cylinder engines. Because of this variation in what the manufacturer considers to be a normal vacuum reading, it is important that you refer to the factory shop manual of the engine to be tested for the basic specifications.

THE VACUUM GAUGE

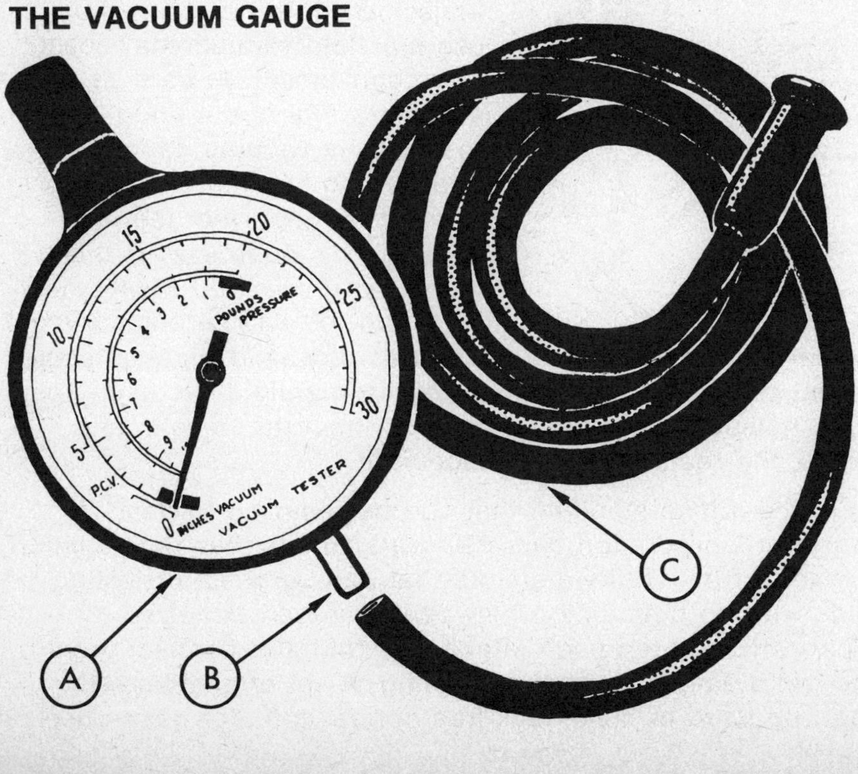

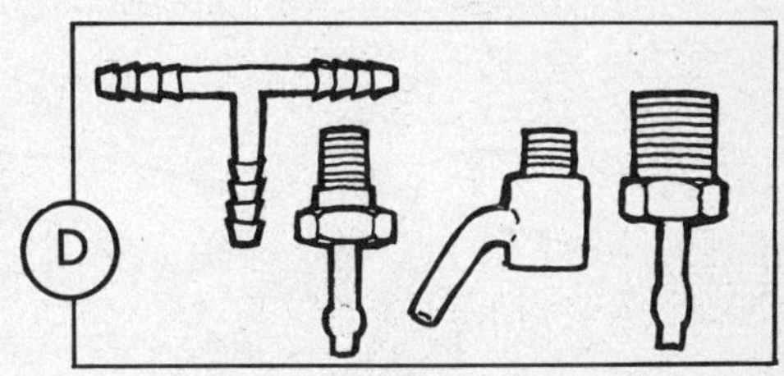

A vacuum gauge is used to measure the difference between atmospheric pressure outside the engine and that present in the intake manifold. This simple test instrument consists of a gauge with a dial and pointer (A), a hose fitting on the gauge (B) and a hose (C) used to connect the unit to the engine. A variety of adapters (D) are usually included with the gauge to help you connect it to the engine.

VACUUM TESTER GAUGE

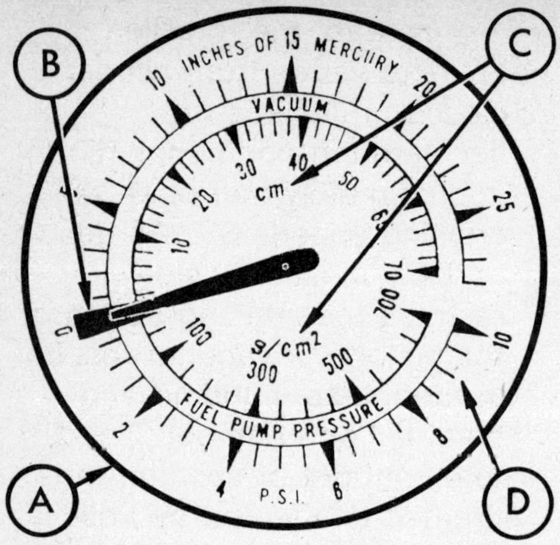

The dial scale (A) is usually calibrated to read 0 to 30 inches of vacuum (B) and 0 to 10 or 0 to 15 pounds per square inch (psi) of pressure (D) for measuring fuel pump pressure. This range exceeds the vacuum and pressure limits of modern automobile engines. Since foreign car manufacturers specify vacuum and pressure measurements in centimeters and grams per cubic centimeter instead of inches and psi, the gauge dial may also include separate, direct-reading scales for these (C).

Inclusion of such scales makes the use of conversion tables and formulas unnecessary when working on imported cars, but once again you must refer to the engine's shop manual to determine what the basic reading should be. If your gauge dial is not marked in centimeters of mercury (cm Hg) or grams per cubic centimeter (g/cm^2) and no conversion tables are handy, the following rule of thumb can be used:

- Four inches of mercury equal 10 cm Hg (vacuum).
- One pound per square inch equals 70 g/cm^2 (pressure).

Atmospheric pressure changes with altitude and weather conditions. Basic readings are usually given for sea level. Atmospheric pressure is specified as 30 inches of mercury at sea level and decreases at a rate of one inch per 1,000 feet of increase in elevation. Thus at 5,000 feet above sea level, atmospheric pressure is 25 inches of mercury. While the exact vacuum reading in barometric inches will provide a general or overall indication of the engine's condition and performance, the *amount* and *speed* of pointer movement on the gauge scale is far more useful in troubleshooting or testing the engine.

CONNECTING THE VACUUM GAUGE

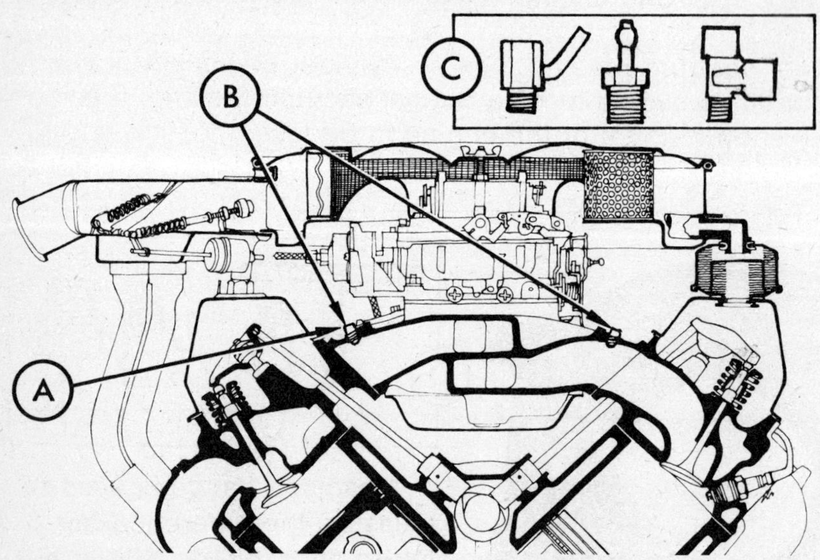

To assure a true vacuum reading, vacuum gauges must always be connected to a source of full manifold vacuum. Connecting the gauge into the power brake system, the transmission modifier or those emission control systems that use manifold vacuum may seem more convenient on cars so equipped, but it is a good idea to hook the vacuum gauge directly into the intake manifold to avoid an incorrect tap (connection). It's quite easy to make a bad tap unless you really know what's in the engine compartment and exactly how it works. On older cars that use vacuum-operated windshield wipers instead of the electrically controlled ones, you can tap into the wiper line for manifold vacuum. Most such systems use a vacuum booster, however, so unless you disconnect the booster and connect the gauge to the engine side, the reading will be inaccurate.

Many engines have a plug (A) in the intake manifold that can be removed for test purposes. Late model (1973 on) Chrysler Corporation cars using an exhaust gas recirculation (EGR) system also have a fitting on top of the intake manifold that can be used for testing. Engines having two intake manifolds with no connection between them can be tested individually (B). Alternatively, you can drill and tap each intake manifold and connect them with an ordinary Y-shaped fitting to get a single, combined reading. A set of adapters (C) is included with your gauge and can be used to make leak-free connections for gauge use.

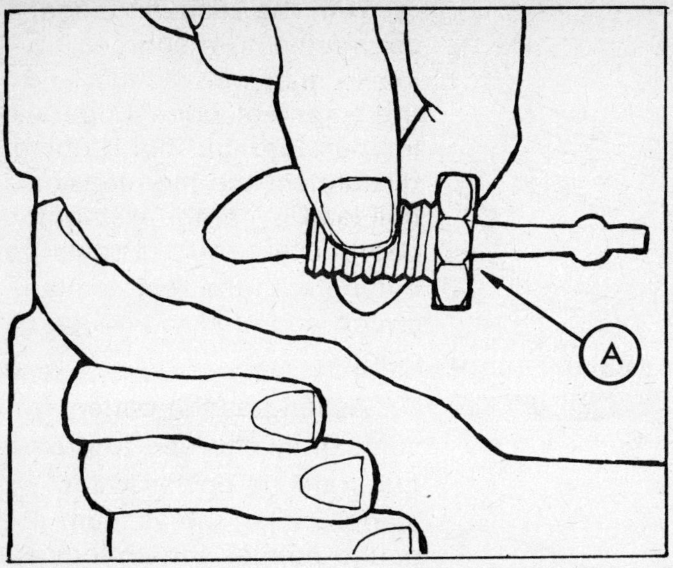

Attach the vacuum gauge by removing the plug from the intake manifold and replacing it with the correct adapter (A).

CAUTION: BE CAREFUL WHEN WORKING ON AN ENGINE THAT IS RUNNING OR HAS RECENTLY BEEN RUNNING. METAL PARTS WILL BE HOT.

As some manifolds use plugs that do not actually provide access into the manifold, make sure that suction is present by placing your fingertip over the hole while the engine is running. If the manifold has no separate plug for this purpose, there are usually several other direct vacuum lines that can be tapped if necessary. This is especially true on post-1968 cars equipped with one or more of the emission control systems such as Transmission Controlled Spark (TCS) or Exhaust Gas Recirculation (EGR). If you find it necessary to use this method of tapping into a vacuum line, make certain that you are connected to the vacuum source.

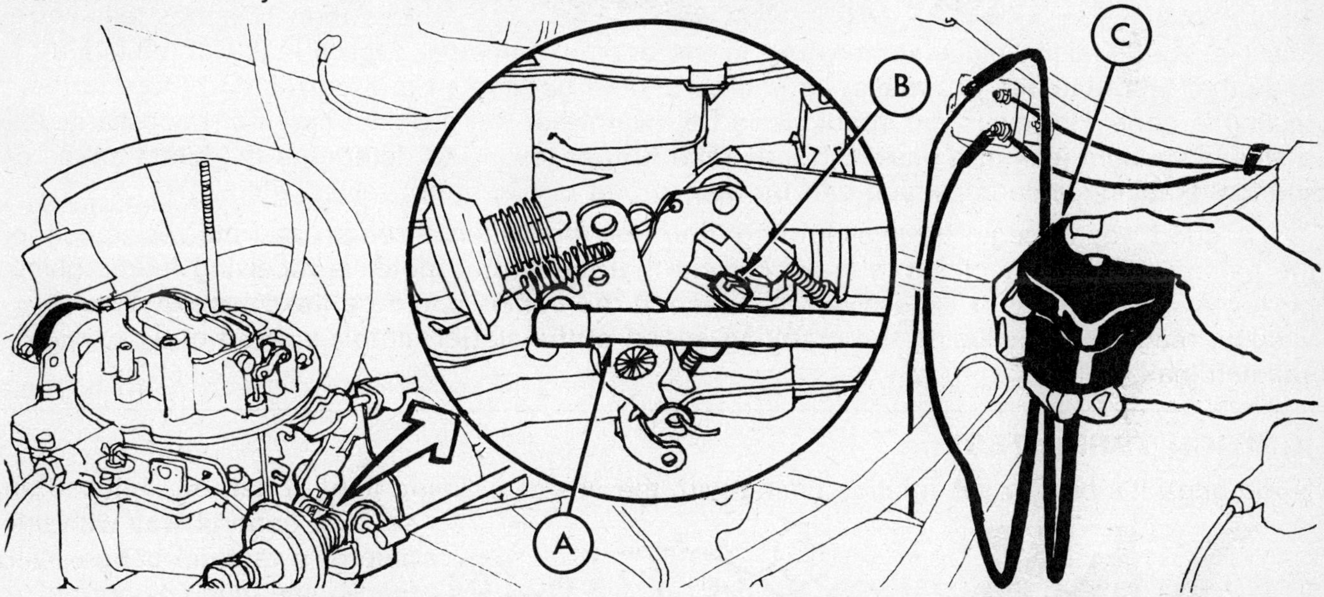

Fit one end of the vacuum gauge hose over the adapter and twist downward to seat the hose completely in position. Connect the other end of the hose to the vacuum gauge. Connect a remote start switch (C) to let you crank the engine from the engine compartment. Start the engine. When the engine has warmed up to normal operating temperature, check the idle speed with a tachometer (see dwell tachometer chapter) to make certain that it is within factory specifications. If not, adjust the throttle linkage (A) or set the carburetor idle screw (B) to get a smooth idle at the correct rpm (revolutions per minute). When adjusting the idle speed, it's well to check the manufacturer's specifications, as there are various ways in which idle speed can be set (lean-drop method, ¼-turn rich method, etc.) and it is important that you use the specified method on emission-equipped cars.

If the engine is operating correctly at sea level, the vacuum gauge pointer should remain steady at the vacuum reading specified for the engine. This can range between 12 and 22 inches on the scale. Subtract one inch per 1,000 feet of elevation from the specified sea level figure to obtain a correct reading at other elevations. Late-model engines using emission control systems will usually idle with much lower vacuum than those with no smog controls. Stop the engine. You can now begin the test sequence.

VACUUM TESTER GAUGE

CRANKING VACUUM TEST

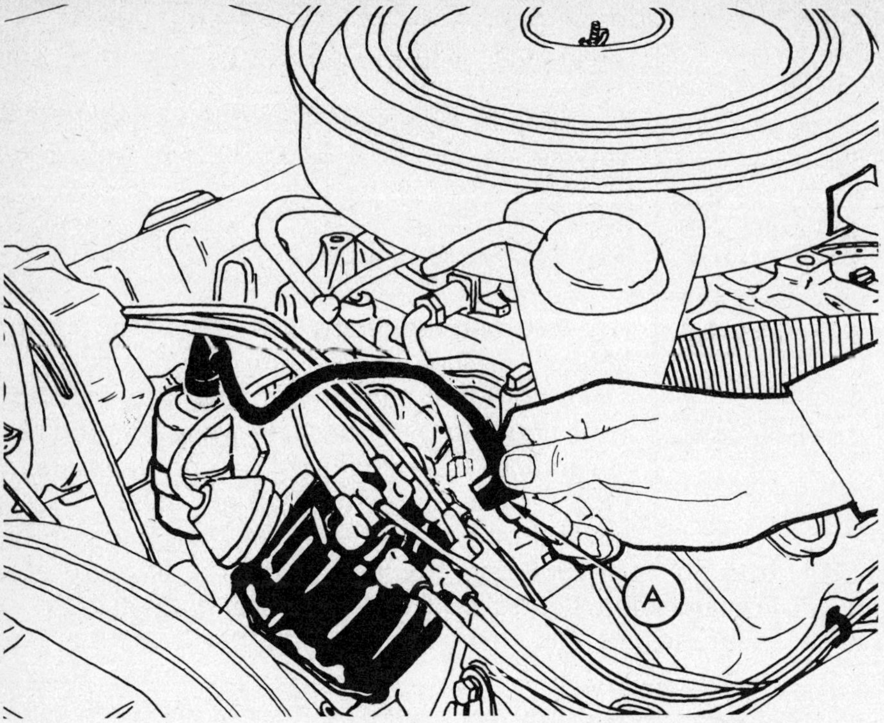

With the vacuum gauge connected to a source of intake manifold vacuum and the engine at normal operating temperature but not running, disconnect the high-tension coil lead (A) at the distributor cap and ground it. Crank the engine with the remote start switch and read the gauge pointer.

As long as the battery is sufficiently charged to provide full cranking power to the starter motor, the vacuum gauge pointer should indicate a steady vacuum reading of approximately 5 inches or more, depending upon the engine. Any reading of less than 5 inches tells you that leakage is taking place somewhere in the engine. Because such a leak will influence any other test readings you make, its source should be located and corrected before further testing is done. However, you should also be aware that the problem may not necessarily be a simple vacuum leak at a gasket. It may take further testing to determine the exact cause of the low cranking vacuum, which can be due to bad piston rings, valves, etc.

A good way to begin troubleshooting is to test the battery strength and the resistance in the battery cable connections to make certain that the starter motor is receiving full cranking power. A weak battery or excessive resistance in the cables could be the cause of a low vacuum reading (by reducing the cranking speed of the starter motor) instead of a vacuum leak.

IGNITION TIMING TEST

Although it's best to set ignition timing with the use of a timing light, it can be checked and set in a general way with the help of a vacuum gauge. With the engine idling at normal operating temperature, the gauge pointer should read at the manufacturer's vacuum specification (C). If the pointer holds a steady reading 2 or 3 inches lower than specified (B), ignition timing is late. Timing that is too early will result in a steady reading 2 or 3 inches higher than normal. (D).

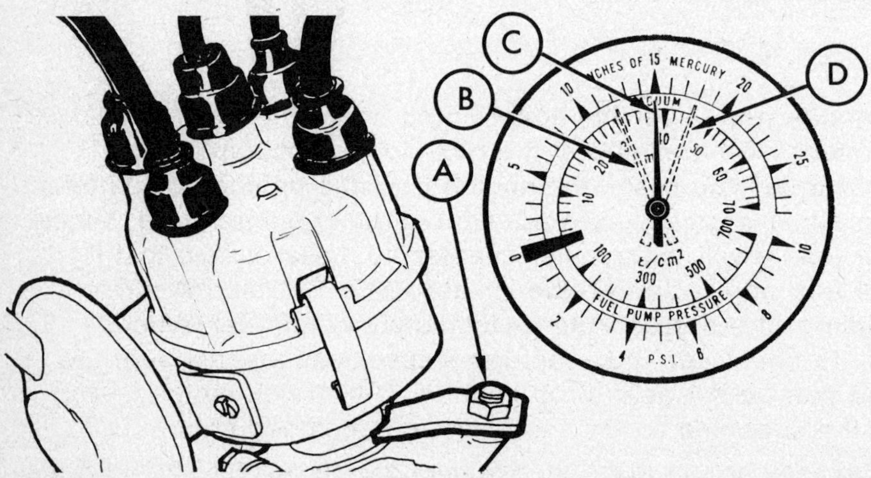

Ignition timing can be set to approximate correctness by loosening the distributor hold-down bolt (A) and turning the distributor slowly until the highest reading on the gauge is noted. At this point, turn the distributor back slowly until the reading drops about one inch. Then tighten the hold-down bolt. This method of setting ignition timing should only be used when a timing light is not available.

VACUUM LEAKAGE TEST

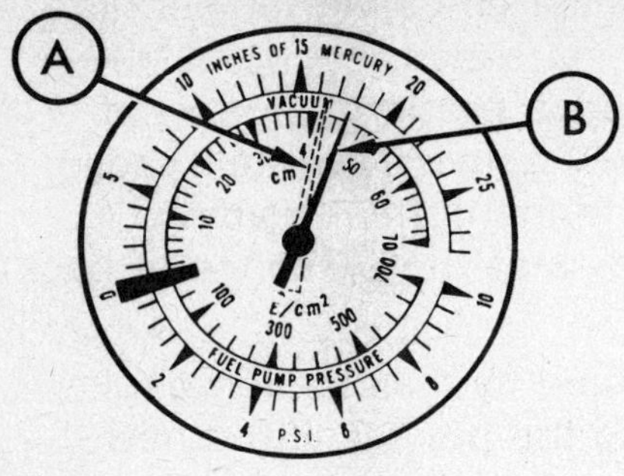

The vacuum gauge can be used to locate leaks in the valve system, cylinder head gasket or intake system. With the engine idling at normal operating temperature, the gauge pointer should hold a steady, normal reading (B). If the pointer drops an inch or more (A), returns to its original position and then drops again, one or more engine valves are closing but not seating properly, causing a vacuum leak. The vacuum drop takes the form of periodic movement of the gauge pointer. Each time the leaking valve fails to seat correctly, the pointer fluctuates. Locate the faulty valve(s) by a compression gauge test.

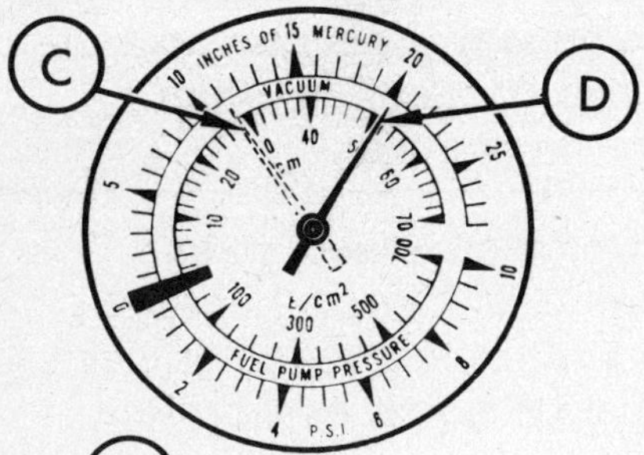

A leaking cylinder head gasket will cause the vacuum gauge pointer to drop *sharply* from the normal steady reading (D). If the pointer drops several inches or more (C), returns to its original position and then drops again, the head gasket is leaking. The vacuum drop will be considerably greater if the leak is located between two cylinders. A compression gauge test will determine the location of the leak, and if nearly identical low readings are given by two cylinders, the leak is between them.

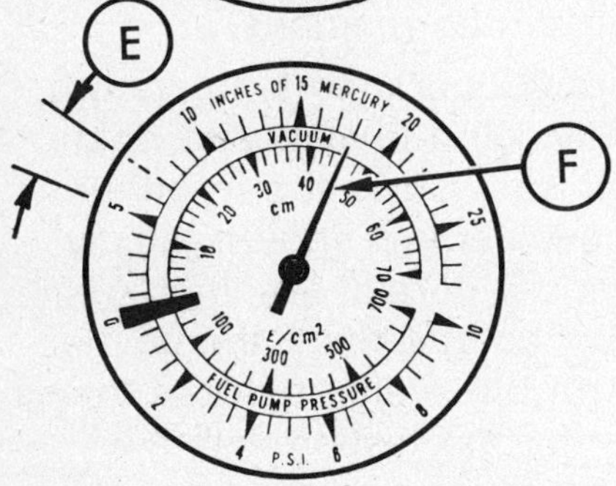

When the vacuum gauge pointer fluctuates constantly at a reading several inches (usually 3 to 8) below normal (F), it indicates intake system leakage. The leak can be located in manifold gaskets, the vacuum starting switch, vacuum transmission control or almost anyplace in the intake system. If the pointer bottoms out to a reading of 5 to 7 inches (E), check the carburetor throttle valve for malfunctioning that may be preventing its closing. Look for a burned heat riser tube, too.

VALVE ACTION TEST

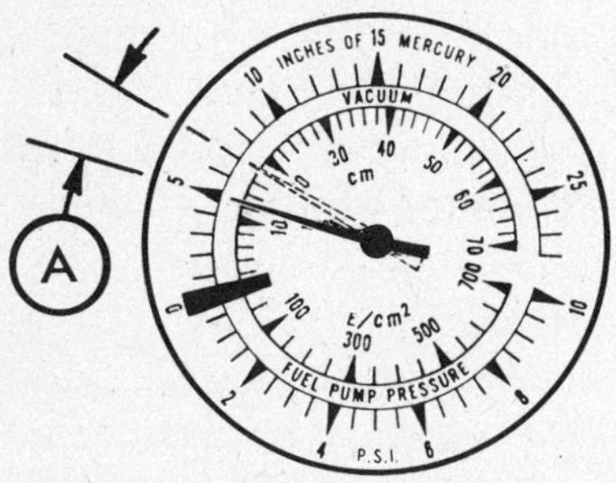

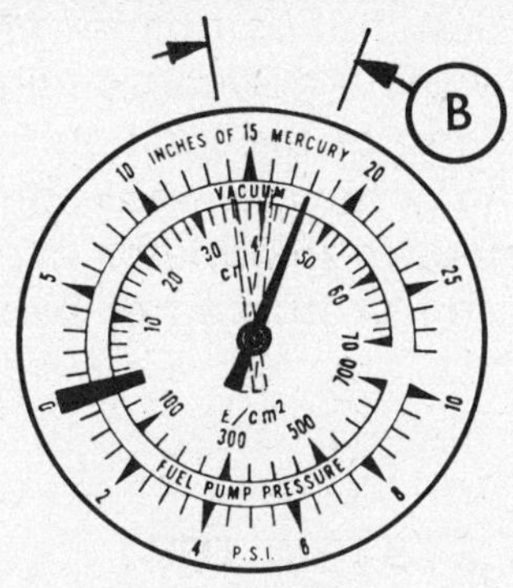

Incorrect valve timing, sticking valves or faulty valve springs are other engine conditions that can be pinpointed using a vacuum gauge. If the gauge pointer fluctuates between 5 and 7 inches on the scale (A) with the engine idling, the cause is most likely excessively late valve timing. As the vacuum reading will usually be a normal one when valve timing is only slightly late, this test works most accurately for bad cases of late valve timing.

When the pointer drops irregularly from a normal reading, one or more valves are sticking. Pointer movement on the scale (B) will be similar to that of a leaking valve, but improper valve seating shows up consistently when the engine is running, and a valve will stick only intermittently. Therefore pointer movement will be irregular instead of consistent. A sticking valve problem can be double-checked by applying penetrating oil to the valves. Penetrating oil will temporarily clear up a sticking valve but will have no effect on one that leaks due to improper seating.

Check for defective valve springs by increasing engine idle to about 2,000 rpm while watching the pointer. If it fluctuates quickly, with the fluctuations becoming faster as engine speed reaches 2,000 rpm, the valve springs are probably weak. A rapid but irregular fluctuation of the pointer will occur every time a valve with a broken spring tries to close.

VACUUM TESTER GAUGE

EXHAUST BACK PRESSURE TEST

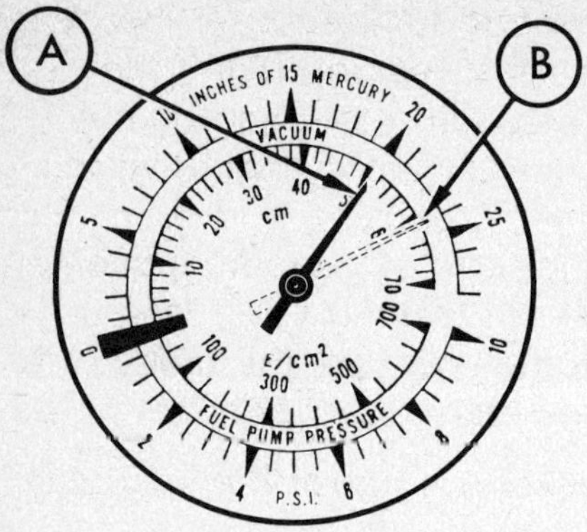

With the engine idling at normal operating temperature, *slowly* open the throttle until engine speed reaches 2,000 rpm, then close it *quickly*. If no back pressure is present, the pointer should jump rapidly (B) above a normal reading and smoothly return just as fast to the normal reading (A). A very slow or uneven return indicates the presence of back pressure.

This test can also be performed by maintaining engine speed at 2,000 rpm and watching the pointer. If the new reading gradually decreases instead of holding steady, the exhaust system is restricted in some way. This can take the form of a defective heat riser valve or shaft or other obstruction in the exhaust pipe or muffler.

PISTON RING TEST

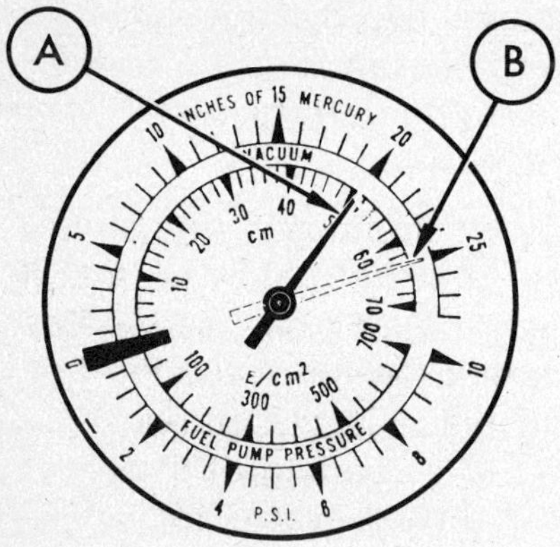

A variation of the procedure for the exhaust back pressure test will indicate piston ring efficiency, but the reading will be accurate only if all other test readings have been normal. Also, engine oil must be good, as worn-out or otherwise diluted oil will influence the readings.

With the engine idling at normal operating temperature, move the throttle to a fully open position *quickly* and hold until engine speed reaches 2,000 rpm. Close the throttle *quickly* and watch the pointer movement—it should rapidly move 5 inches or more (B) above the normal reading (A) at idle and then return to normal just as quickly. If the pointer movement is less than 5 inches on the scale, it shows a loss of compression and the need for further testing with a compression gauge.

IGNITION SYSTEM TEST

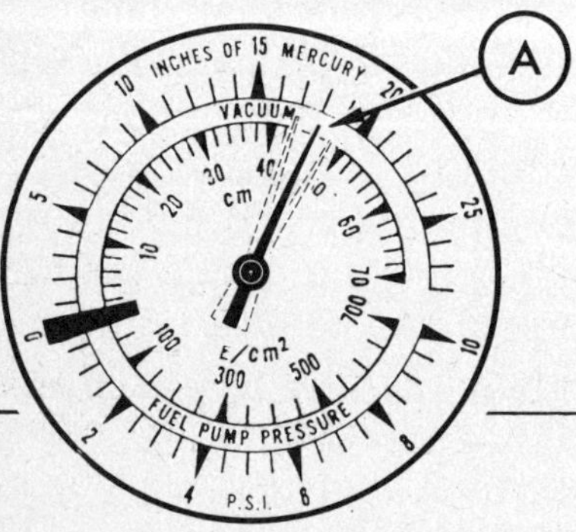

Continuous pointer movement an inch or so above or below (A) a normal reading indicates the need for further checking of the ignition system with electrical test equipment. You may have a weak ignition coil, high-tension cable leaks, a defective distributor cap or spark plug, incorrect spark plug gapping or burned or badly adjusted breaker points.

PCV SYSTEM TEST

With the engine idling at normal operating temperature, remove the air cleaner hose from the rocker arm cover (B) or closed oil filler cap (A). Attach the vacuum gauge in its place, using an adapter if necessary. The gauge pointer should move between 3 and 5 inches in a few seconds. If it does not, the PCV valve is clogged or there are one or more air leaks into the crankcase.

WINDSHIELD WIPER TEST

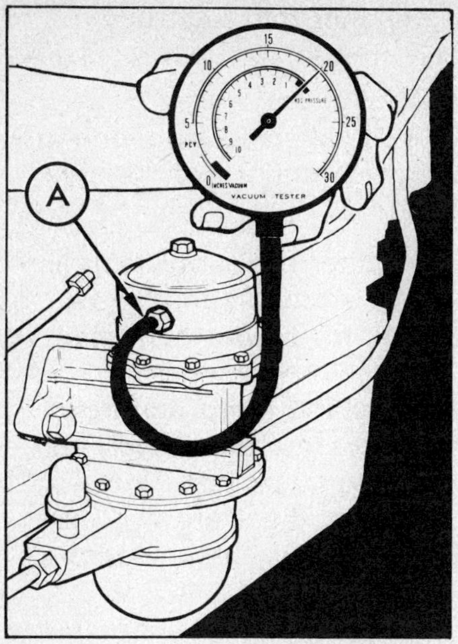

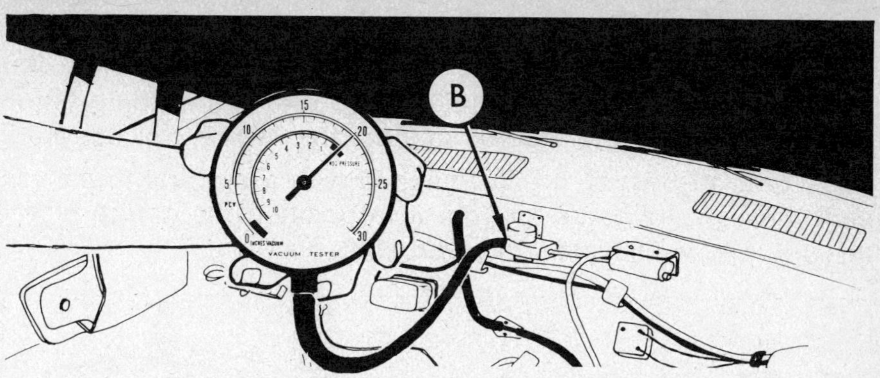

The efficiency of vacuum-operated windshield wipers can also be determined with the vacuum gauge. If the car is not equipped with a vacuum booster, connect the gauge at the wiper connection (B); for those with a vacuum booster, connect the gauge to the booster's vacuum side (A). The gauge reading should remain constant as you alternately idle and accelerate the engine. If it does not, there is a leak in the line or a defective booster.

CARBURETOR ADJUSTMENT—PRE-EMISSION CARS

Provided that the engine is in proper operating condition, it is possible to use the vacuum gauge to adjust the idle mixture on cars not equipped with emission control systems. With the engine idling at normal operating temperature and the vacuum gauge connected, a slow floating motion of the gauge pointer above or below a normal reading indicates that carburetor adjustment is required.

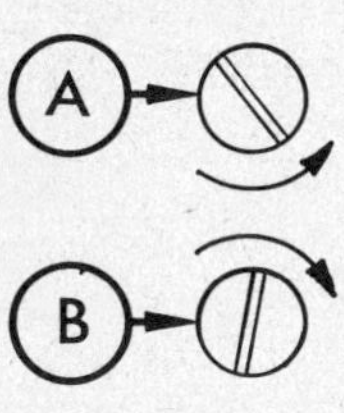

Turn the idle speed screw on the carburetor clockwise (B) or counter-clockwise (A—lean or rich) until the gauge pointer reaches the highest possible steady reading. If the carburetor has more than one idle speed screw, each must be adjusted. Turn both an equal amount in the same direction and watch the action of the pointer; you may have to readjust the idle speed screw when the idle mixture adjustment is completed.

EMISSION CONTROL TESTING

Any emission control systems that use manifold vacuum can be checked for proper operation with the vacuum gauge. Before you tap the line for a reading, though, you must know exactly how the system works. For example, the Transmission-Controlled Spark (TCS) system used on GM cars has a switch in the transmission connected to a solenoid that controls the vacuum applied to the distributor diaphragm. This results in the elimination of vacuum spark advance in all but high gears (third and fourth gear in four-speed manual transmissions). Whether or not the distributor vacuum advance actually shuts off or comes on when it is supposed to can be determined with a vacuum gauge. There are many variations of the system whereby vacuum is shut off only in certain gears or at certain speeds, however, so you must know exactly what you are looking for or you will be unable to interpret the gauge reading correctly when you find vacuum or the lack of it.

To test the system described above, jack the rear wheels of the car off the ground and support them with jack stands (A).

CAUTION: BEFORE PROCEEDING, BE SURE THAT THE JACK STANDS ARE PROPERLY POSITIONED AND WILL HOLD THE CAR SAFELY.

VACUUM TESTER GAUGE

With the engine idling at normal operating temperature and the gear selector lever in drive, open the throttle wide open and watch the gauge pointer. When the shift into high gear occurs, you should get a reading. If you have a reading before the shift takes place or if you get no reading at the time of shifting, the TCS is not operating correctly.

If you are testing a car equipped with a manual transmission, you will need someone to shift through the gears for you or to position the gauge high enough so that it can be seen through the windshield while you shift.

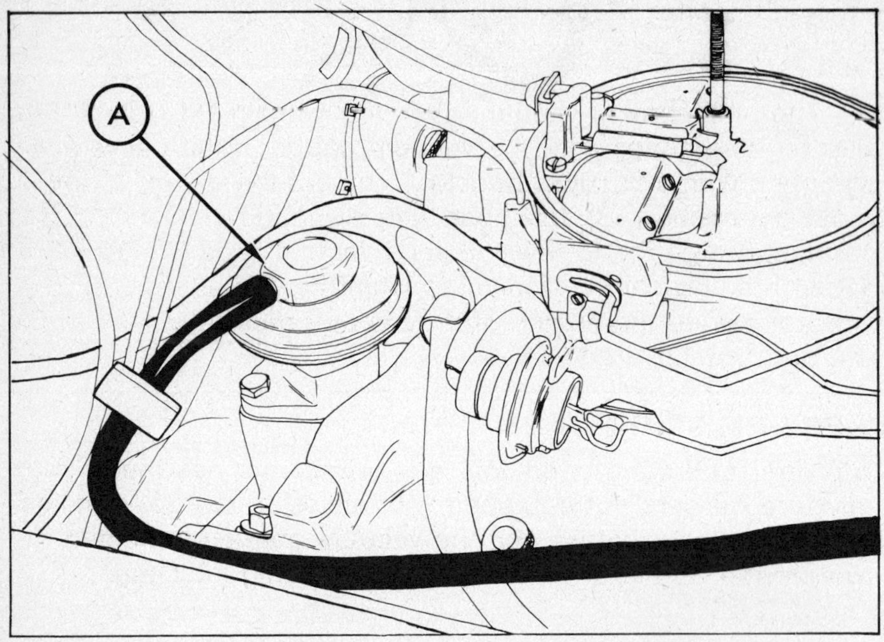

Other emission controls using vacuum can also be tested for proper operation. But again, you must understand when and how they operate before you can run a vacuum test. For example, there are two basic Exhaust Gas Recirculation (EGR) systems used on 1973–74 American cars, but the manufacturers have modified the basic systems to some extent to fit operating characteristics of their engines. Chrysler Corporation cars use three variations: floor jet, ported vacuum and venturi vacuum systems. As each differs in the manner and timing of vacuum application, it is absolutely necessary that you know when vacuum should be present and when it should not, as well as understanding the interrelation of the temperature control valve, temperature sensing device, thermal delay switch, etc. to the application of vacuum. Actually running the test is very simple; just disconnect the vacuum hose at the EGR valve (A) and check for the availability of vacuum with the engine running at 1,500–2,000 rpm. (EGR systems do not operate at idle speeds.)

FUEL PUMP PRESSURE

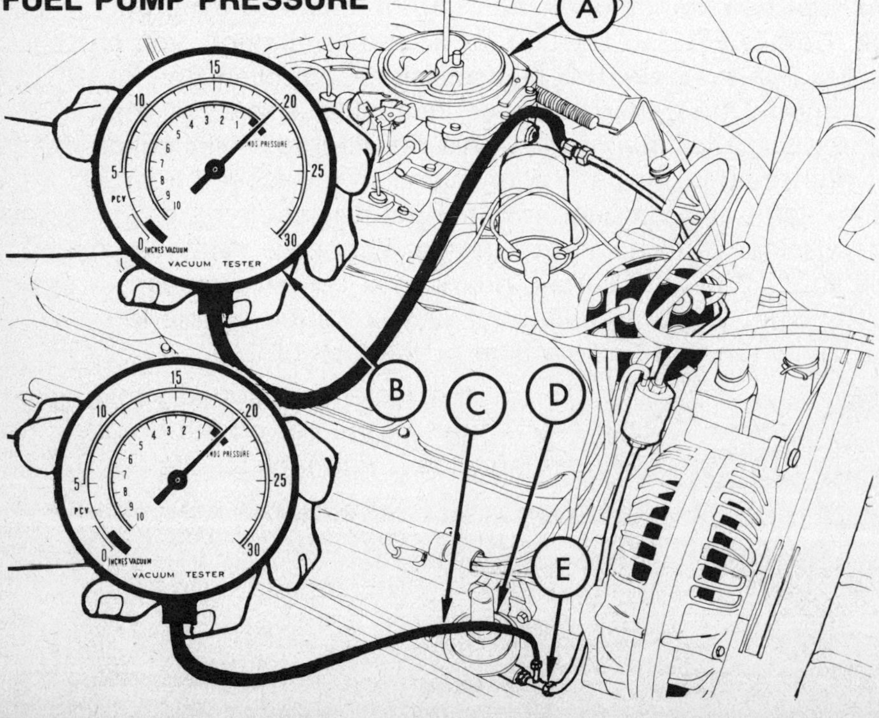

If the vacuum gauge has a pressure scale on its face, you can perform this test by simply disconnecting the fuel line at the carburetor (A) and attaching it to the vacuum gauge (B), but as there is no fuel flow with this method, it is a static (non-moving) test and less accurate.

It also can be used to check fuel pump pressure by connecting the gauge hose (C) between the fuel pump (D) and the carburetor (A), using a T-adapter (E).

Start the engine and let it idle. Hold the gauge at the carburetor height and check its reading. Fuel pump pressure should fall within the range specified by the manufacturer, a figure which is available in the factory shop manual. If it does not, the pump is faulty.

Now stop the engine and watch the gauge pointer. Pressure should remain for several minutes, and the pointer should also hold steady during that time before gradually dropping to zero. If the pressure does not hold, it indicates that the pump is defective or that it contains a bleed hole to prevent static pressure. Late-model cars all use fuel pumps containing a bleed hole, and all replacement fuel pumps sold for several years now have them, so this test may be useful only on cars eight to 10 years old which still have their original pump.

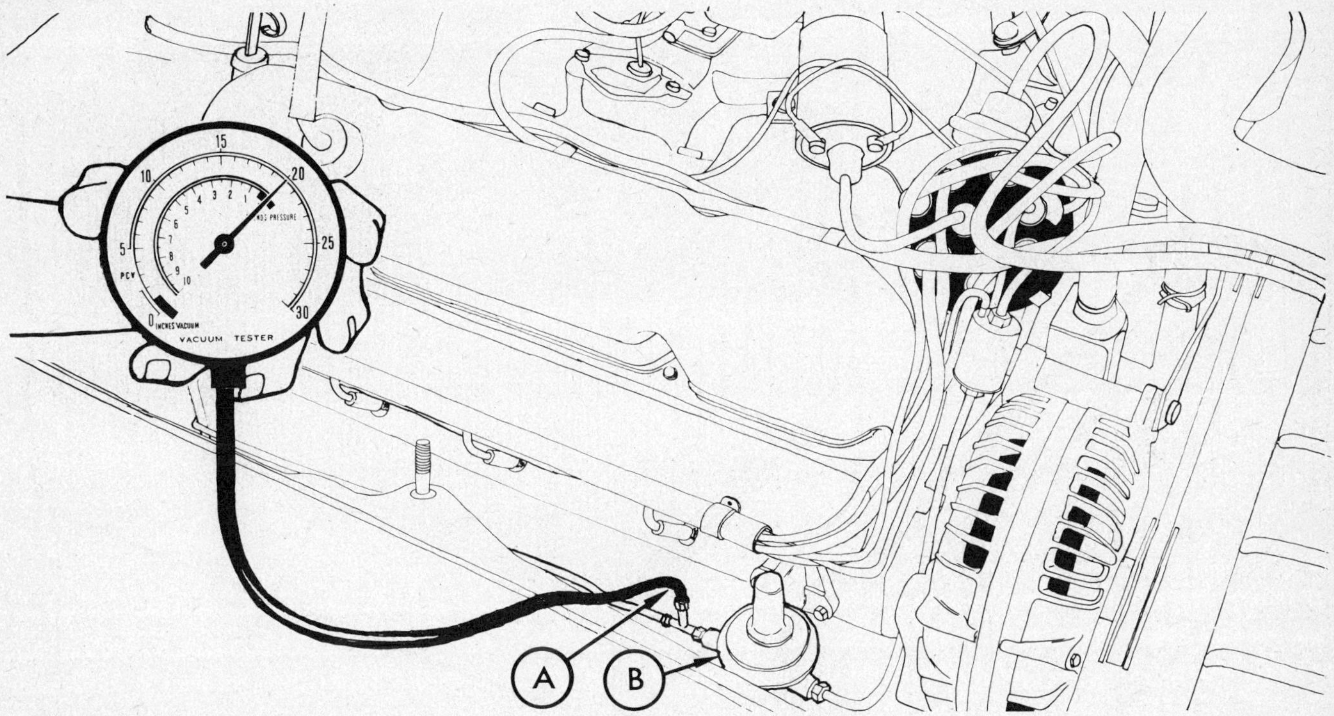

In addition to the pressure test, you can check fuel pump vacuum by connecting the gauge hose (A) to the suction side (B) of the pump. With the engine idling, you should get a reading of approximately 10 inches on the scale. If the reading differs very much from this figure, the pump is weak and may need replacement.

You should also make a flow test at this time, as pressure and vacuum may be adequate while flow is not. This can happen when there is an obstruction in the fuel pump or pump-to-carburetor line or when a fuel filter is clogged. The pump is actually working as it should, but delivery of the fuel to the carburetor is halted.

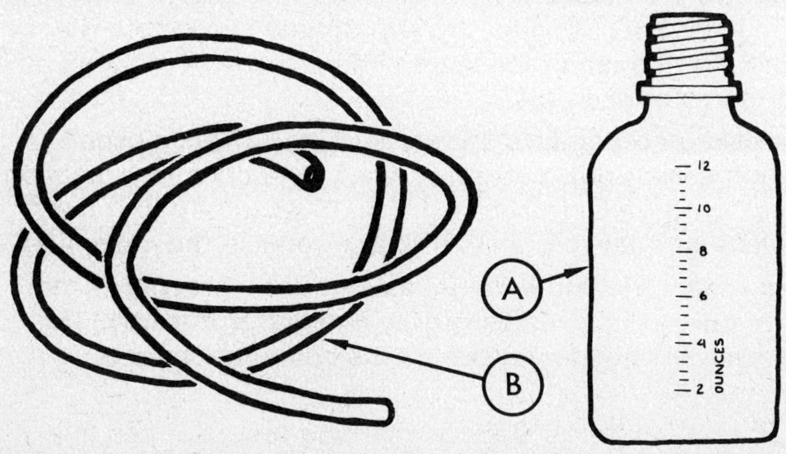

The vacuum gauge cannot be used for a flow test, but all you need is a container marked in fluid ounces (A) and a 3/8-inch flexible hose (B). Remove the vacuum gauge hose from the fuel line and replace it with the 3/8-inch hose. Place the other end of the hose in the container. Then run the engine on the fuel remaining in the carburetor for 30 seconds. Stop the engine and reconnect the fuel line. Then compare the amount of fuel collected with the manufacturer's specifications for the exact delivery rate.

BATTERY HYDROMETER

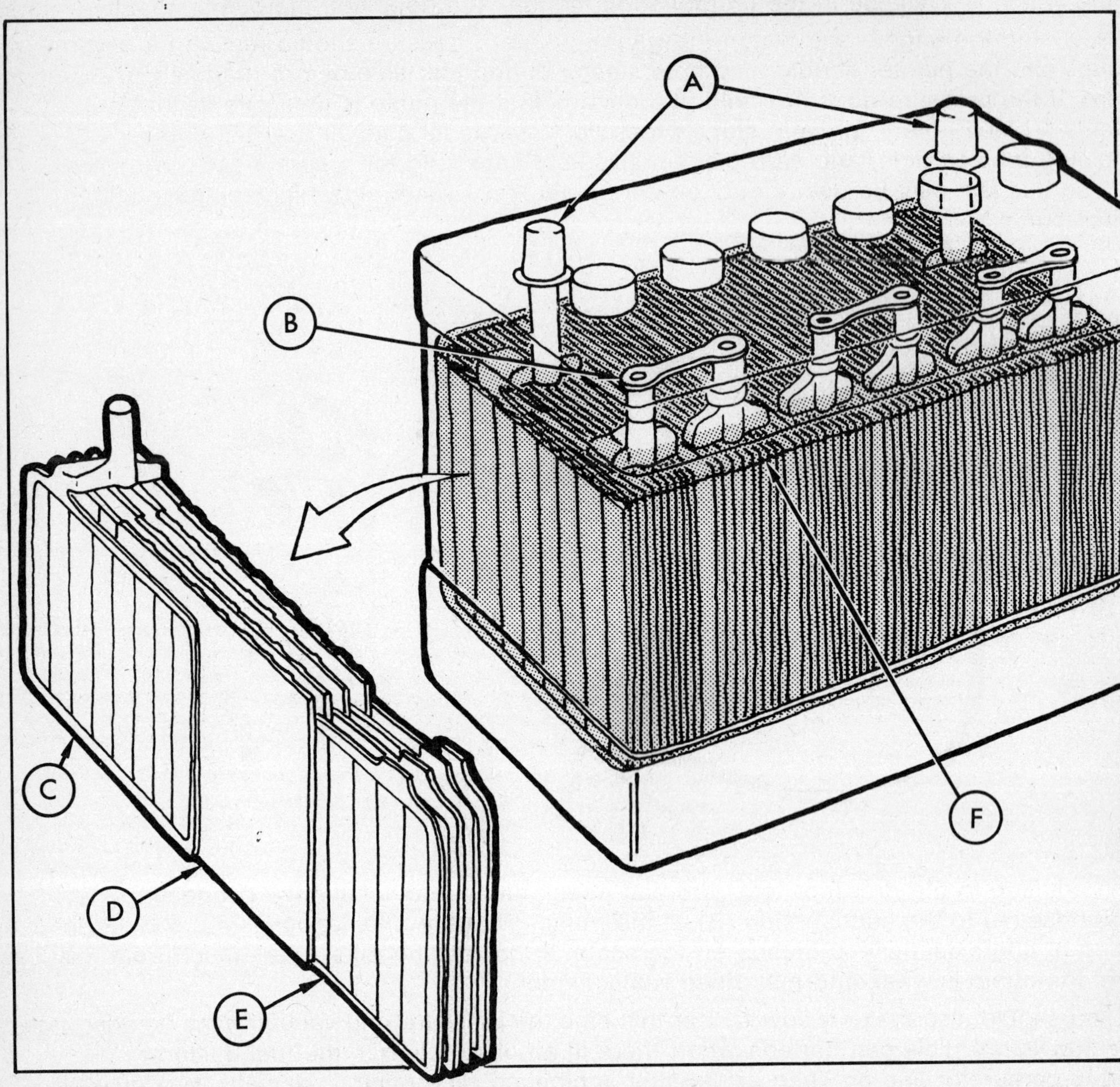

The lead/acid battery used in automobiles contains a number of positive plates of lead peroxide (E) and negative plates of spongy lead (C). Divided by appropriate separators (D) and attached in series by cell connectors (B), the lead plates are submerged in a solution of sulfuric acid and water, which is called the electrolyte (F).

When a circuit connected to the positive and negative battery terminals (A) is closed (which happens when you turn the ignition switch on), current created by a chemical reaction flows through the circuit. The sulfuric acid in the electrolyte is gradually absorbed into the lead plates, changing them into lead sulfate and leaving only ordinary water in the cells.

A battery's state of charge can be quickly determined by measuring the percentage of sulfuric acid in the battery electrolyte in terms of its specific gravity (weight or density). This is possible because the electrolyte's specific gravity decreases as the sulfuric acid enters the plates during a period of discharge.

Recharging a battery with a generator, alternator or battery charging unit reverses this chemical process. The plates return to their original state as the sulfuric acid reunites with the water, *increasing* the specific gravity of the electrolyte during a period of charge.

BATTERY HYDROMETER DESIGN

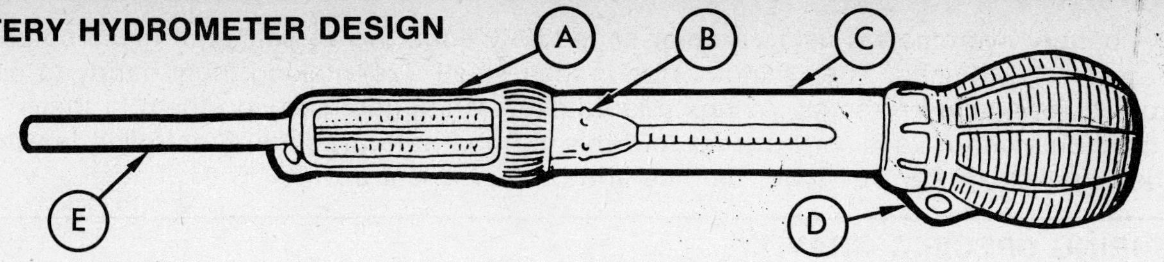

Measuring the concentration of the electrolyte is done with a battery tester called a battery hydrometer (A). This simple test instrument consists of a weighted and calibrated float (B) inside a glass barrel (C). When the rubber bulb (D) at the top of the barrel is squeezed, liquid is drawn into the tester through a syringe tip or tubing at its bottom (E).

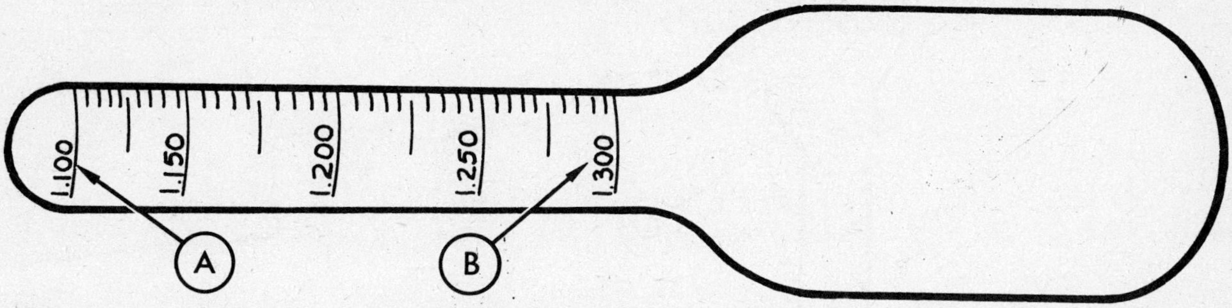

The weighted float or hydrometer is calibrated to indicate the comparative weight or density of the electrolyte sample drawn into the hydrometer barrel from the battery cell. Since the electrolyte in a fully charged battery has a specific gravity of approximately 1.280 at 80° F, float calibration usually extends from 1.100 (A—discharged state) to 1.300 (B—fully charged).

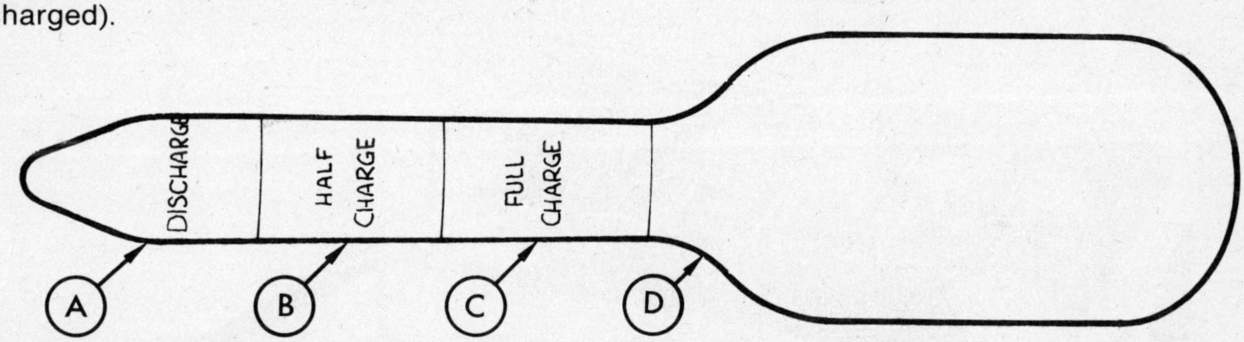

Inexpensive battery hydrometers often use a color-coded float (D) instead of one that is calibrated to specific gravity. A reading in the yellow or green zone (C) indicates a full charge; one in the white or mid-zone (B) indicates a half charge. If the reading is in the red zone (A), the battery is considered to be dead. A reading from such floats—not calibrated in terms of specific gravity—is a generalized one and not very useful beyond a yes/no indication of battery cell condition.

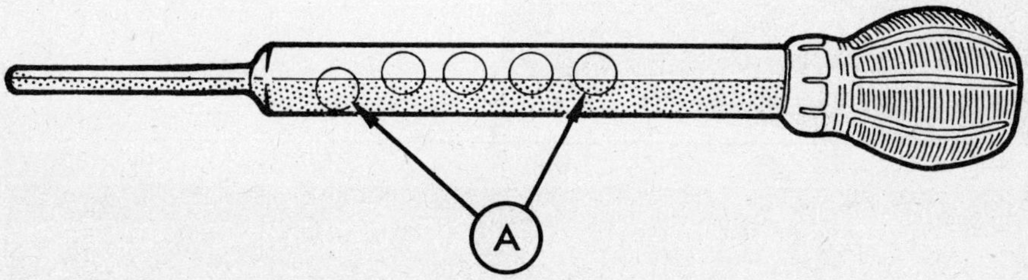

NO. OF BALLS FLOATING	STATE OF CHARGE
5	Overcharged
4	100%
3	75%
2	50%
1	25%
0	None

Still other inexpensive battery hydrometers use a series of five color-coded balls (A) instead of a weighted float. When battery electrolyte is drawn into the barrel, one or more of the tiny balls will float. The number of balls that float depends on the state of charge, as shown in this table.

BATTERY HYDROMETER

Like battery hydrometers using a color-coded-only float, the versatility of these so-called "power testers" is limited. Their major virtue is their small size, making them handy to carry in a glove compartment or small toolbox. Manufacturers suggest carrying them in their protective container in your shirt pocket, but this is not a safe practice. Electrolyte leaking from the tester barrel can ruin your clothes and cause painful burns.

MEASURING SPECIFIC GRAVITY

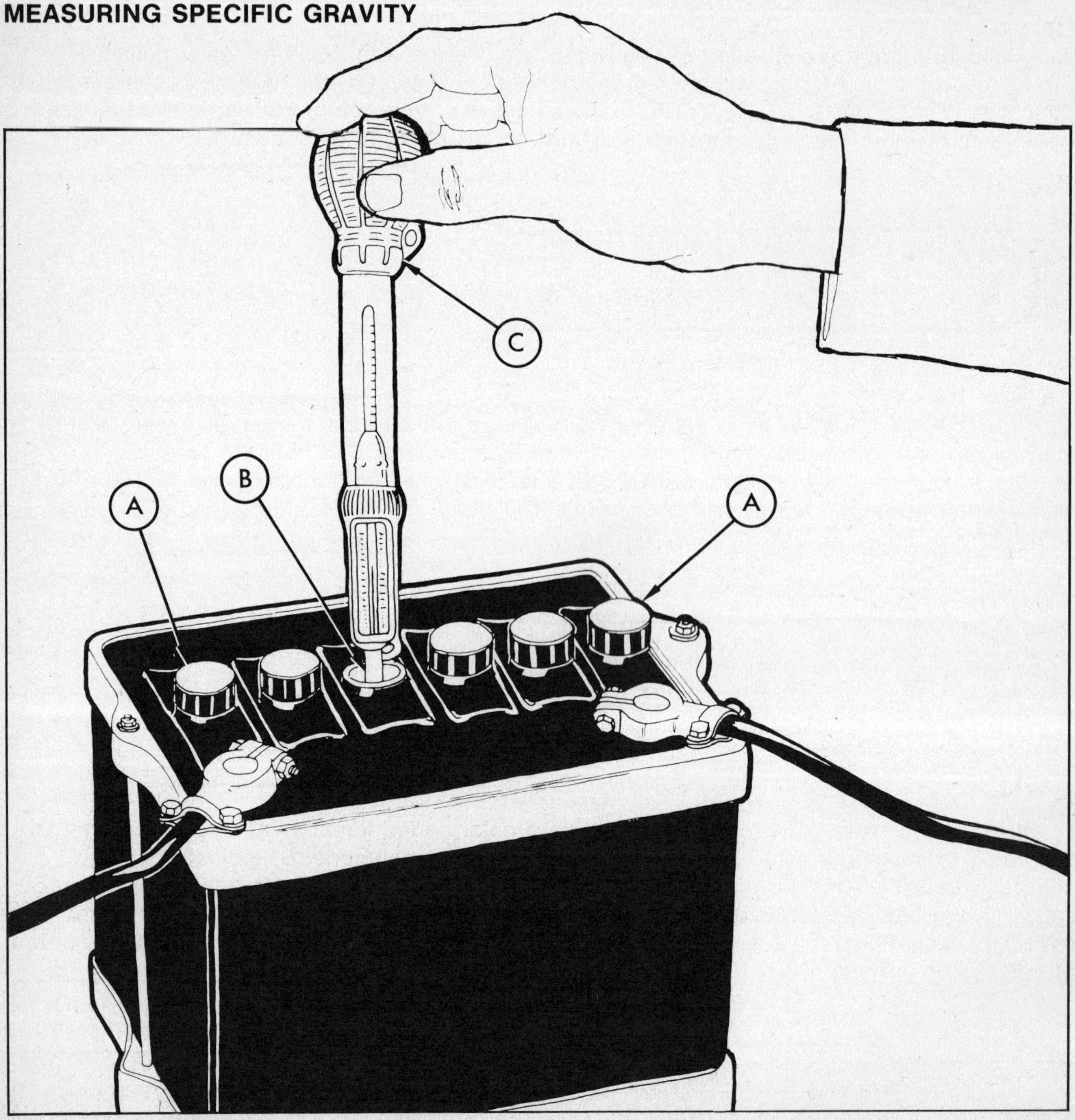

WARNING: DO NOT GET ANY BATTERY FLUID ON SKIN OR CLOTHING. SEVERE BURNS AND DAMAGE TO CLOTHES WILL RESULT.

To use a battery hydrometer, remove the filler/vent cap from all battery cells (A). Insert the tube at the bottom of the tester (B) into each cell in turn. Holding the tester vertically, squeeze the rubber bulb (C) gently and release it. This should draw sufficient electrolyte into the tester barrel to float the hydrometer freely.

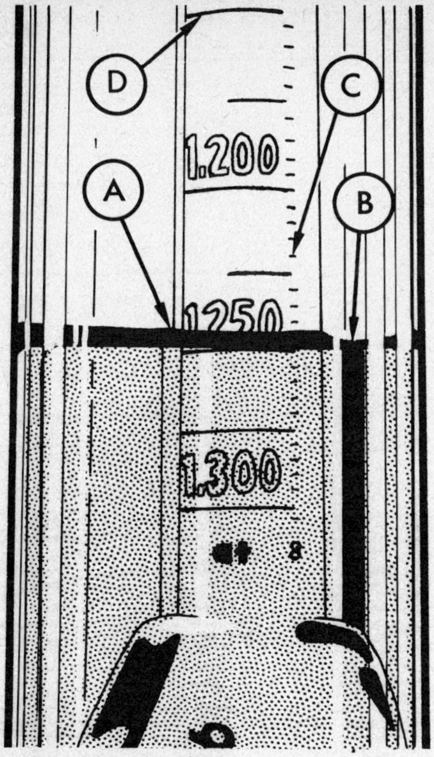

SPECIFIC GRAVITY.	CHARGE
1.260—1.280	100%
1.230—1.250	75%
1.200—1.220	50%
1.170—1.190	25%
1.140—1.160	Almost Useless
1.110—1.130	Discharged

Withdraw the tester from the cell. Hold it in a vertical position at eye level so that the float stands free of the barrel sides. Then take a reading where the surface of the liquid meets the float scale (A). With some battery hydrometer designs, surface tension will cause a curvature of the liquid (B) against the float, but if the hydrometer scale is read to include this curvature, the resulting figure will be incorrect—disregard any such curvature. Compare the reading obtained with the specific gravity table provided here.

WHAT THE READING MEANS

Any reading below 1.220 (C) indicates a poor charge condition, and one below 1.150 (D) means that the cell is dead for all practical purposes. If the reading of any one cell is lower than the others by 0.50 or more, that particular cell is shorted and the battery will have to be replaced.

When all cells test in the 1.220 or below range (C), the battery should be recharged with a battery charging unit (see that chapter). The cells can be tested during charging to determine when the process is complete. When the specific gravity remains the same in three successive readings an hour apart and all cells are gassing, the battery has accepted as much of a charge as it can hold. To determine the actual specific gravity, however, it is necessary to let the battery sit off-charge for an hour and then take a battery hydrometer reading. Readings taken while the battery is gassing will not be correct. If charging does not bring the battery cells up to at least a 50% charge, the battery will have to be replaced.

TEMPERATURE CORRECTION

While the test procedure and interpretation of results just given are correct, there is another factor involved in measuring the specific gravity of the electrolyte—its temperature. As temperature increases, the electrolyte expands, reducing the specific gravity; as temperature decreases, the electrolyte contracts, increasing the specific gravity.

Battery hydrometer readings are considered to be correct when the electrolyte temperature is 80° F. For each 10° above 80° F, you must add .004 (also known as four "points of gravity") to the original reading; for each 10° below 80°, you must subtract .004 from the reading.

Unless these variations in specific gravity and temperature are taken into consideration, the hydrometer reading will give only an approximate indication of the amount of acid in the electrolyte when its temperature is other than 80° F. Here's why.

Suppose that testing your battery cells results in a specific gravity reading of 1.230; this would lead you to believe that the battery is in fair condition. But your reading was obtained when the electrolyte temperature was 20° F. Once the reading has been temperature-corrected according to the previously mentioned formula, the actual specific gravity (or ASG) of 1.206 (6 x 4 = 24; 1.230 — .024 = 1.206) reveals that the battery is really very low and absolutely in need of charging.

Temperature correction can be made by taking the temperature of the electrolyte with a thermometer before using the battery hydrometer or, more conveniently, by the use of a thermo battery tester.

BATTERY HYDROMETER

THERMO BATTERY TESTERS

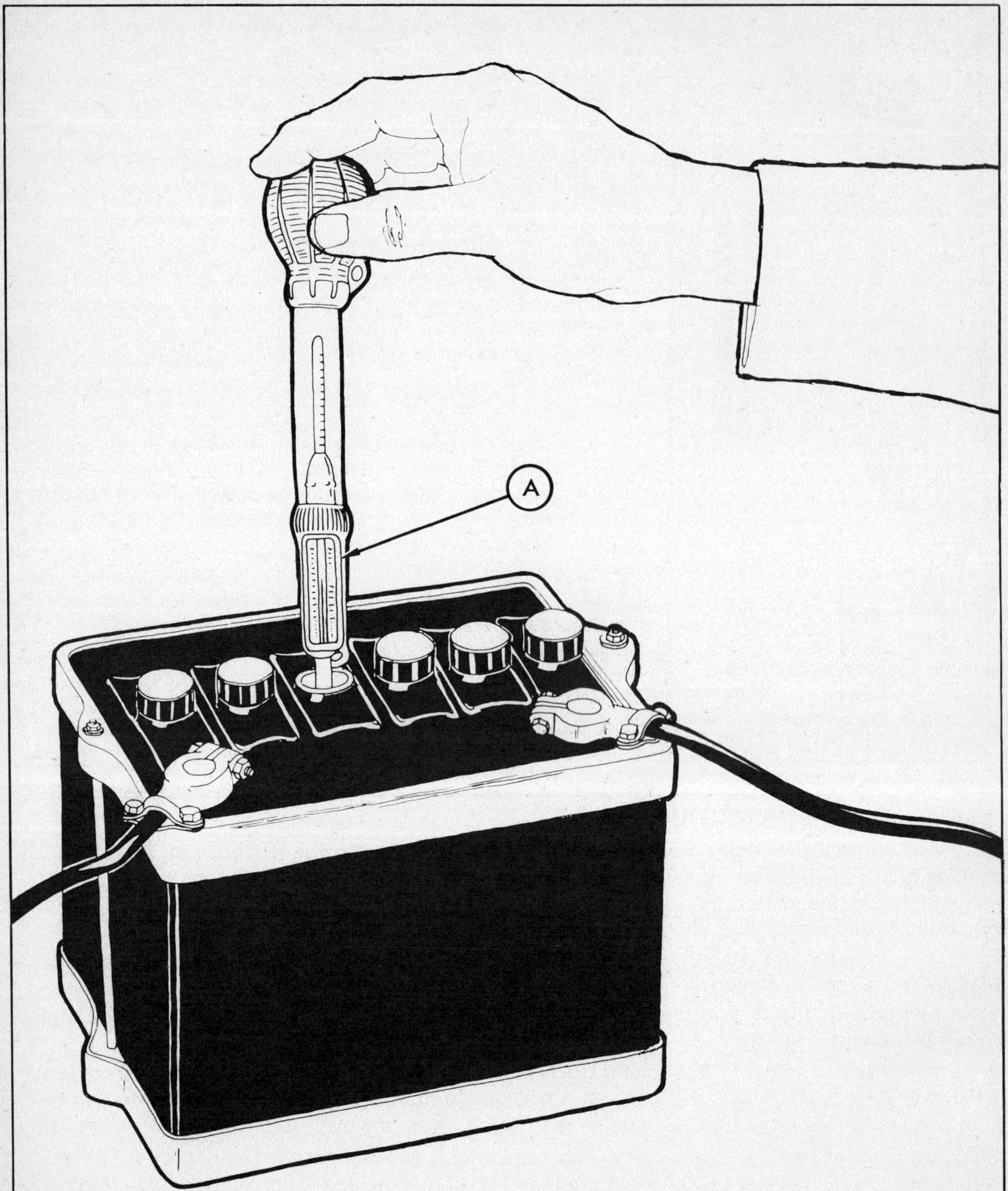

Temperature correction is easily performed using a battery hydrometer containing a built-in temperature and correction scale (A), often called a thermo battery tester. To arrive at the actual specific gravity (ASG) of the electrolyte, take the hydrometer float reading as before, but also note the level of the mercury in the thermometer.

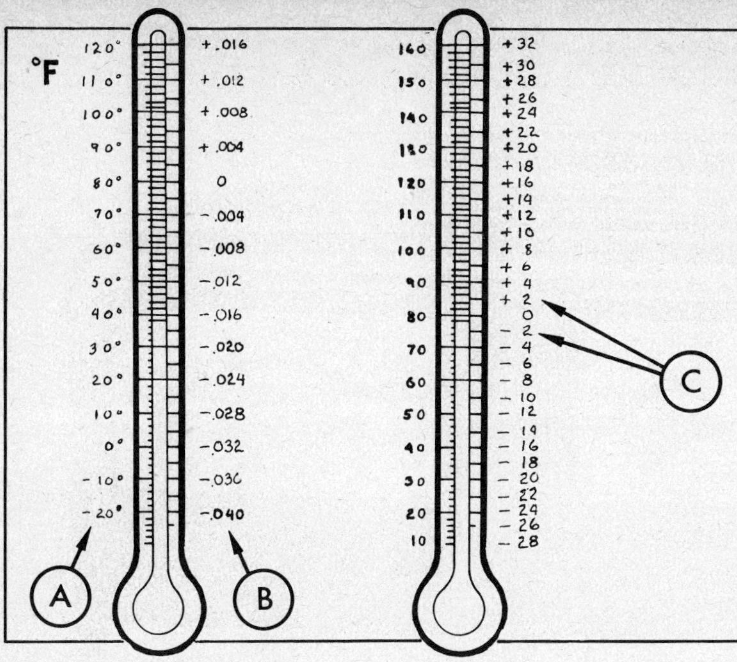

On some thermo battery testers, the thermometer tells the temperature in degrees Fahrenheit at one side (A) and the correction in "points of gravity" at the other (B). Add or subtract the specified "points of gravity" to the float reading to arrive at the actual specific gravity (1.230 — .01 = 1.218 ASG).

A variation of this correction scale reads in plus (+) or minus (—) points of gravity (C).

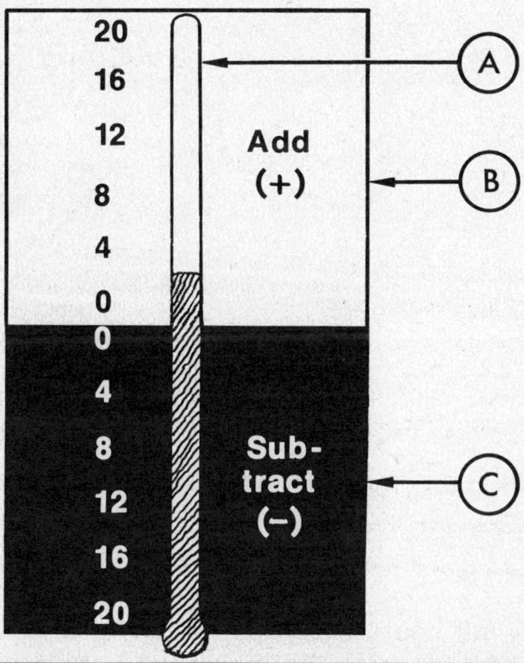

Other thermo battery testers read "corrective points of gravity" from 0 to 20 directly on a color-coded scale (A), and no thermometer reading in degrees is provided. The numerical reading in the white range (B) is *added* to the hydrometer float reading (1.230 + 8 = 1.238 ASG), while one in the black range (C) is *subtracted* from the float reading (1.230 — 8 = 1.222 ASG).

PRECAUTIONS IN BATTERY HYDROMETER USE

Certain precautions should be taken when using a battery hydrometer to assure accurate readings and insure personal safety.

1. The instrument must be clean, inside and out, for an accurate reading. *Do not* use the battery hydrometer to test coolant and *do not* test the battery with a coolant hydrometer.

2. *Do not* take readings immediately after adding water to the cells. If you do, the water will not be thoroughly mixed with the electrolyte and the readings will be incorrect.

3. *Do not* take readings immediately after returning from a long or high-speed trip, because battery condition will not then be "average."

4. When using a thermo battery tester, draw and expel the electrolyte several times before taking the reading to insure a correct temperature.

5. Use care when removing the hydrometer from one cell and inserting it in the next. Electrolyte is extremely corrosive; any that spatters on the car or your clothes should be washed off at once with a neutralizer solution like baking soda and water to prevent damage.

BATTERY-STARTER TESTER

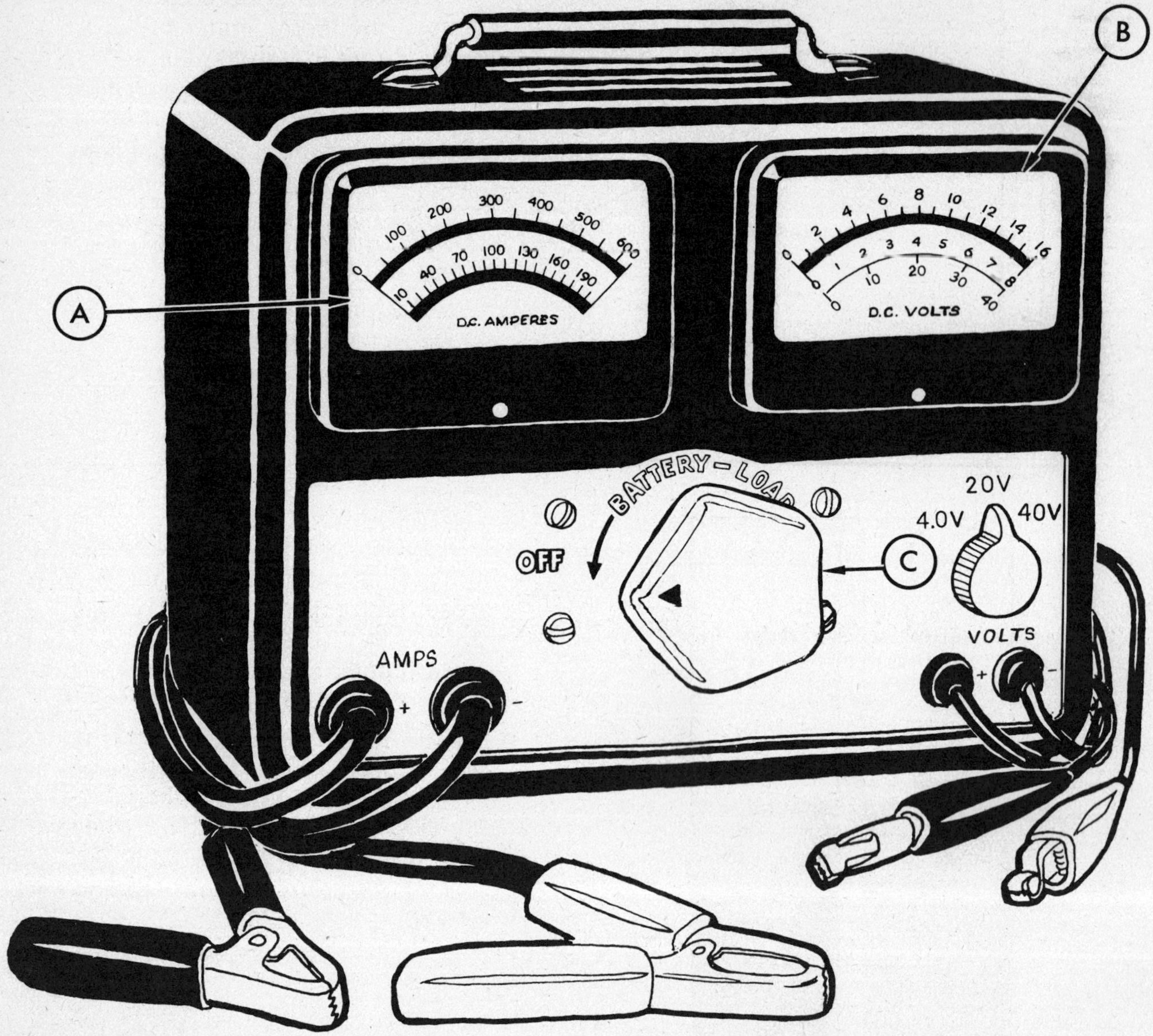

A storage battery at full charge (or nearly full charge) can be checked for internal defects by performing a capacity test. This test duplicates the maximum battery effort needed to crank a cold engine. It requires the use of a battery-starter tester, which can uncover defects in both the battery and starting system that might otherwise go undetected. The battery-starter tester consists of three units housed in a single case: an ammeter (A) and voltmeter (B) for current and voltage measurement and an adjustable resistor or high-capacity carbon pile (C) to control the amount of current passing through the circuit.

BATTERY-STARTER TESTER CONTROLS

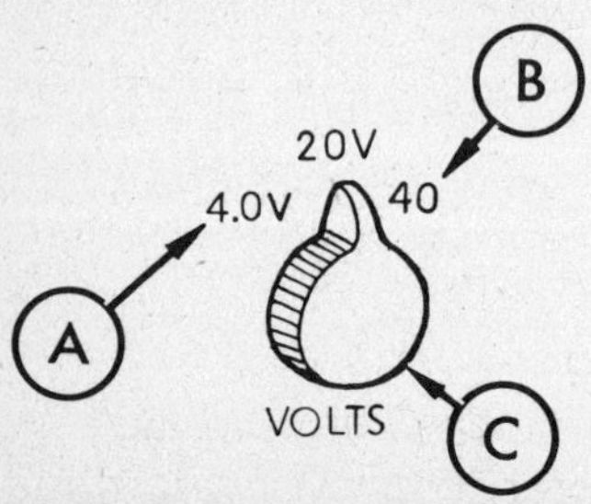

Battery-starter testers are equipped with a voltage selector switch (C), usually marked with at least three different voltage ranges. While the specific ranges differ according to unit manufacturer, there is a minimum voltage (A) range (usually 4 volts) and a maximum (B) range (usually 40 volts). Regardless of how the ranges are marked (4-8-16-40 or 4-20-40 etc.), you can test all six, 12, 24 or 32-volt systems by setting the selector switch accordingly.

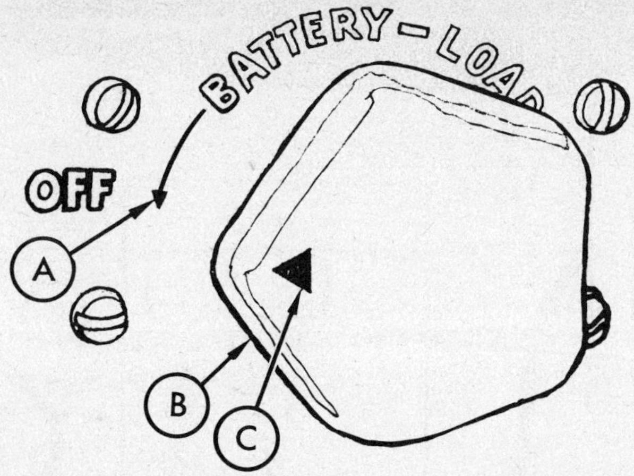

A load control knob (B) is located at the bottom center of the tester face and may be marked OFF/LOAD or DECREASE/INCREASE. Arrows are provided on the knob (C), tester face (A) or both to indicate direction.

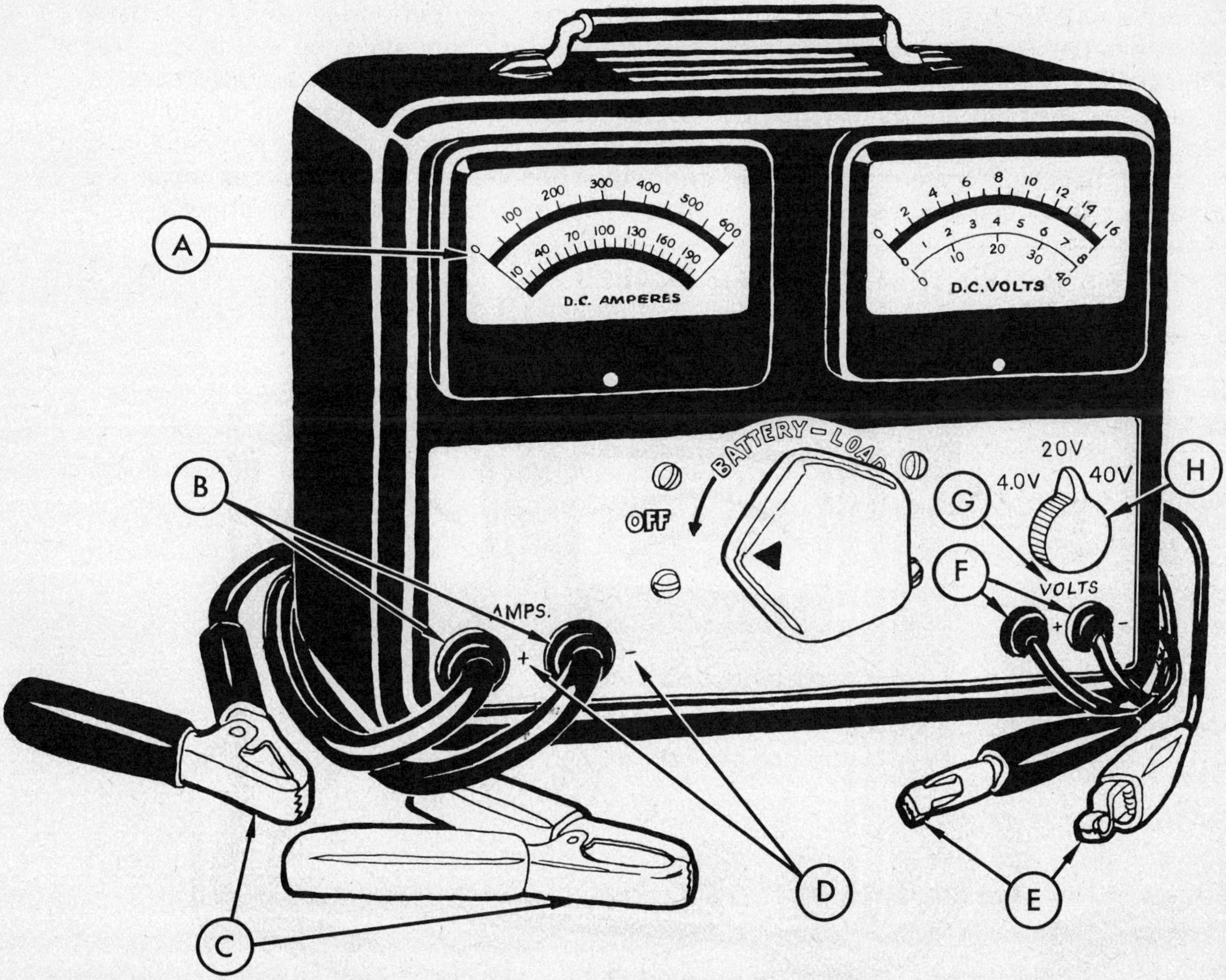

Test leads are permanently connected to each side of the tester. One set is used for ammeter readings and can be found at the left side of the tester face (B) beneath the ammeter scale (A). Voltmeter leads are located at the right side (F) beneath the voltage selector switch (H). Each set of leads is marked according to function (G). Positive and negative leads are color-coded (red for positive and black for negative), and they are also identified by positive and negative symbols (D) on the tester face. Additionally, voltmeter leads usually terminate in small, alligator-type clips (E), while ammeter leads are equipped with larger, scissor-type clips (C).

BATTERY-STARTER TESTER

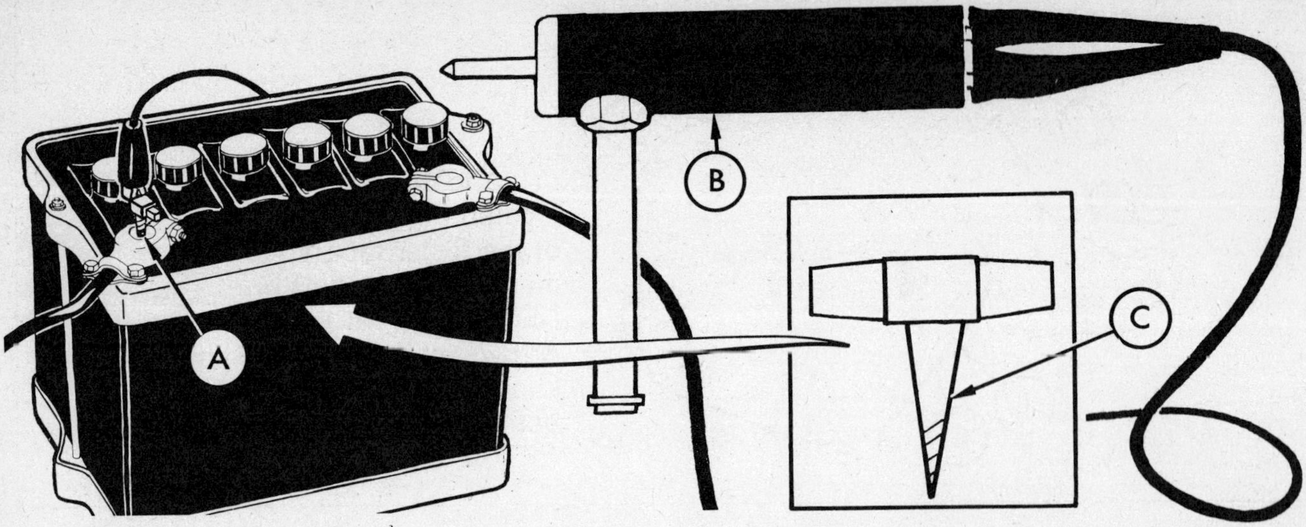

Battery-starter testers are also equipped with accessory pin connectors (C) or battery cell probes (B). Pin connectors are most useful today in connecting the test leads directly to the battery terminals (A), but in the days of soft-top batteries, they were also used to connect test leads to the battery cell connector straps for voltage readings of individual cells. Today's hard-top batteries, with connecting straps that run *through* the cells instead of across their tops, have made the pin connector obsolete for individual cell tests. The cadmium battery cell probes are now used to make individual cell voltage checks.

BATTERY-STARTER TESTER METER SCALES

Separate meter scales are provided—one for amperage readings (A) and one for voltage readings (E). The amp meter is traditionally placed at the left side of the tester face above the ammeter leads (B) and the voltmeter at the right above the voltage selector switch (D), but each will be clearly marked (C) as to function.

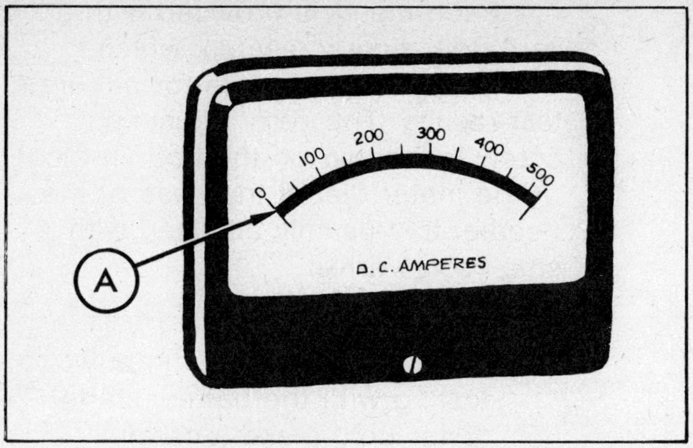

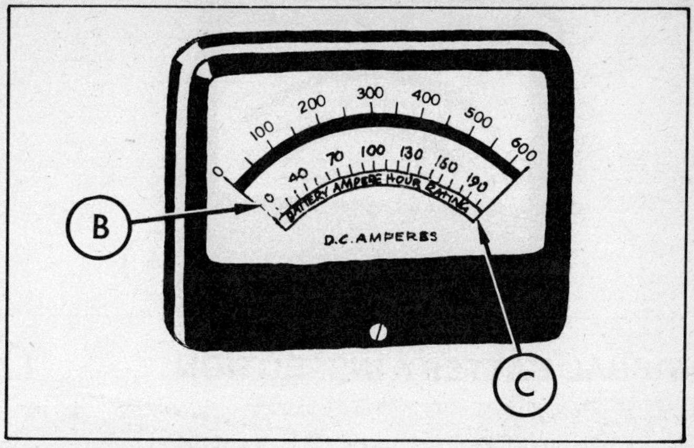

The ammeter scale will show at least a 0-to-500-amp range (A) in 10-amp divisions, but some meters may be marked to 600 amps or more. You may even find a second amp scale (B), reading 0 to approximately 200 amps in 5-amp divisions. This scale will be marked BATTERY AMPERE-HOUR RATING (C). It allows the tester load to be adjusted exactly to the battery ampere-hour rating. On tester units without this scale, you must calculate load to three times the ampere-hour rating when performing certain tests.

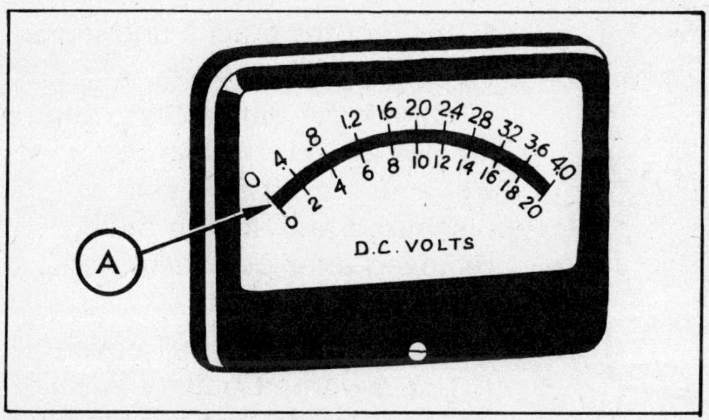

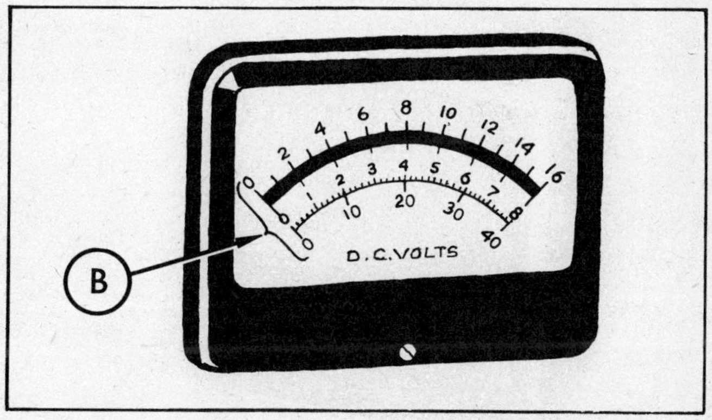

The voltmeter will have at least two (A) and as many as four (B) scales, depending upon meter design and the number of voltage ranges provided by the voltage selector switch. While the usual arrangement is to include one scale for each voltage range, some meters use a double-purpose scale. As an example, the 40-volt range may be read on the 4-volt scale by multiplying the meter reading by a factor of 10 ($2.4 \times 10 = 24$ volts).

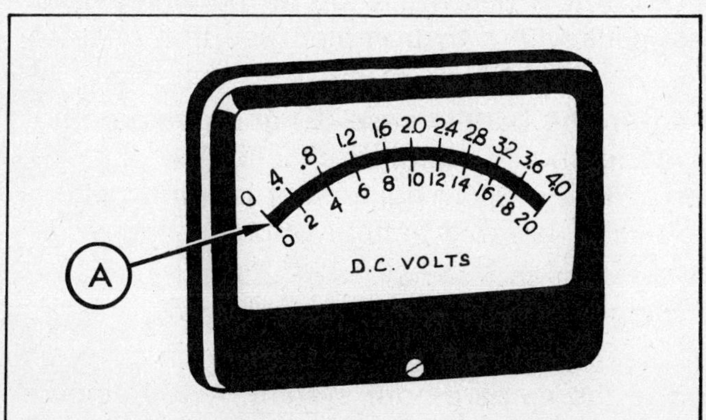

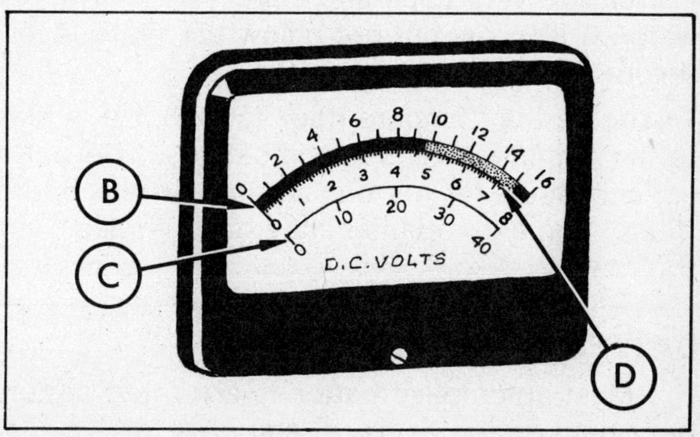

A 4-volt scale is usually provided with 0.1-volt divisions (A), while a 40-volt scale is marked in 1.0-volt divisions (C). Voltage scales in between these two (8, 16, 20, etc.) are most often found with 0.2-volt divisions. In addition, some meter scales are provided with color-coded bands (B) to simplify interpretation of voltage readings when testing the battery. The color-coded band(s) may indicate the 9.6-15.5-volt range for 12-volt batteries and/or the 4.8-7.75-volt range for six-volt batteries (D).

BATTERY-STARTER TESTER

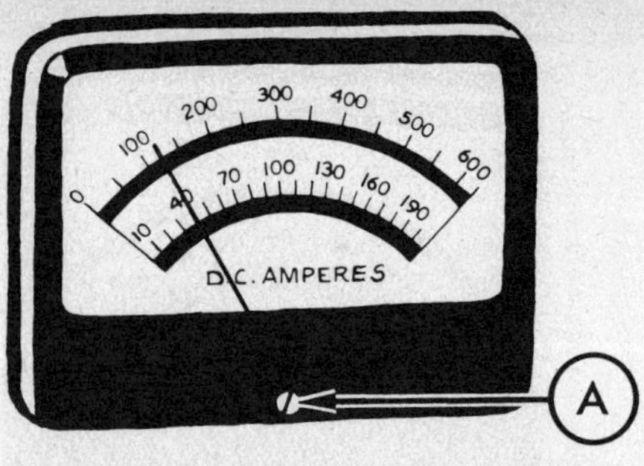

Each meter is provided with a zero adjusting screw (A), which zeroes the meter needle for accurate test results. The zero adjustment screw is located on the bottom face of the meter dial at the base of the needle. It is usually adjusted with a small screwdriver.

VISUAL BATTERY INSPECTION

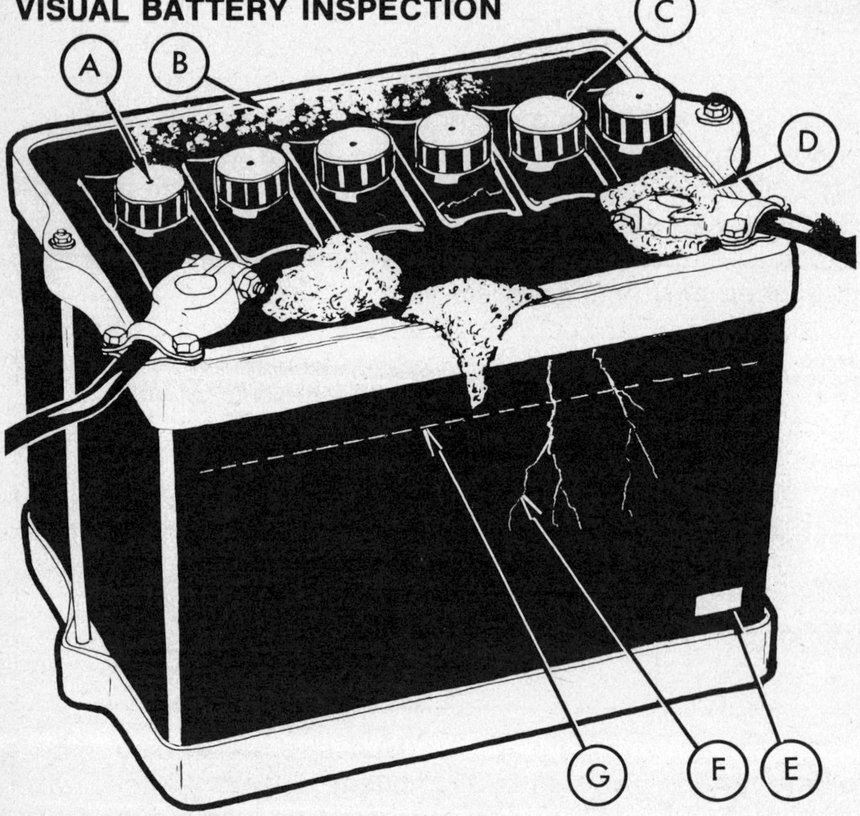

Before undertaking any testing with the battery-starter tester unit, make a careful visual inspection of the battery. What you see can be very useful as well as important when analyzing test readings. Follow the inspection procedure outlined below.

1. Inspect for cracks and leaks (F) in the battery case.

2. Check the battery case top for dirt and electrolyte (B); this will cause excessive self-discharge and should be removed after performing a battery leakage test.

3. Look for raised cell covers (C) or a warped battery case. These indicate that the battery has been overcharged or overheated.

4. Make sure that the filler-vent cap holes are not plugged (A).

5. Check battery posts, clamps and cables (D) for loose connections, corrosion, etc.

6. Remove vent caps and check electrolyte level (G). If it is below the top of the plates, add water. If it is low but not below the plate level, test cells with a hydrometer (see that chapter) before adding water.

7. Locate the "ampere-hour rating" if it is stamped on the battery case (E) and compare it to the manufacturer's specifications. The battery rating should equal or exceed the specifications for normal starting system operation. While original equipment batteries will do so, some companies may have installed an "on-sale" bargain battery whose rating is too low.

BATTERY SPECIFICATIONS

Most automotive batteries are rated according to the ampere-hour system. A 100-ampere-hour battery can deliver 5 amperes for a period of 20 hours. Although the 20-hour figure is not mentioned in the rating, it is generally understood when dealing with automotive batteries. The greater the plate area in a battery, the higher the ampere-hour rating. Thus there is a close relationship between the number of plates found in each cell and the rating figure applied to the battery. For example, batteries rated as 48-ampere-hour units have four positive plates per cell, while those with a 70-ampere-hour rating have six positive plates per cell.

Some manufacturers have also started to rate batteries by their reserve capacity, which gives the number of minutes that the battery will sustain a 25-amp discharge at 80° F and still remain above 10.5 volts. It is expected that this rating will gradually replace the ampere-hour rating. However, since all current battery-starter testers are designed to work with the ampere-hour rating specification, and as the latter can be determined either from the battery case or the battery manufacturer's specifications, we will use ampere-hour ratings in the test sequences that follow.

CONNECTING THE BATTERY-STARTER TESTER

Test unit connections depend upon the test being performed and the system to be checked. Connection patterns for each individual test are given under that section and are for negative ground systems. If the car being tested uses positive ground, reverse the leads from the pattern specified.

Ground can be determined by locating the battery cable which is connected to the engine or frame. The battery terminal post to which the other end of this cable connects should be marked with a + (positive) or − (negative) sign. The marking may appear on top of the battery post (A) or on the battery case beside the post (B). Generally speaking, all American cars are negative ground.

BATTERY CAPACITY TEST

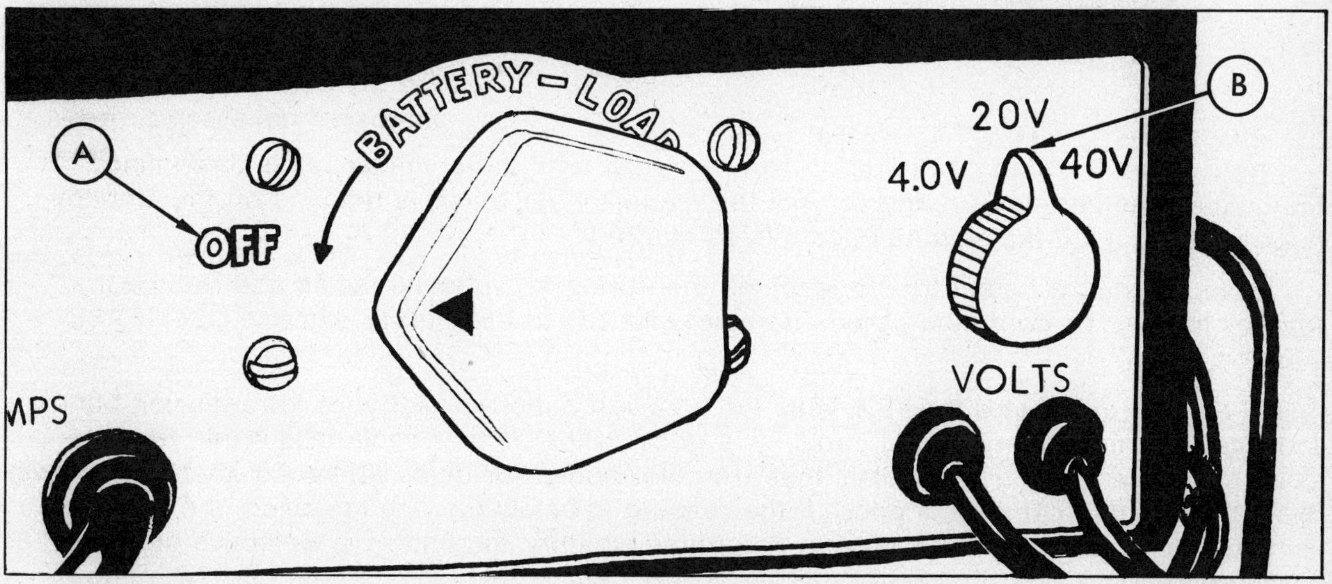

Also known as a "load" test, this check of a battery's ability to provide current and maintain the minimum required voltage determines its overall condition. Begin by turning the load control knob to the OFF or extreme DECREASE position (A) and set the voltmeter selector switch to the closest range (B) appropriate for the battery being tested. On some units, this will be a 20-volt position, while others may have an 8-volt range for use with six-volt batteries and a 16-volt range for 12-volt batteries.

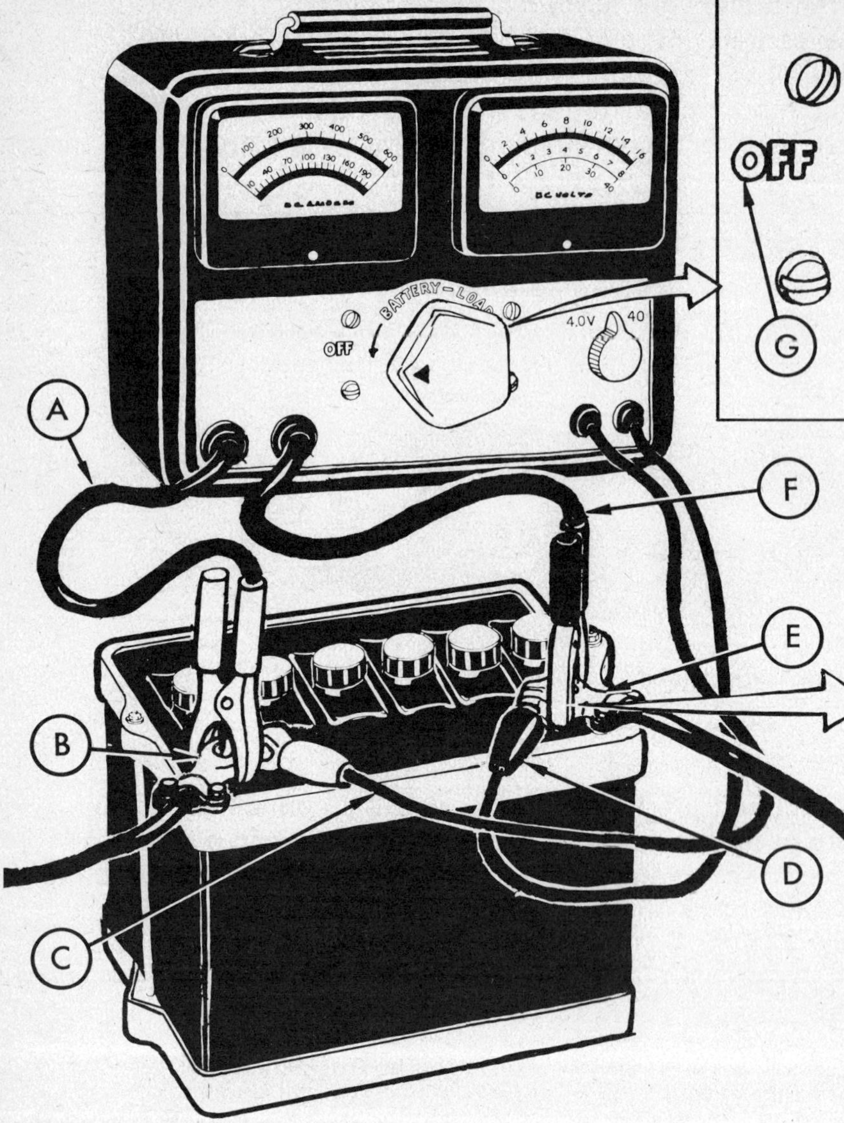

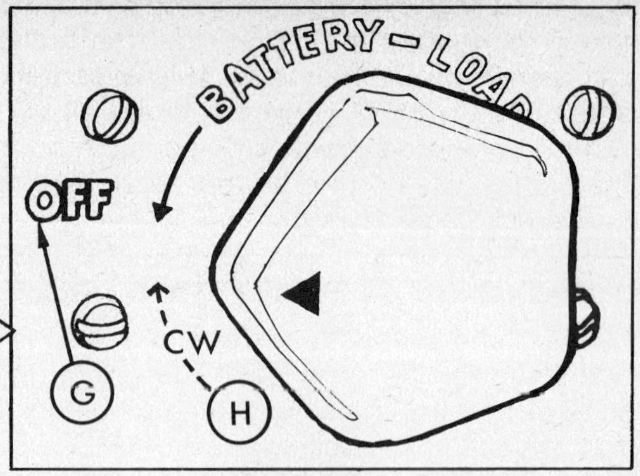

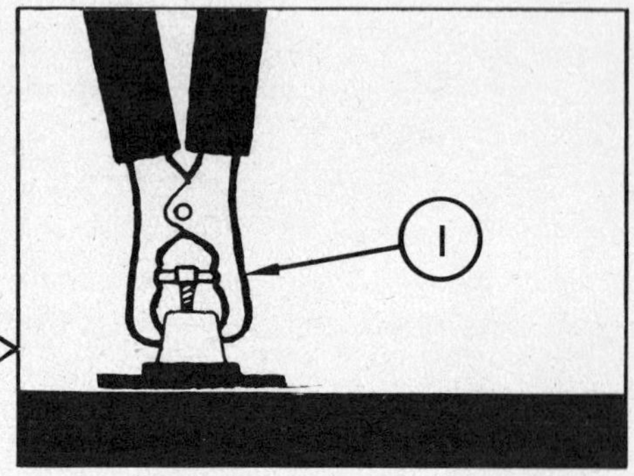

Connect the red ammeter (A) and voltmeter (C) leads to the positive battery terminal post (B) and the black ammeter (F) and voltmeter (D) leads to the negative battery terminal post (E). Voltmeter clips must contact the battery posts rather than the cable clamps. Use pin connectors (I) for this purpose if necessary.

Now adjust the load control knob clockwise (H) until the ammeter reads three times the battery's ampere-hour rating or until the ampere-hour rating is reached on the ampere-hour rating scale (if the ammeter scale is so marked).

Hold the load constant for 15 seconds, then read the voltmeter scale and reduce the load by turning the control knob counterclockwise (G) to the OFF or extreme DECREASE position.

A reading of not less than 9.6 volts for a 12-volt battery or 4.8 volts for a six-volt battery indicates that the battery output capacity is good. If the meter needle falls within the color-coded band(s) on those dials that are color-coded for quick reference, that also shows the battery output capacity is good. If the reading is below the one specified, the three-minute charge test should be used to determine whether the battery is defective or simply discharged.

CAUTION: A BATTERY HAS A LOWER DISCHARGE CAPACITY WHEN COLD, SO IT MAY FAIL THE CAPACITY TEST IF PERFORMED IN COLD WEATHER. A CAPACITY TEST SHOULD ALWAYS BE PERFORMED WHEN THE BATTERY IS AT A NORMAL (80° F) TEMPERATURE.

INDIVIDUAL CELL VOLTAGE TEST

Even though the battery passes the capacity test, the voltage of the individual cells should be checked while the battery is under load. Cadmium battery cell probes are required for this test.

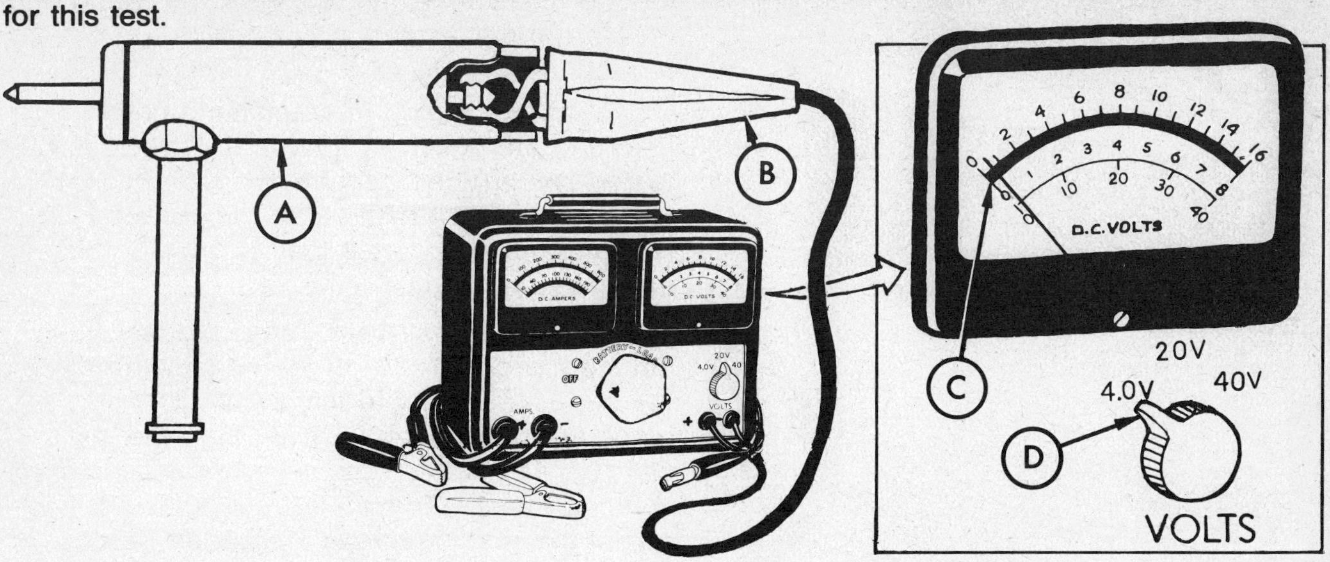

To use the battery cell probes, remove both ammeter and voltmeter leads from the battery posts, then connect the red positive probe (A) to the positive voltmeter lead (B) (red-to-red) and the negative probe to the negative voltmeter lead (black-to-black). Set the selector switch to the 4-volt range (D) and make certain that the meter needle is zeroed (C). If it is not, adjust it to zero.

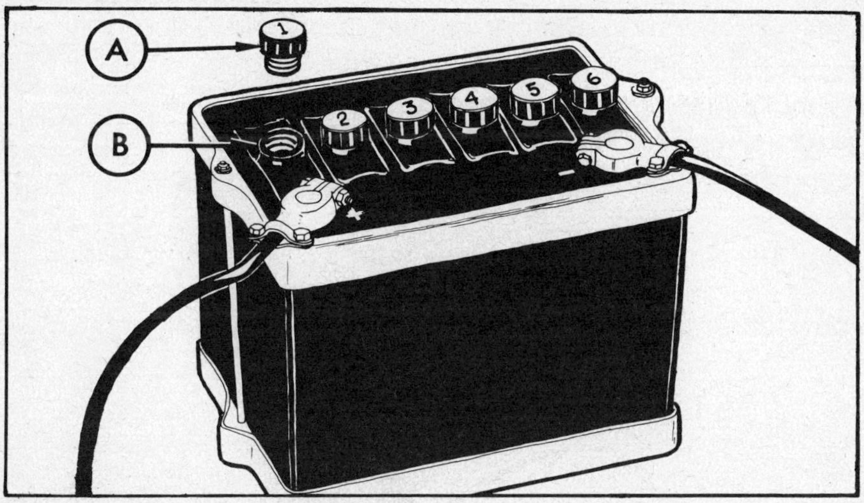

Remove the filler vent caps (A) from each cell (B) and begin your test with the cell nearest the positive terminal end of the battery. We'll consider this as cell No. 1 and number the remaining cells in order (2, 3, 4, 5, 6).

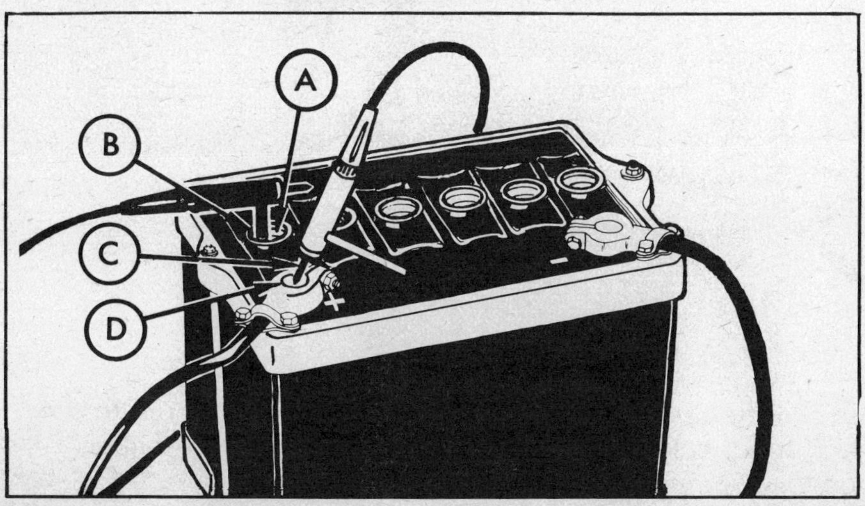

Hold the steel prod end (C) of the positive probe on the positive battery terminal post (D) and place the negative probe end (B) in cell No. 1 (A). Adjust the load control to three times the ampere-hour rating and record the meter reading for cell No. 1. This figure will be revised slightly as a final step in obtaining the exact voltage of cell No. 1. Reduce the load to zero and remove the probes.

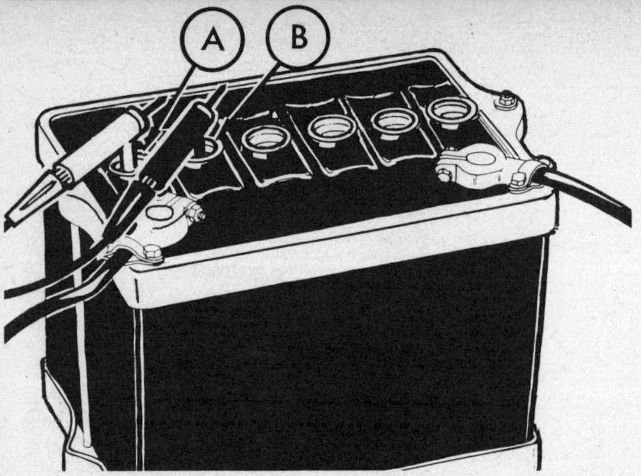

Test cell No. 2 by inserting the positive probe in cell No. 1 (A—previously tested) and the negative probe in cell No. 2 (B). Adjust the load to three times the ampere-hour rating and record the meter reading for cell No. 2. Then reduce the load to zero.

Continue testing the remaining cells by inserting the positive probe in the previously-tested cell and the negative probe in the cell to be tested. Adjust the load, record the reading and reduce the load to zero in this same sequence each time.

After testing all cells in this manner, hold the steel prod end (D) of the negative probe on the negative battery terminal post (C) and place the positive probe end (A) in the last cell (B). Adjust the load as before and read the voltmeter scale. This meter reading is used to adjust the reading already recorded for cell No. 1.

If the meter needle reads above zero, *add* the reading to that obtained from cell No. 1 to get its adjusted correct voltage. If the meter needle reads less than zero, reverse the position of the probes and repeat, *subtracting* the reading from that of cell No. 1 to obtain its adjusted correct voltage.

Individual cells should not vary more than 0.1 volt on this test. If they do, replace the battery.

THREE-MINUTE CHARGE TEST

If battery voltage obtained during the capacity test was less than 9.6 volts for a 12-volt battery (4.8 volts for a six-volt battery) or if the meter needle did not reach the colored band on coded dials, the three-minute charge test will determine whether the battery is just discharged or defective.

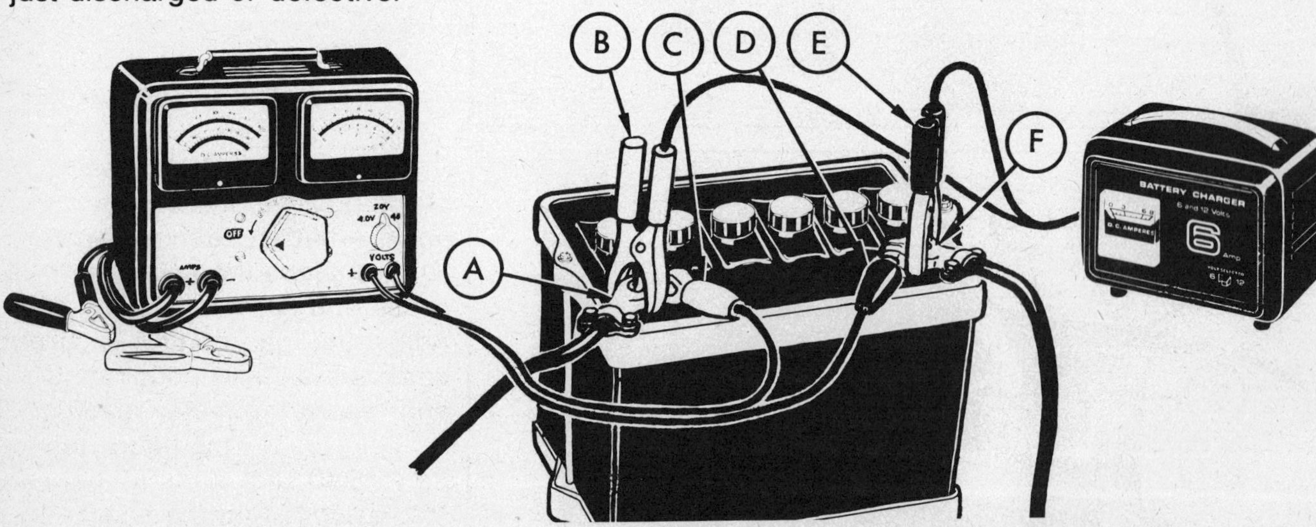

Connect the red voltmeter leads of the tester (C) and a battery charging unit (B) to the positive battery terminal post (A) and the black leads of the voltmeter (D) and the battery charging unit (E) to the negative battery terminal post (F).

Adjust the voltage selector switch (if necessary) and turn on the battery charging unit, setting it to produce a fast-charging rate of 40 amps for a 12-volt battery or 75 amps for a six-volt battery. Wait three minutes; then, with the battery still on fast charge, read the voltmeter scale.

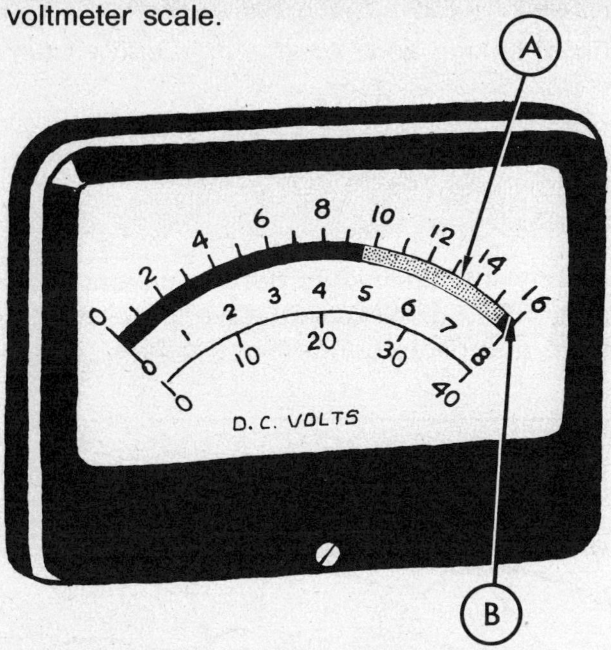

A reading of more than 15.5 volts for a 12-volt battery (B) or 7.75 volts for a six-volt battery indicates that the battery is probably sulphated or worn out and must be replaced.

If the reading falls within the 9.6-15.5-volt range (A—4.8-7.75 for a six-volt battery), set the voltage selector switch to the 4-volt position and perform the individual cell voltage test with the battery still on fast charge. Individual cells should not vary more than 0.1 volt on this test; if they do, the battery has one or more defective cells and must be replaced.

When individual cells test within 0.1 volt of each other and the total voltage falls within the above specifications, the battery is discharged but not defective. Test its specific gravity (see the battery hydrometer chapter) and charge as required.

BATTERY LEAKAGE TEST

When the battery top is covered with dirt and/or electrolyte, the contamination can result in a continuous battery discharge, because it produces a path which battery current can follow. Due to its higher voltage, a 12-volt battery is more likely to lose energy in this manner than a six-volt battery.

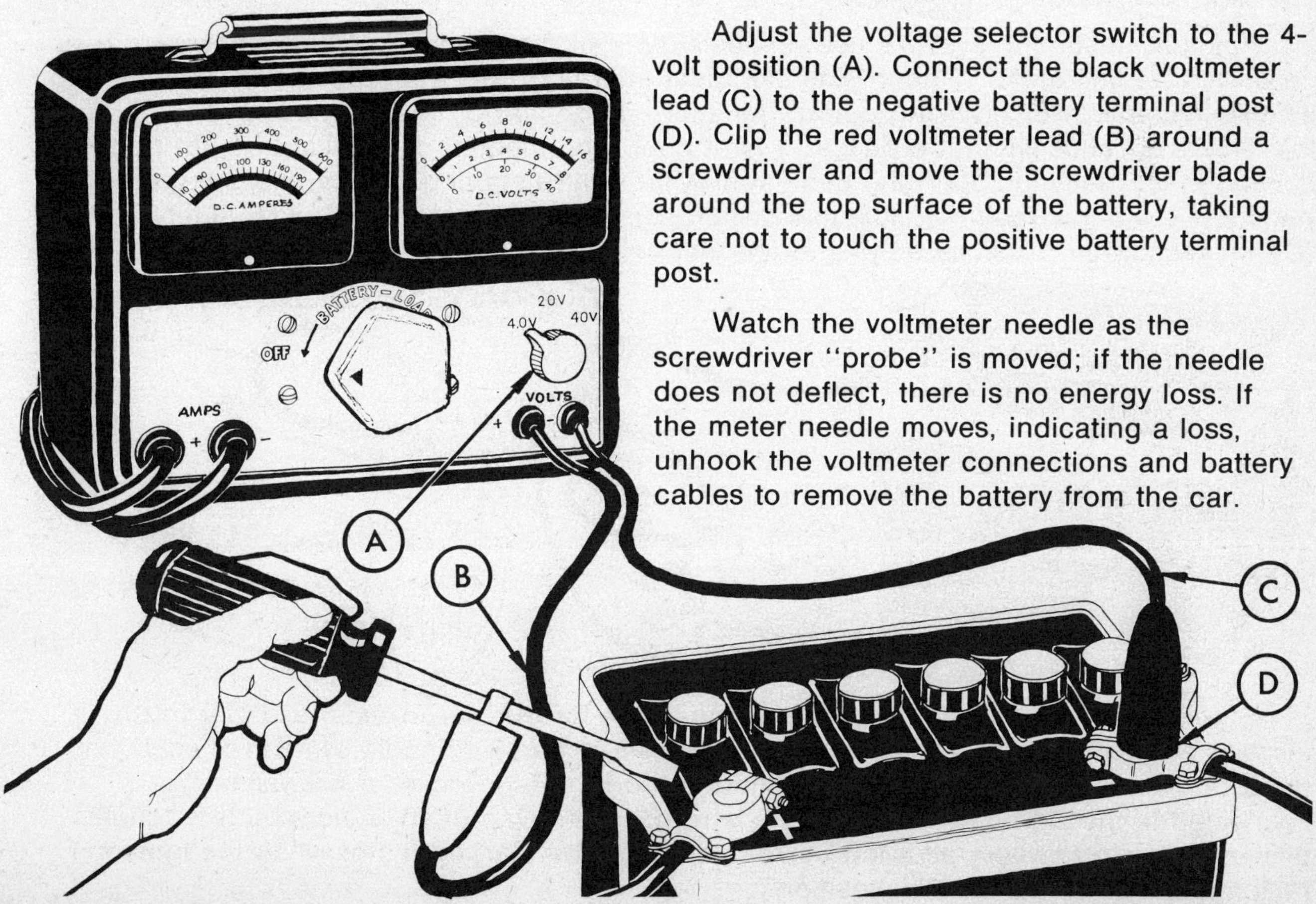

Adjust the voltage selector switch to the 4-volt position (A). Connect the black voltmeter lead (C) to the negative battery terminal post (D). Clip the red voltmeter lead (B) around a screwdriver and move the screwdriver blade around the top surface of the battery, taking care not to touch the positive battery terminal post.

Watch the voltmeter needle as the screwdriver "probe" is moved; if the needle does not deflect, there is no energy loss. If the meter needle moves, indicating a loss, unhook the voltmeter connections and battery cables to remove the battery from the car.

WARNING: BE SURE TO UNHOOK THE NEGATIVE OR GROUND CABLE FIRST, THEN THE POSITIVE OR HOT CABLE TO PREVENT POSSIBLE ELECTRIC SHOCK.

Clean the battery case thoroughly with a solution of baking soda or ammonia and water and dry it completely before reinstalling.

CAUTION: BE CAREFUL TO KEEP THE CLEANING SOLUTION OUT OF THE CELLS. IF SOME GETS INTO THE CELLS, IT WILL NEUTRALIZE THE ACID IN THE BATTERY.

A small piece of tape over each filler/vent cap hole will usually keep out the cleaning solution, but don't forget to remove it when cleaning is finished. While you have the battery out of the car, also clean the battery carrier, hold-down and cable connections before replacing the battery.

THE AUTOMOTIVE STARTING SYSTEM

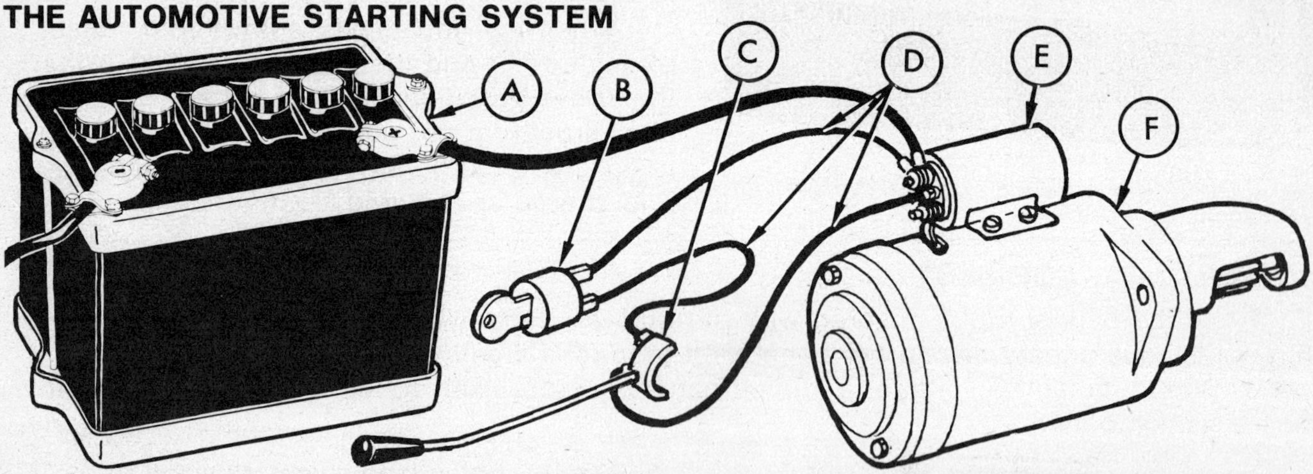

Automotive starting systems use a special type of motor designed for intermittent use only. This starting motor (F) converts electrical energy drawn from the storage battery into mechanical energy or cranking output. As such, it should not be operated for periods longer than 15-20 seconds without a pause to let it cool. The amount of cranking output from the starting motor depends upon the condition of the entire starting system, including the battery (A), wiring circuit (D), solenoid switch (E), ignition switch (B), neutral safety switch (C–automatic transmissions only) and the engine's cranking requirements.

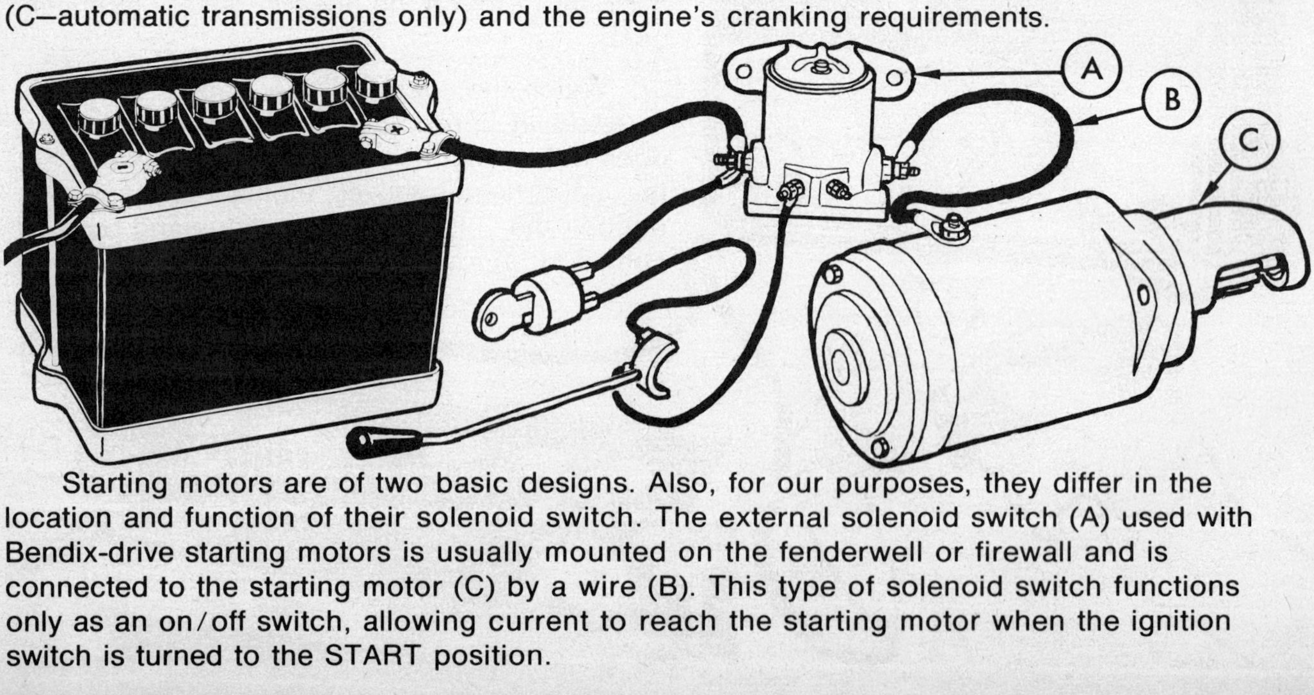

Starting motors are of two basic designs. Also, for our purposes, they differ in the location and function of their solenoid switch. The external solenoid switch (A) used with Bendix-drive starting motors is usually mounted on the fenderwell or firewall and is connected to the starting motor (C) by a wire (B). This type of solenoid switch functions only as an on/off switch, allowing current to reach the starting motor when the ignition switch is turned to the START position.

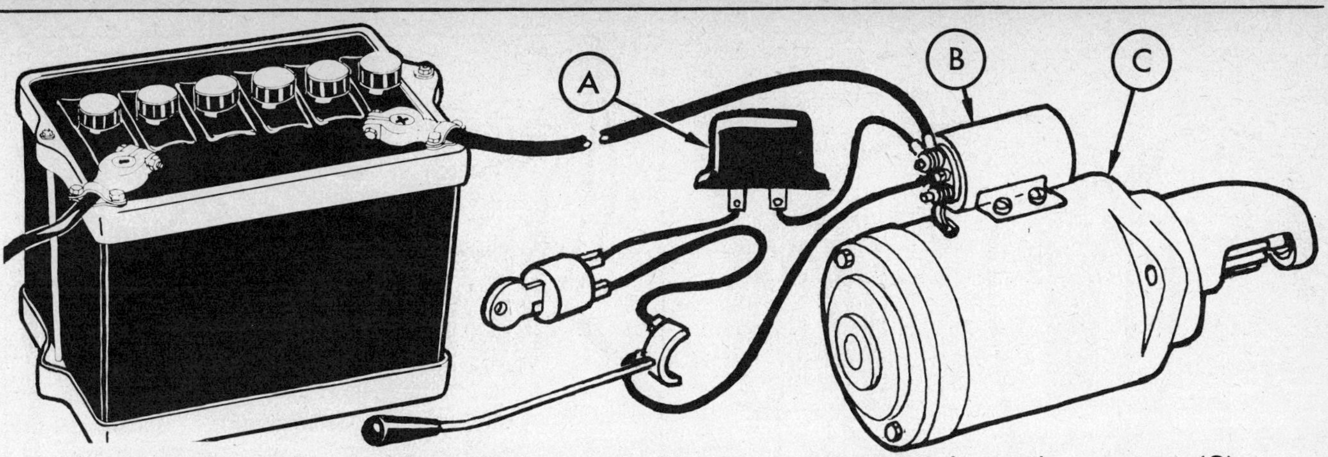

The integral solenoid switch (B) is used with overrunning-clutch starting motors (C). In addition to controlling the current flow, it moves the starter pinion gear into engagement with the engine's flywheel. Some Chrysler Corporation ignition systems also incorporate a relay (A) in the ignition switch-to-solenoid switch wiring. When the ignition switch is turned to the START position, the relay is activated, and this in turn activates the solenoid switch mounted on the starting motor.

STARTING SYSTEM TESTS—CRANKING VOLTAGE

This is a quick and easy overall test of the entire starting system and will only indicate whether problems do or do not exist. If cranking voltage is low or cranking speed sluggish, further tests must be made to locate and determine the exact cause.

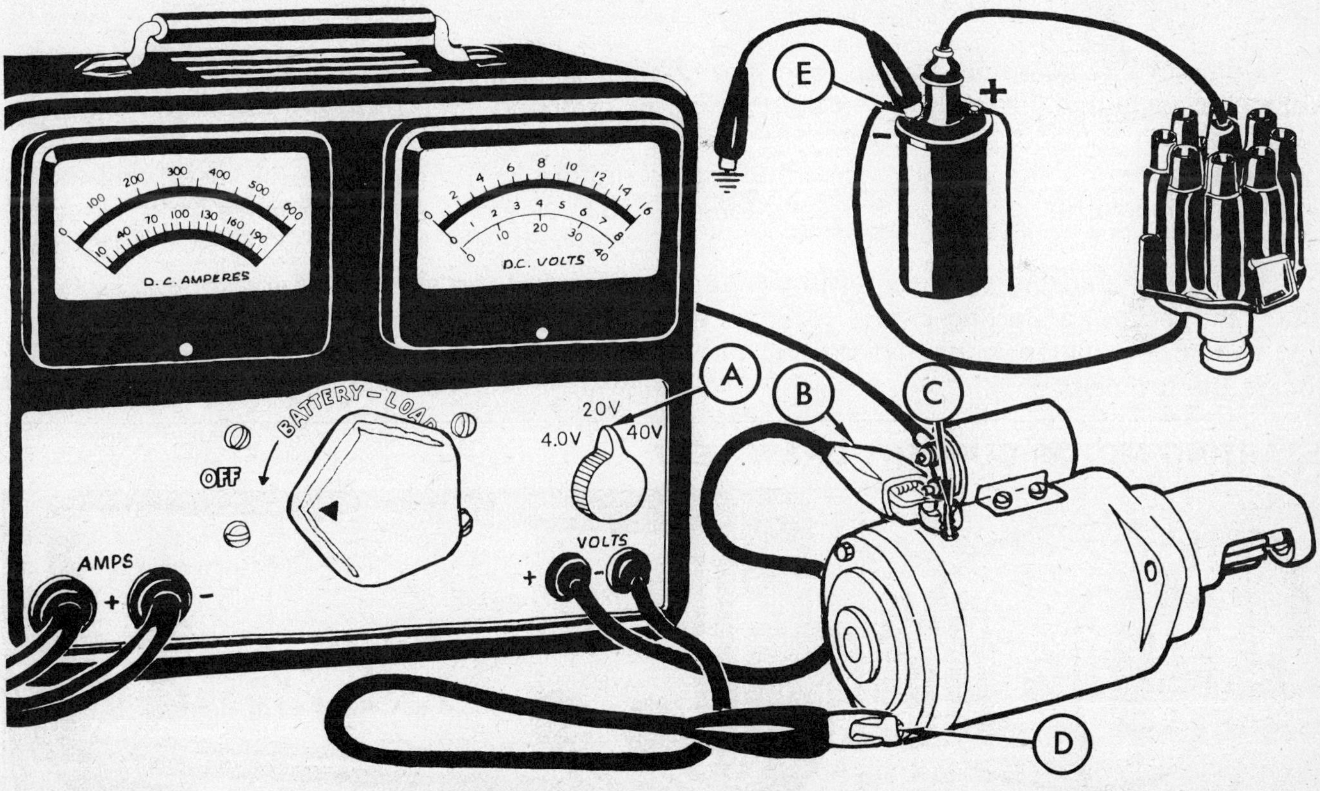

Adjust the voltage selector switch to the closest range appropriate for the system being tested (A). Connect the red voltmeter lead (B) to the starter terminal (C) on the starting motor and the black voltmeter lead to ground (D) on the starting motor frame.

CAUTION: GROUND OR DISCONNECT THE COIL DISTRIBUTOR LEAD (E—TERMINAL MARKED −) TO PREVENT THE ENGINE FROM STARTING WHEN IT IS CRANKED.

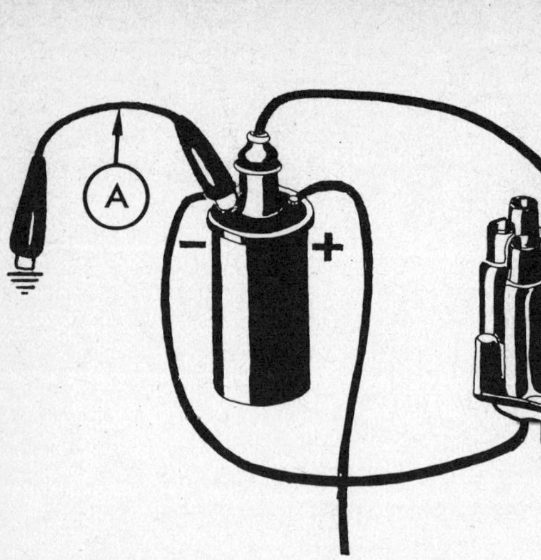

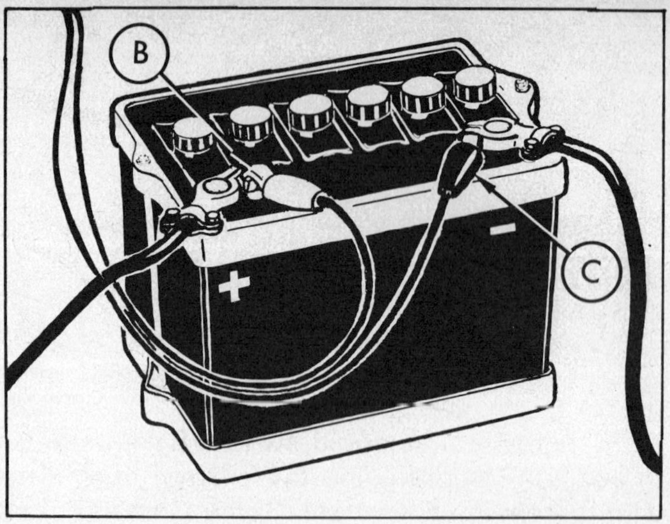

If you have difficulty in using this connection pattern because of the location of the starting motor, cranking voltage can also be measured at the battery. Connect the red voltmeter lead (B) to the positive battery terminal post and the black voltmeter lead (C) to the negative battery terminal post.

CAUTION: DISCONNECT THE DISTRIBUTOR PRIMARY LEAD AT THE COIL (TERMINAL MARKED –) OR USE A JUMPER LEAD (A) TO GROUND IT AND PREVENT THE ENGINE FROM STARTING WHEN CRANKED.

Connect a remote start switch (see that chapter) and crank the engine while reading the voltmeter scale and listening to the starting motor cranking speed. *Do not* crank the engine for more than 15 seconds at a time. The starting system can be considered to be in good condition if the starting motor cranks the engine uniformly at a good rate of speed and the voltmeter needle reads at least 9.6 volts for a 12-volt system; 4.8 volts or more for a six-volt system.

When the cranking voltage is within specifications but the starting motor sounds sluggish, perform a starting circuit resistance test. If cranking voltage is below specifications, you should test battery capacity, starting circuits and the starting motor cranking current to locate the problem.

STARTING MOTOR CURRENT DRAW TEST

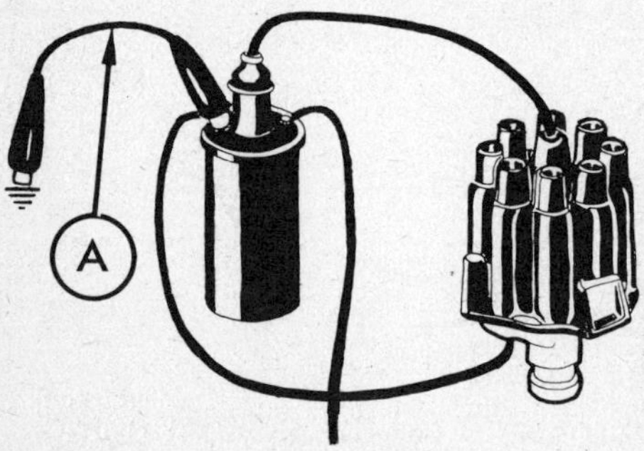

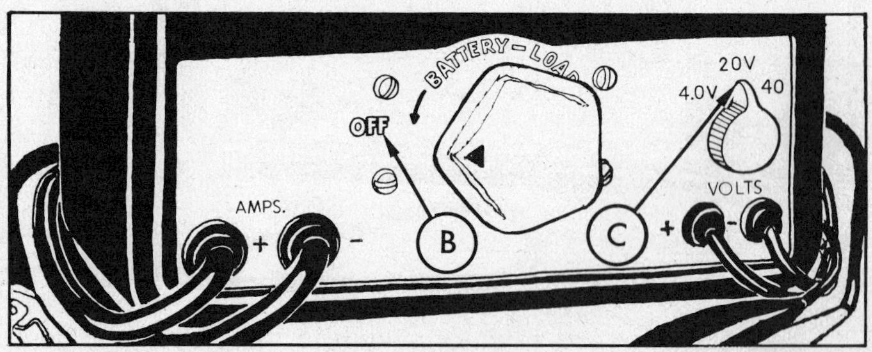

With the load control knob in the OFF or extreme DECREASE position (B), adjust the voltage selector switch to the closest range appropriate for the system being tested (C).

CAUTION: GROUND OR DISCONNECT THE COIL DISTRIBUTOR LEAD (A—TERMINAL MARKED –) TO PREVENT THE ENGINE FROM STARTING WHEN CRANKED.

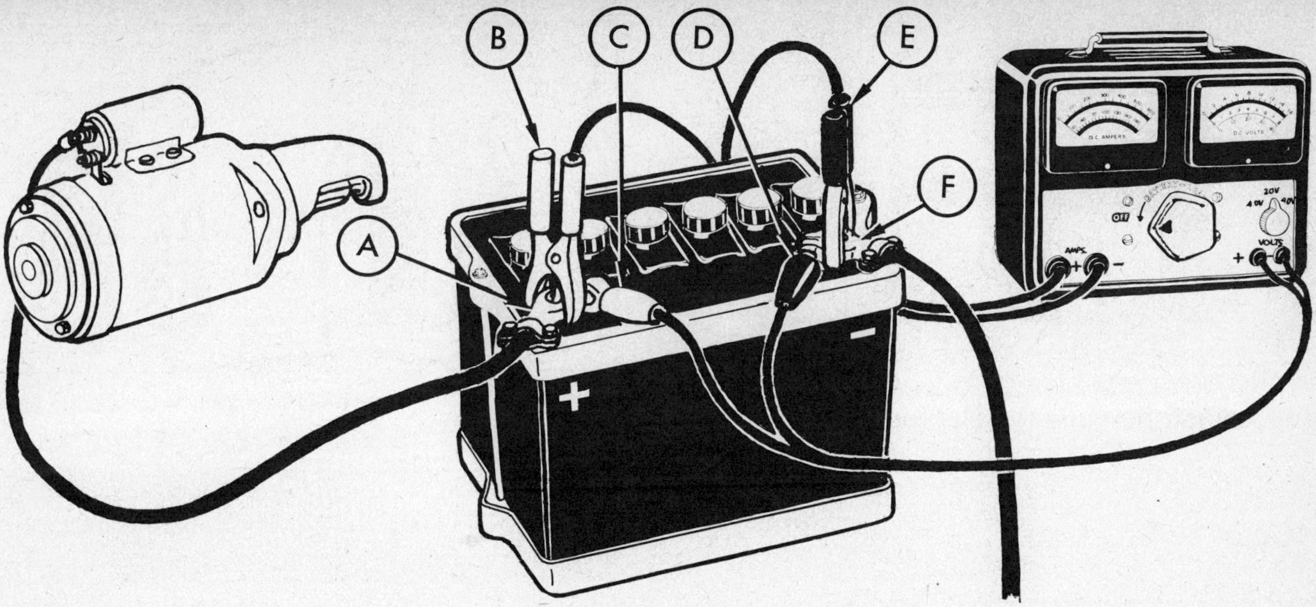

Now connect the red ammeter (B) and voltmeter (C) leads to the positive battery terminal post (A) and the black ammeter (E) and voltmeter (D) leads to the negative battery terminal post (F).

Crank the engine, using a remote start switch, and listen to the starting motor cranking speed. Do not crank the engine for more than 15 seconds at a time. Read the voltmeter scale at its final, steady reading and remember it or jot it down on a piece of paper.

Without cranking the engine, adjust the load control knob in a clockwise direction until the voltmeter reading is exactly the same as that obtained while the engine was cranking. The ammeter needle will now indicate the current drawn by the starting motor while it is cranking the engine. As battery voltage is proportional to amperage flow, the battery-starter tester duplicates the voltage reading during cranking and eliminates having to insert the ammeter in the circuit during cranking. The same amount of current (amperage) flows through the tester that was formerly flowing through the cranking motor, since the voltage is the same in each case. Reduce the load to OFF and compare this reading against that specified by the manufacturer.

A normal current draw accompanied by a low cranking speed means there is excessive resistance in the starting circuit. Run a complete resistance check of the insulated and ground circuit. If resistance is not excessive in the circuit, the starting motor must be removed and bench-checked for poor brush or commutator conditions and/or loose or high-resistance connections in the armature and field circuits.

A low starting motor current draw coupled with low cranking speed or none at all (total failure to crank the engine) requires the same troubleshooting procedure.

A high starting motor current indicates trouble in the starting circuit, but can also occasionally be due to an overheated or overly tight engine. If the engine has been recently overhauled or if it runs hot, let it cool for 30 minutes and then repeat the current draw test.

If current draw is still above specifications, you may have either a mechanical drag in the starting motor or the engine or a short or excessive resistance in the starting circuit wiring. Check the possibility of shorts and excessive resistance in the wiring as a first step; if neither is present, the starting motor must be removed for a free-running or "no-load" test to find out whether it or the engine is at fault.

STARTER INSULATED CIRCUIT RESISTANCE TEST

This measures the resistance of cables and switches that carry current to the starting motor. The heavy current required by the starting motor produces a voltage drop in the wiring that can be measured by this test as an indicator of resistance.

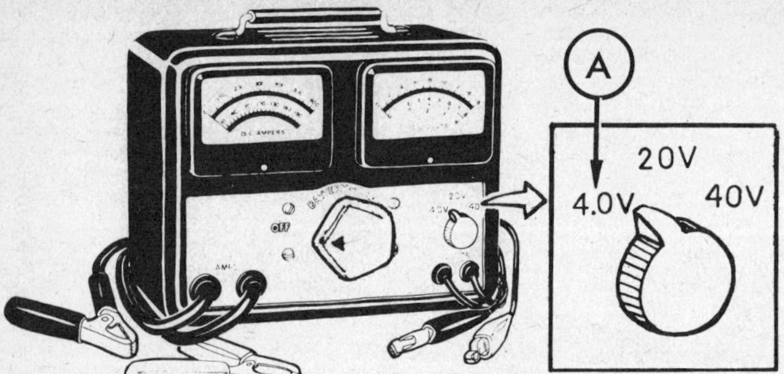

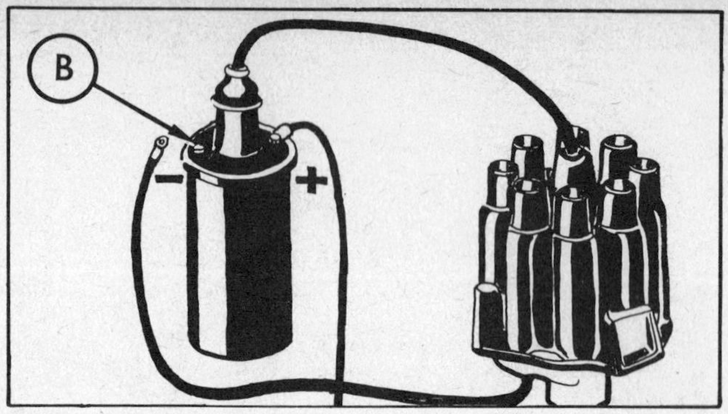

Adjust the voltage selector switch to the 4-volt position (A).

CAUTION: DISCONNECT THE DISTRIBUTOR LEAD AT THE COIL (B—TERMINAL MARKED −) TO PREVENT THE ENGINE FROM STARTING WHILE IT IS BEING CRANKED.

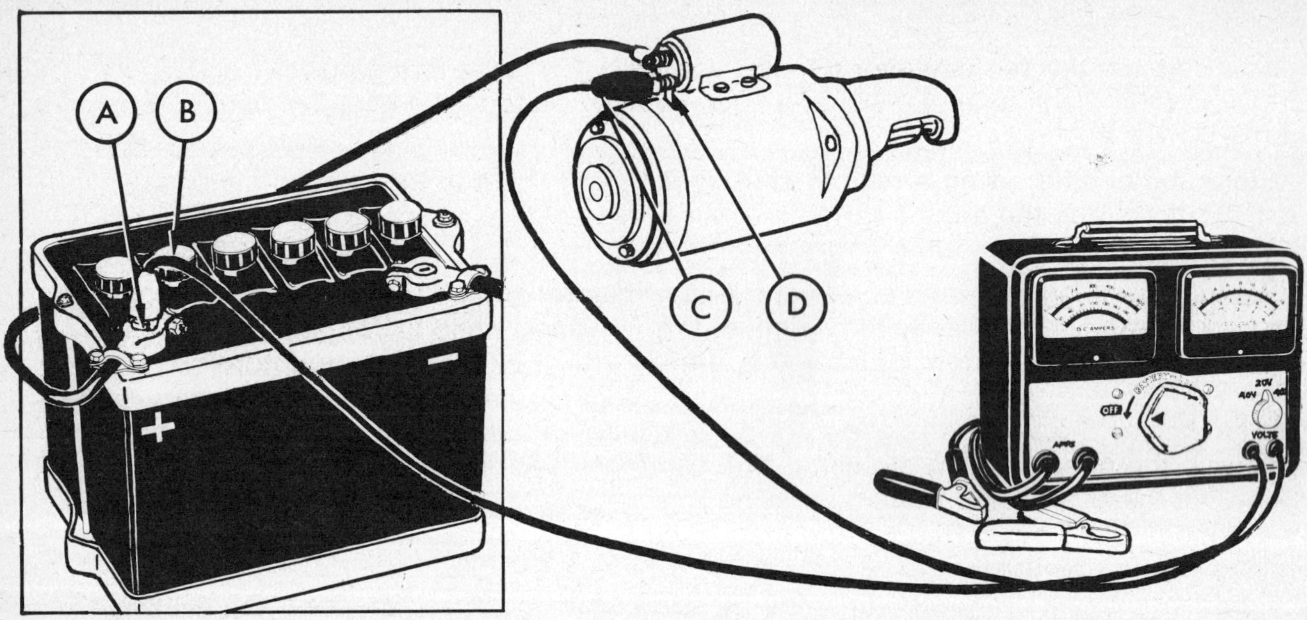

Connect the red voltmeter lead (B) to the battery positive terminal post (A)—not the cable clamp—and the black voltmeter lead (C) to the starting motor terminal (D). The voltmeter will now read off-scale to the right until the starting motor is operated.

Crank the engine while watching the voltmeter needle. A voltage reading of 0.3 volt or less usually indicates that voltage drop in the cables, connections and starting relay solenoid is normal. However, it is best to check the manufacturer's specifications for the exact voltage loss permitted, since some systems do allow as much as 0.6 volt.

Voltmeter readings above those specified by the manufacturer indicate high resistance in the starter insulated circuit. To trace the exact point of excessive resistance, disconnect the black voltmeter lead and connect it to each connection in sequence, repeating the test with the engine cranking.

	12-VOLT BATTERY	SIX-VOLT BATTERY
BATTERY INSULATED CABLE	0.2 Volt	0.1 Volt
SOLENOID SWITCH	0.1 Volt	0.1 Volt
EACH CONNECTION	0.0 Volt	0.0 Volt

Unless otherwise specified by the car's manufacturer, the maximum allowable voltage loss across various parts of the insulated circuit should not exceed the amounts in the chart.

GROUND CIRCUIT RESISTANCE TEST

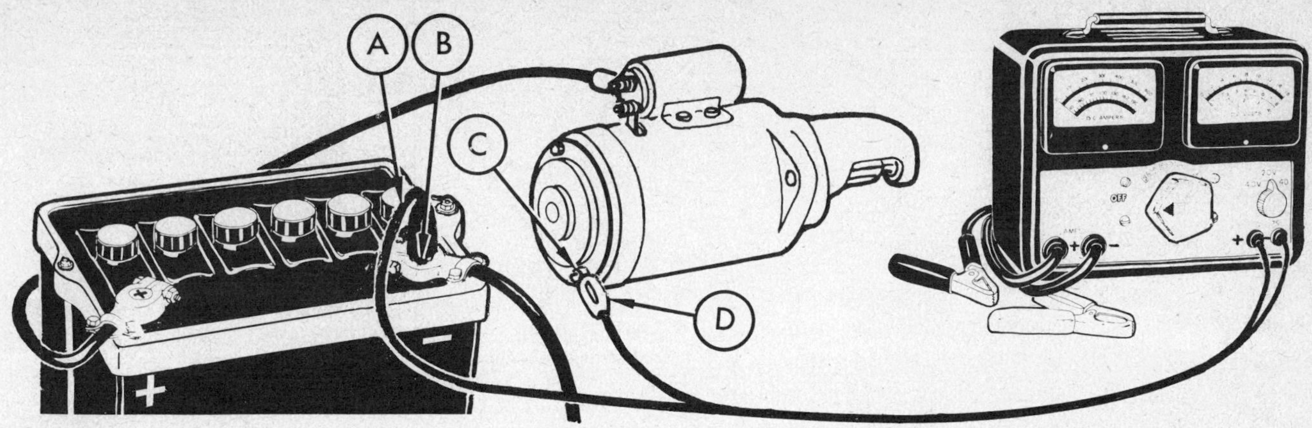

With the voltmeter selector switch set to the 4-volt position and the distributor primary lead disconnected from the coil (terminal marked —) as in the insulated circuit resistance test, attach the red voltmeter lead (D) to the starting motor case (C) and the black voltmeter lead (A) to the battery negative terminal post (B)—*not* the cable clamp.

Crank the engine while watching the voltmeter needle. A voltage reading of 0.2 volt or less usually indicates that voltage drop in the cables is normal, but it is best to check the manufacturer's specifications for the exact voltage loss. Some systems do allow as much as 0.3 volt.

Voltmeter readings above those specified by the manufacturer indicate high resistance in the starting ground circuit. To trace the exact point of excessive resistance, remove the voltmeter leads and touch them across each connection and cable, reading the voltmeter scale while cranking the engine. Readings across each connection should be zero; on ground cables shorter than 24 inches, 0.1 volt or less; on ground cables 24 inches or longer, 0.2 volt or less.

STARTING SOLENOID SWITCH RESISTANCE TEST

High resistance in the solenoid switch circuit cuts the current flowing through the solenoid windings. This in turn causes the solenoid to work incorrectly or may even prevent it from functioning at all. It can also burn the main switch contacts, which will reduce the current flow in the starting motor circuit.

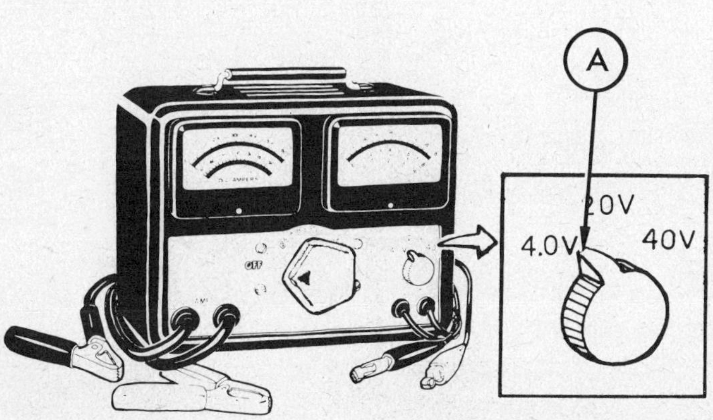

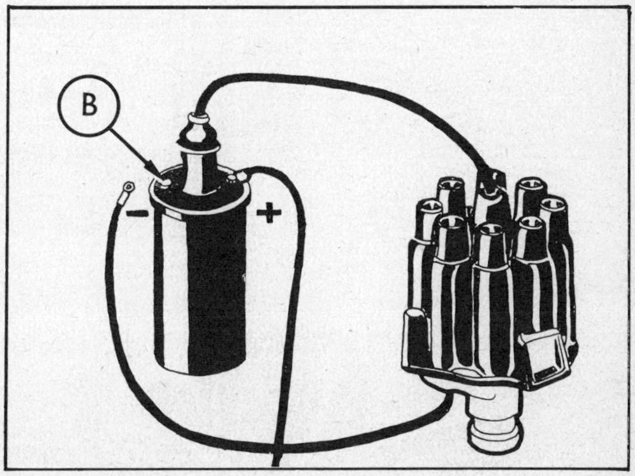

Adjust the voltage selector switch to the 4-volt position (A).

CAUTION: DISCONNECT THE DISTRIBUTOR PRIMARY LEAD FROM THE COIL (B—TERMINAL MARKED —) TO PREVENT THE ENGINE FROM STARTING WHEN CRANKED.

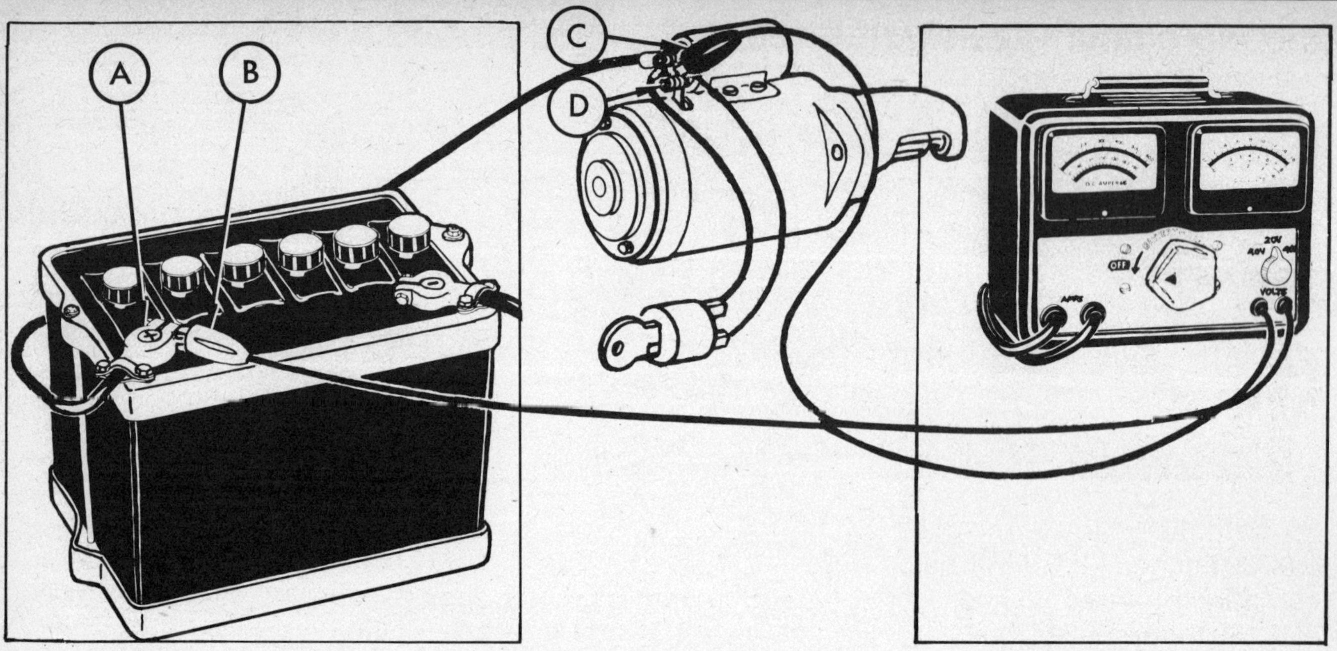

Connect the red voltmeter lead (B) to the positive battery terminal post (A) and the black voltmeter lead (C) to the solenoid ignition switch terminal (D). The voltmeter lead should be clipped *directly* to this terminal and not to the end of the ignition switch wire.

Turn the ignition switch ON and crank the engine (do not use the remote start switch in this test) while watching the voltmeter scale. Compare this reading against that specified by the manufacturer; if the voltage loss is higher than specified, the solenoid starting switch should be replaced.

FREE-RUNNING STARTER CURRENT DRAW TEST

When starter current draw is above normal but circuit resistance is within specifications, the starting motor must be removed from the car and tested for its free-running current draw. Also known as the "no-load" test, this should be done before disassembling the starting motor and is a bench test. It cannot be done with the starting motor mounted on the car.

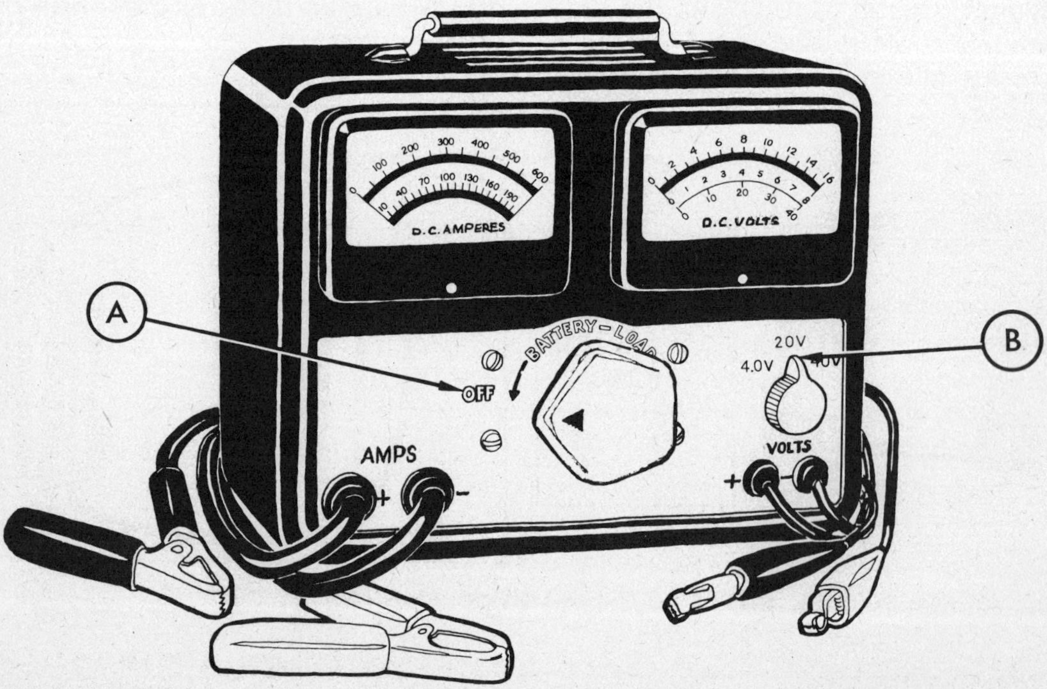

With the load control knob in the OFF or extreme DECREASE position (A), adjust the voltage selector switch to the closest range appropriate for the system being tested (B).

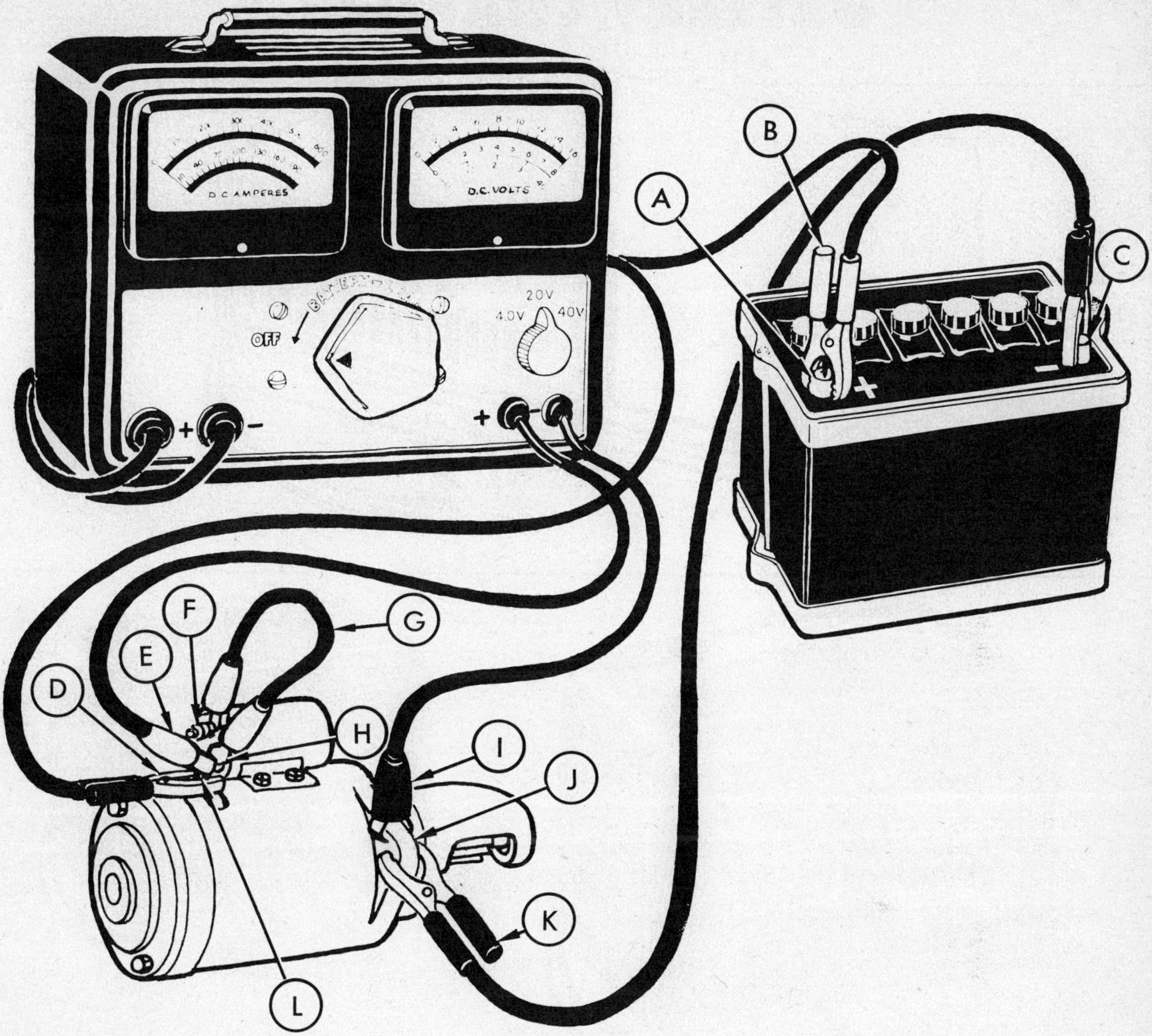

Connect the red ammeter lead (B) to the positive battery terminal post (A) and the black ammeter lead (D) to the starting motor terminal post (L). Connect the red voltmeter lead (E) to the starting motor terminal post (H) and the black voltmeter lead (I) to the starting motor frame (J). Run one jumper lead (K) from the starting motor frame (J) to the negative battery terminal post (C). If the starting motor being tested has an attached solenoid, then run a second jumper lead (G) from the solenoid's input terminal (F) to its control terminal (H).

The ammeter needle will indicate the starting motor free-running current draw when you turn the load control knob clockwise until the free-running test voltage is set on the voltmeter scale. Unless the current draw is measured at the exact voltage as specified, though, the ammeter reading will be meaningless.

If the current draw is greater than specified, the starting motor may have shorted circuits, a rubbing armature, a bent armature shaft or tight bushings.

A low or zero reading indicates excessive internal resistance. Check for loose or high-resistance connections in the field or armature circuit and for poor brush or commutator conditions.

BATTERY CHARGER

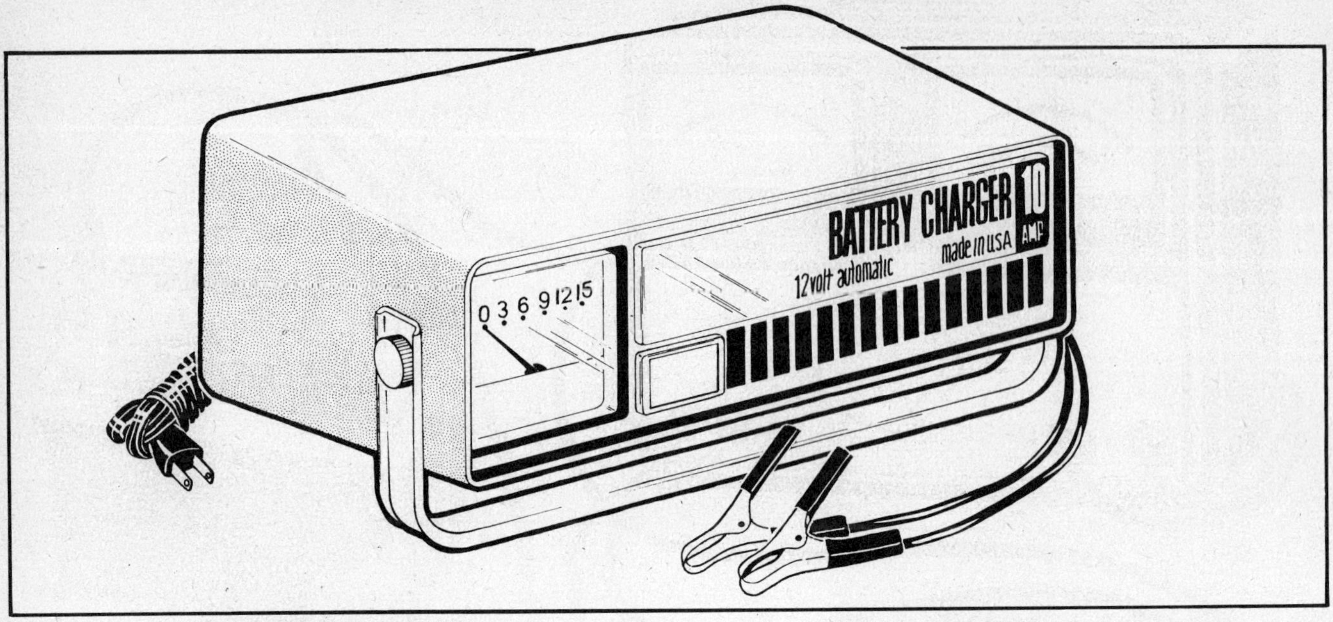

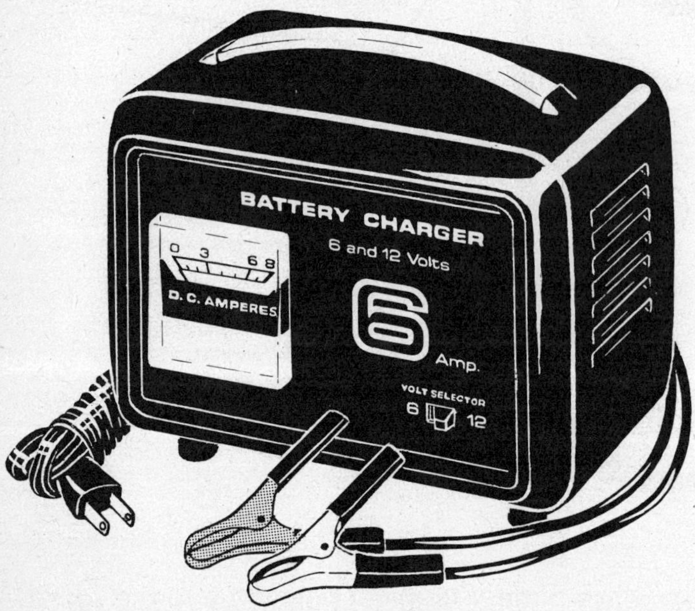

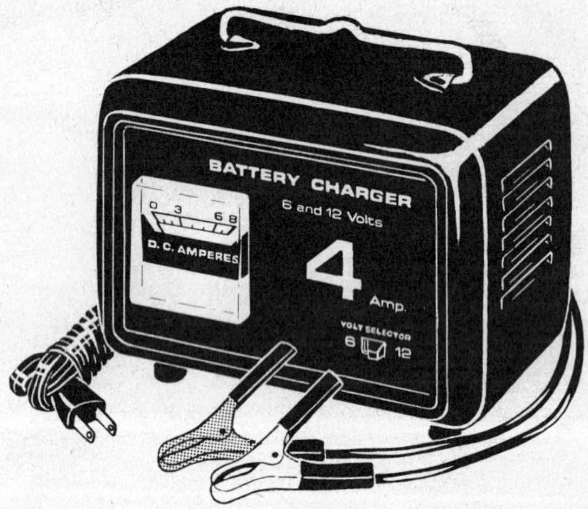

Although the battery charger is not strictly a piece of tune-up/test equipment, its use is required at times in conjunction with a battery hydrometer or battery-starter tester (see those chapters). Home battery chargers are available in a wide variety of amperage ratings and designs, most of which are quite satisfactory as long as they are used for the purpose they are designed for—keeping the battery safely charged under difficult service conditions.

Many discharged batteries, especially those that are sulphated, can be brought back to good condition by a slow charge. Sulphating is a battery condition that takes place when large areas of the plates become covered with heavy deposits of lead sulphate due to inadequate charging or old age. The chemical reaction between the lead plates and electrolyte acid changes some of the material into lead sulphate. If the battery is not charged enough to convert the compound back into usable materials, sulphating gradually takes place. As sulphated areas tend to harden permanently, their chemical convertability can be lost. A long, slow charge is the only means of completely displacing the acid from the sulphated areas and restoring the battery to its full capacity. A lesser charge will not remove the sulphate but it can return the battery to service temporarily.

ELECTROLYTE LEVEL

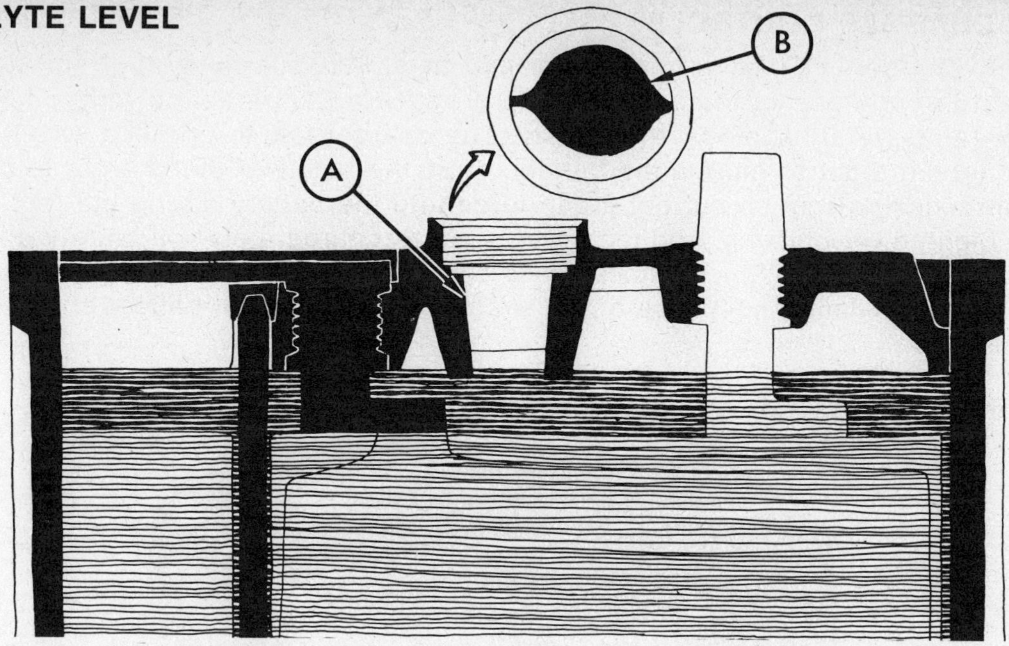

Most batteries are constructed with some kind of electrolyte level indicator. Conventional automotive batteries usually use a tube-shaped filler or vent (A) that extends down into the case. Some have slotted sides or a diamond-shaped bottom to mark the bottom of the vent well. When the electrolyte touches the vent well, its surface will appear distorted (B).

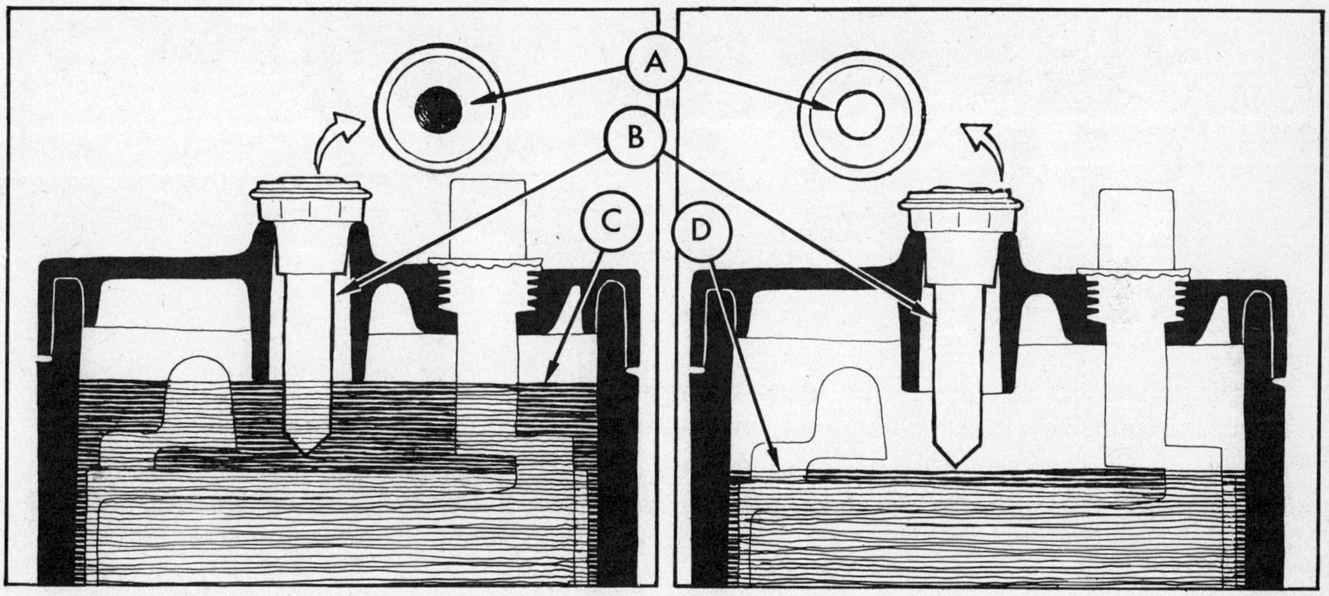

Delco Energizers use what is known as a Delco Eye (A). This specially designed vent plug has a transparent rod (B) extending through its center. When the electrolyte is at its proper level (C), it covers the lower part of the rod, giving the uncovered top part of the rod a dark appearance. If the electrolyte level is low (D), it falls below the tip of the rod, which then seems to glow.

CAUTION: BEFORE CHARGING YOUR BATTERY, MAKE SURE THAT THE ELECTROLYTE COVERS THE PLATES; IF NOT, ADD SUFFICIENT WATER TO DO SO.

Do not bring the level up to normal at this point, as charging tends to raise the electrolyte level, which could cause the battery to overflow. If more water is required, it should be added after the charging process is complete.

BATTERY CHARGER CONTROLS

A battery charger is simply a transformer that takes AC (alternating current) line voltage and passes it through one or more rectifiers to produce a DC (direct current) output at a current flow (amperage) usable by the battery. The charger has to produce a voltage higher than the battery in order to charge the battery. Thus the ability of the charger to produce voltage determines how much current it can force into the battery during the recharging process. A high-power unit will produce seven volts to charge a six-volt battery and 14 volts to charge a 12-volt battery. Since smaller units cannot produce that much voltage, their charging capacity is limited. Because of the simplicity of operation, battery charger controls are minimal.

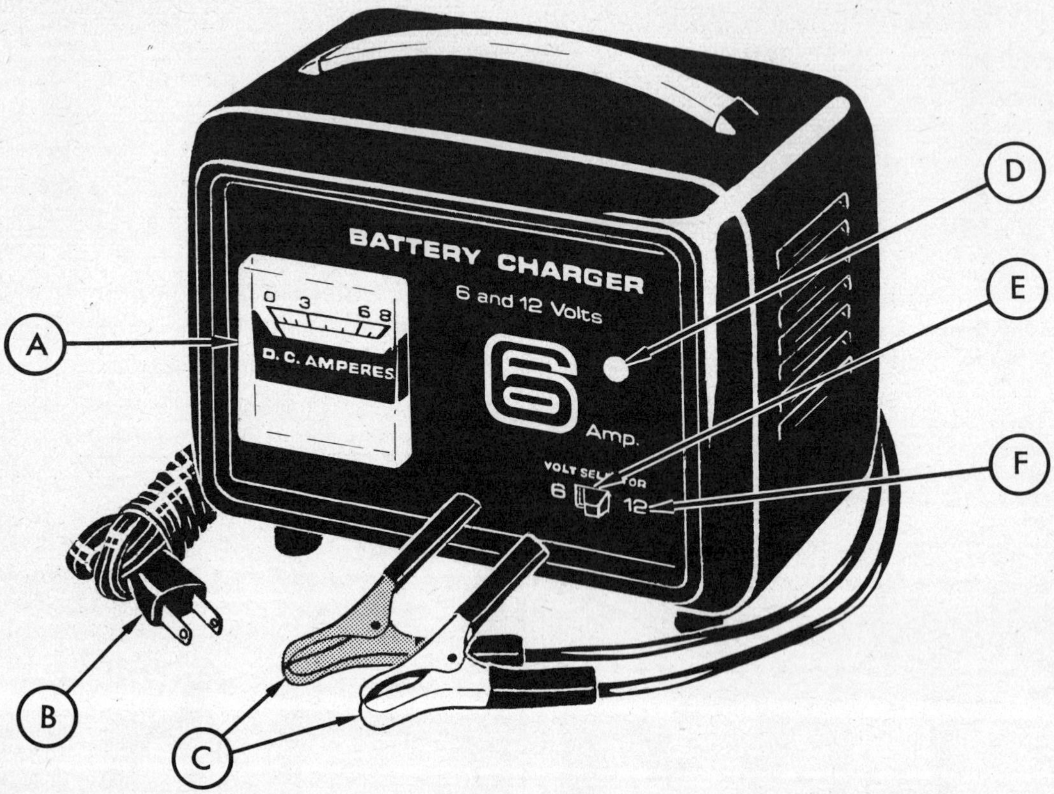

Each charger is equipped with a DC ammeter (A) on its front panel, although some of the very inexpensive "trickle" chargers (three amps or less) may not use one.

Battery chargers that are adjustable for six- or 12-volt operation will have a voltage selector switch (E), with each position clearly marked on the panel (F). The present trend in charger design (other than professional shop equipment) seems to be in the direction of 12-volt-only units.

Recent models may include a reverse polarity light (D) to indicate incorrect charger lead connections to the battery.

CAUTION: IF THE POLARITY LIGHT COMES ON, REVERSE THE LEAD CONNECTIONS. THE LIGHT MUST BE OFF BEFORE THE UNIT CAN BE OPERATED WITHOUT DAMAGE TO THE BATTERY OR CHARGING SYSTEM.

Twin leads (C) connected permanently to the charger are used to connect it to the battery. These are usually equipped with scissor-type spring return clamps and have color-coded (positive = red; negative = black) insulation for ease in making the proper connections.

An attached AC line cord (B) is used for connecting the charger to a power source.

PREPARATION FOR CHARGING

Loosen or remove the filler/vent caps. A six-volt battery has three caps, while a 12-volt battery has six caps.

WARNING: A HIGHLY EXPLOSIVE GAS MIXTURE (HYDROGEN) FORMS IN EACH CELL DURING CHARGING, SO BE SURE TO USE PRECAUTIONS WHEN WORKING AROUND A BATTERY THAT IS UNDER CHARGE. DO NOT SMOKE, CREATE SPARKS OR BRING AN OPEN FLAME NEAR THE BATTERY. TO PREVENT AN ELECTRICAL ARC THAT COULD CAUSE AN EXPLOSION, ALWAYS CONNECT THE AC POWER CORD TO A LINE OUTLET *AFTER* THE CHARGER LEADS ARE CONNECTED, AND DISCONNECT THE POWER CORD *BEFORE* DISCONNECTING THE LEADS.

Check, clean and tighten both battery cables. If the battery is to be charged with its cables removed, clean the terminal posts thoroughly with a wire post cleaner (A). Worn or defective cables should be replaced.

CONNECTING THE BATTERY CHARGER

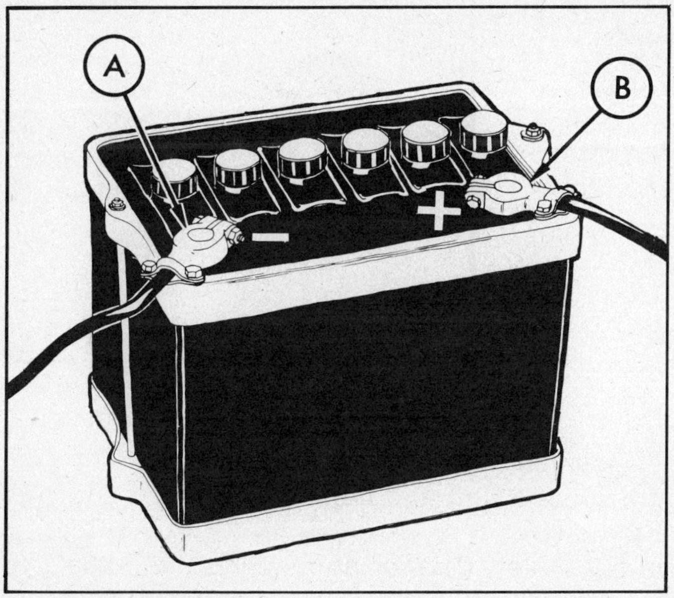

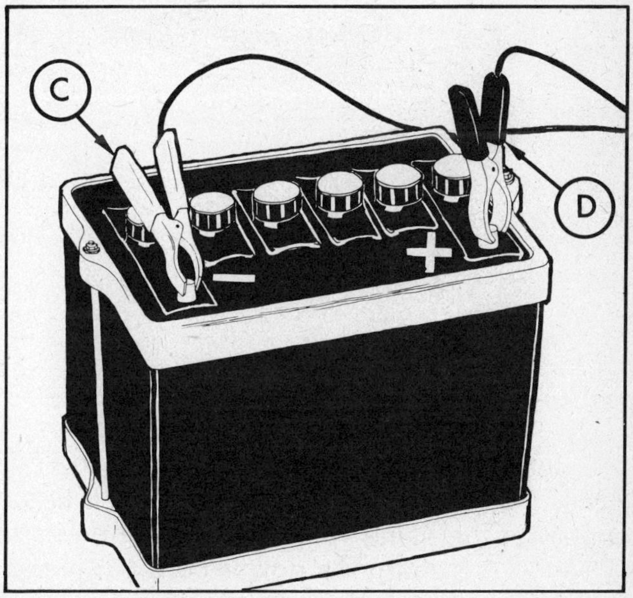

The charger should *always* be connected to the battery *before* it is plugged into an AC outlet. Disconnect the battery cables on alternator-equipped cars before connecting the charger. This will prevent accidental damage to electronic components from reverse polarity.

Remove the positive or "hot" cable (B) first, then the negative or ground cable (A). Now connect the red charger lead (D) to the positive battery terminal and the black lead (C) to the negative battery terminal. Rock the leads back and forth to assure a good connection. Double-check the connection sequence by making certain the reverse polarity light is off. On chargers without this warning light, if the connections are incorrect, sparking may take place when clamping the leads to the battery terminal posts and the ammeter will read upscale.

OPERATING THE CHARGER

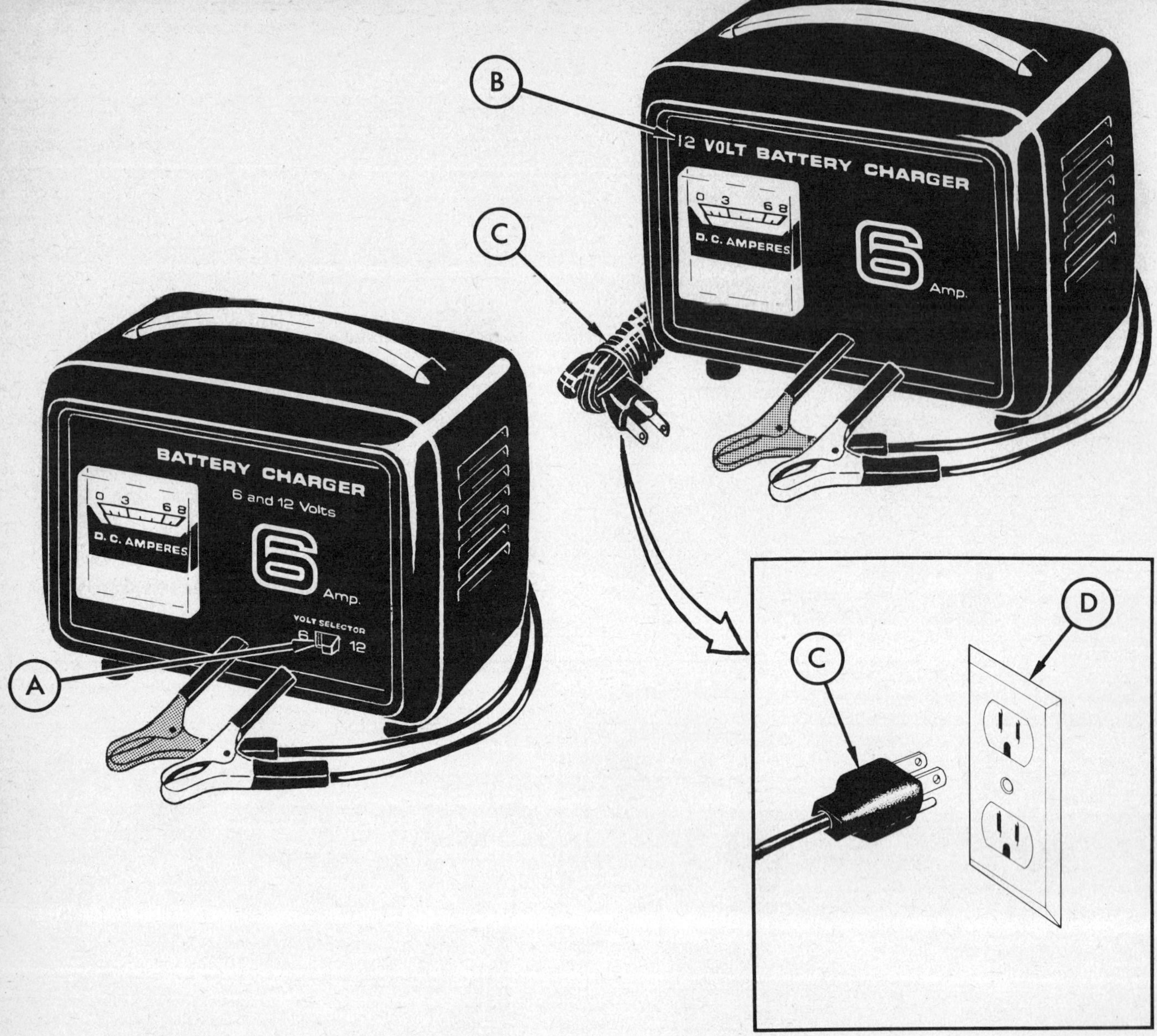

Set the voltage selector switch to the correct position (A) for the battery under charge. If the charger has no selector switch, it is a single-voltage unit and must be connected to a battery of the same voltage specified on its face plate (B).

Plug the charger's power cord (C) into a live AC outlet (D) *after* connecting the leads to the battery. The ammeter on the charger panel should now indicate the amount of charge current. If it does not, rock the charger lead/battery post connections; if the ammeter still does not read, check the AC line outlet to make sure that it is live.

The charging current will automatically adjust its rate to the battery condition. In the case of a warm, discharged battery, the current starts well upscale on the ammeter and decreases as the charge continues; on a cold or partially charged battery, it starts lower on the scale but decreases less. When the battery is fully charged, the charger's automatic sensing circuit will reduce the charging rate to a level sufficient only to replace battery self-losses.

Chargers without this automatic sensing circuit will also reduce their charging rate but not as low, and any prolonged connection to the battery can result in an overcharge and eventual damage to the battery.

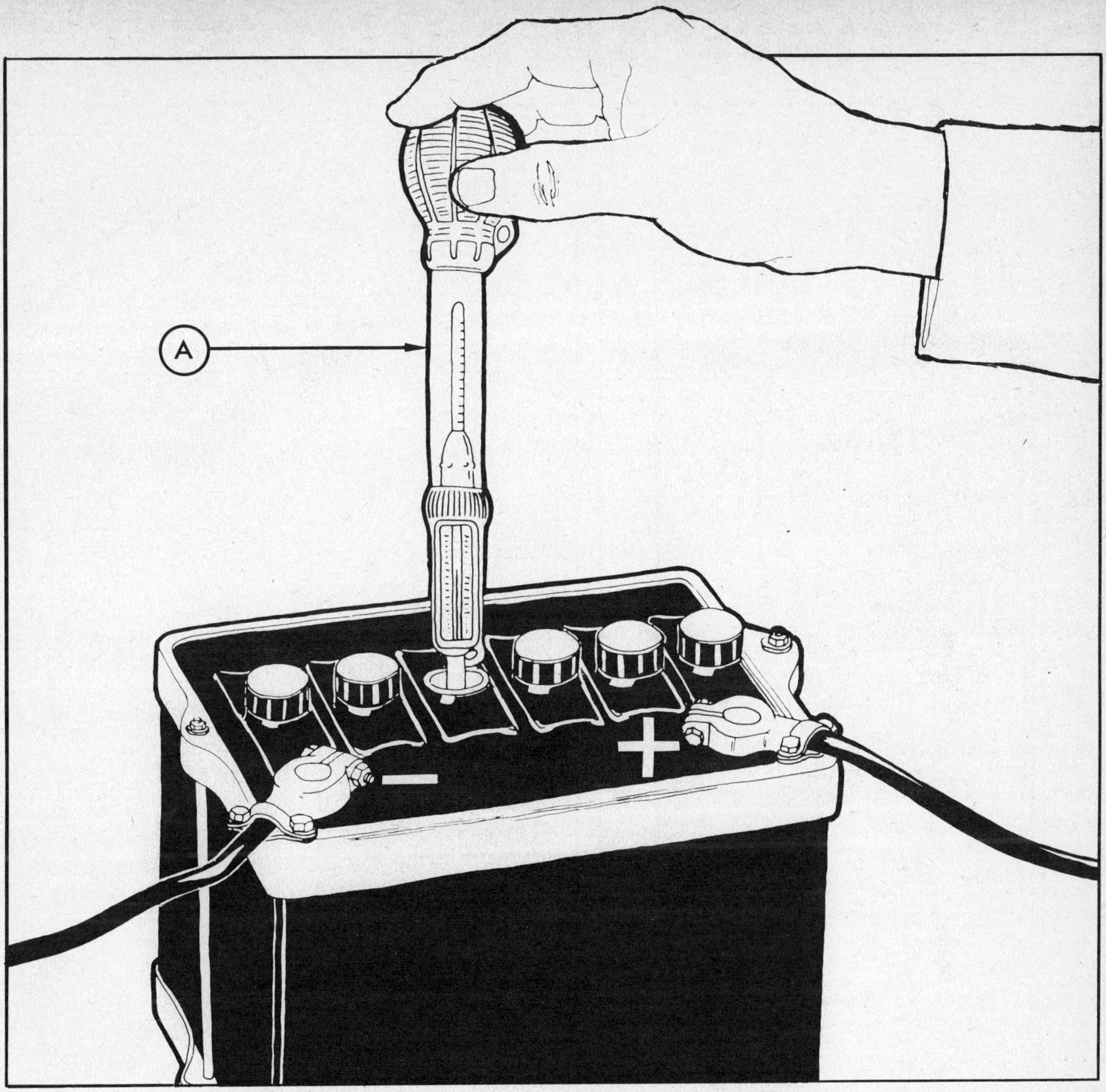

You can determine when a battery is fully charged by taking specific gravity readings (A) with a battery hydrometer (see that chapter). When the specific gravity remains unchanged on three successive readings an hour apart and all cells are gassing, the battery has accepted as much of a charge as it can hold. To determine the actual specific gravity, it is necessary to let the battery sit off-charge for an hour and then take a battery hydrometer reading, as readings taken while the battery is gassing will not be correct. They will only tell you whether or not the battery is charging.

CHARGING TIMES

The length of time required to fully charge a battery depends upon several internal conditions. An old battery will need up to 50% more ampere-hour charging than a relatively new one. A completely discharged battery needs twice the ampere-hour charge of one that is only half discharged. A 70-ampere-hour battery requires twice as much of a charge as a 35-ampere-hour battery, and charging at temperatures below 80° F will require more time. Specific gravity readings taken with the battery off-charge are the only sure way of telling when a battery has accepted a sufficient charge.

EXHAUST GAS ANALYZERS

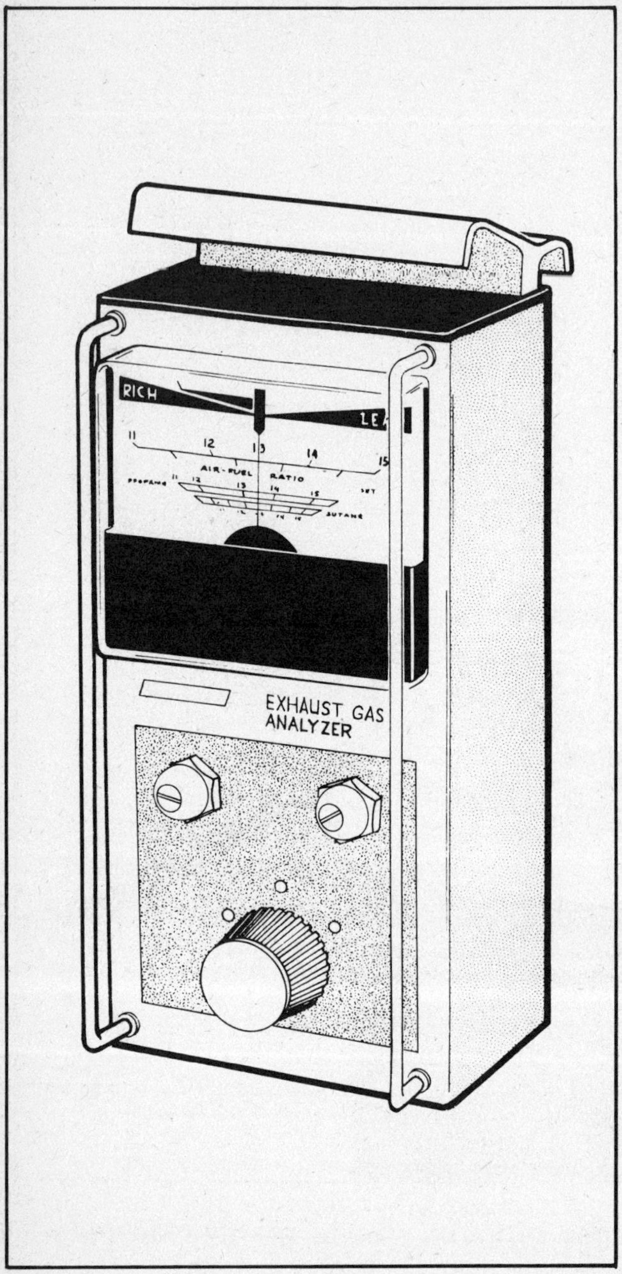

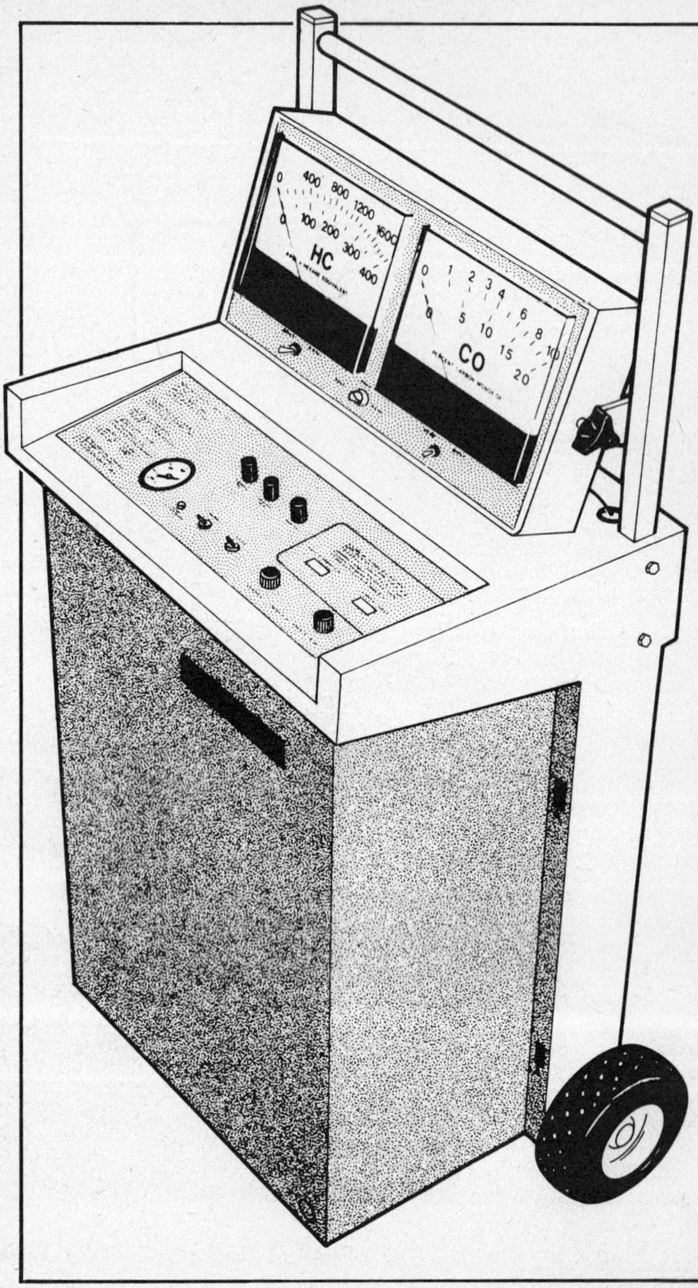

Since 1968, all American and many foreign cars sold in the United States have been equipped with one or more emission control systems to reduce hydrocarbons, carbon monoxide and nitrates of oxide pollutants from their exhaust. As "clean air" legislation requirements grow stronger each year in an attempt to solve the problem of smog and air pollution, what passes through your car's tailpipe and into the atmosphere becomes an important end product of a correct tune-up.

Although an automotive engine can be tuned and timed to run well in several different ways, there is only one correct tune-up method for cars equipped with emission controls, and that is to follow the manufacturer's specifications to the letter. Even if an engine *has* been tuned for top efficiency, the fact that specifications were followed is no real guarantee that it will operate efficiently. There may be an intake manifold air leak, a restricted air filter or some other factor unrelated to a tune-up that would not be discovered while tuning the engine, but which prevents it from working at peak efficiency. You can use an exhaust gas analyzer to discover just how well you have done your job.

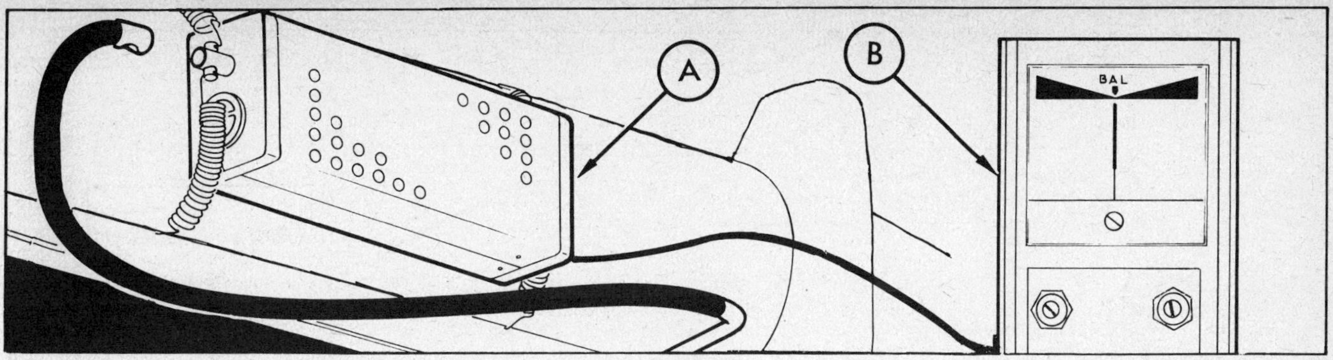

This test equipment uses a thermal conductivity (heat sensing) cell in a unit (A) to sample and measure the exhaust gas coming from the car's tailpipe. It provides a reading of the air/fuel ratio during idle, cruising and acceleration conditions on its meter scale (B). In addition to determining just how efficient the engine's operation really is, the exhaust gas analyzer can check carburetor adjustments and the proper operation of the carburetor's circuits as well. While exhaust gas analysis is usually thought of as a final step in the tune-up process to double-check adjustments and make sure that the engine is tuned to peak efficiency, it can be used before a tune-up to check engine operation and efficiency, determine if fuel is being wasted and if so, locate the cause so it can be corrected.

The most practical use for the exhaust gas analyzer comes in setting the idle air/fuel ratio on those cars for which such a ratio is specified in their factory shop manuals. If the manufacturer does not provide an idle air/fuel ratio setting, he will specify use of the "lean-drop" method of carburetor adjustment, where you lean the mixture with the carburetor mixture screw(s) until the rpm (revolutions per minute) drop a specified amount from the best idle setting. (See tach-dwell meter chapter.) Alternatively, use of the "quarter-turn rich" method may be advised, where you adjust the carburetor mixture screw(s) one-quarter turn richer after reaching the best idle setting.

Use of the specified method is absolutely essential to proper engine operation. The shop manual must be consulted for any changes. For example, the 1973 Chevrolet engines were adjusted by the "quarter-turn rich" method, but all 1974 Chevrolet engines are set by the "lean-drop" method. Adjusting the 1973 models by the lean-drop method will not allow the engine to function correctly, and vice versa.

BASIC ANALYZER DESIGN

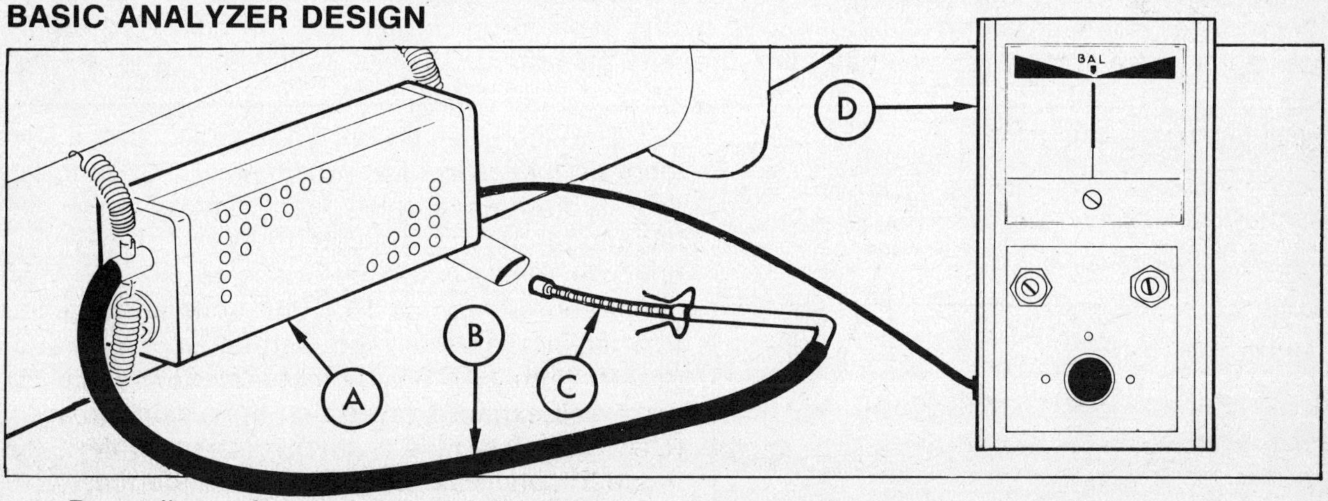

Regardless of manufacturer, all exhaust gas analyzers currently available are composed of three main and separate units: a meter or instrument panel with calibrated scale (D), a sensing unit (A) and a pickup tube (B) with probe (C) attached. Many sensing units contain a number of 1½-volt batteries which provide the electric current necessary to operate the system. Others hook directly to the car's battery or to a separate battery of equal voltage. Some manufacturers store the 1½-volt batteries under the meter's display scale (D) instead of in the sensing unit.

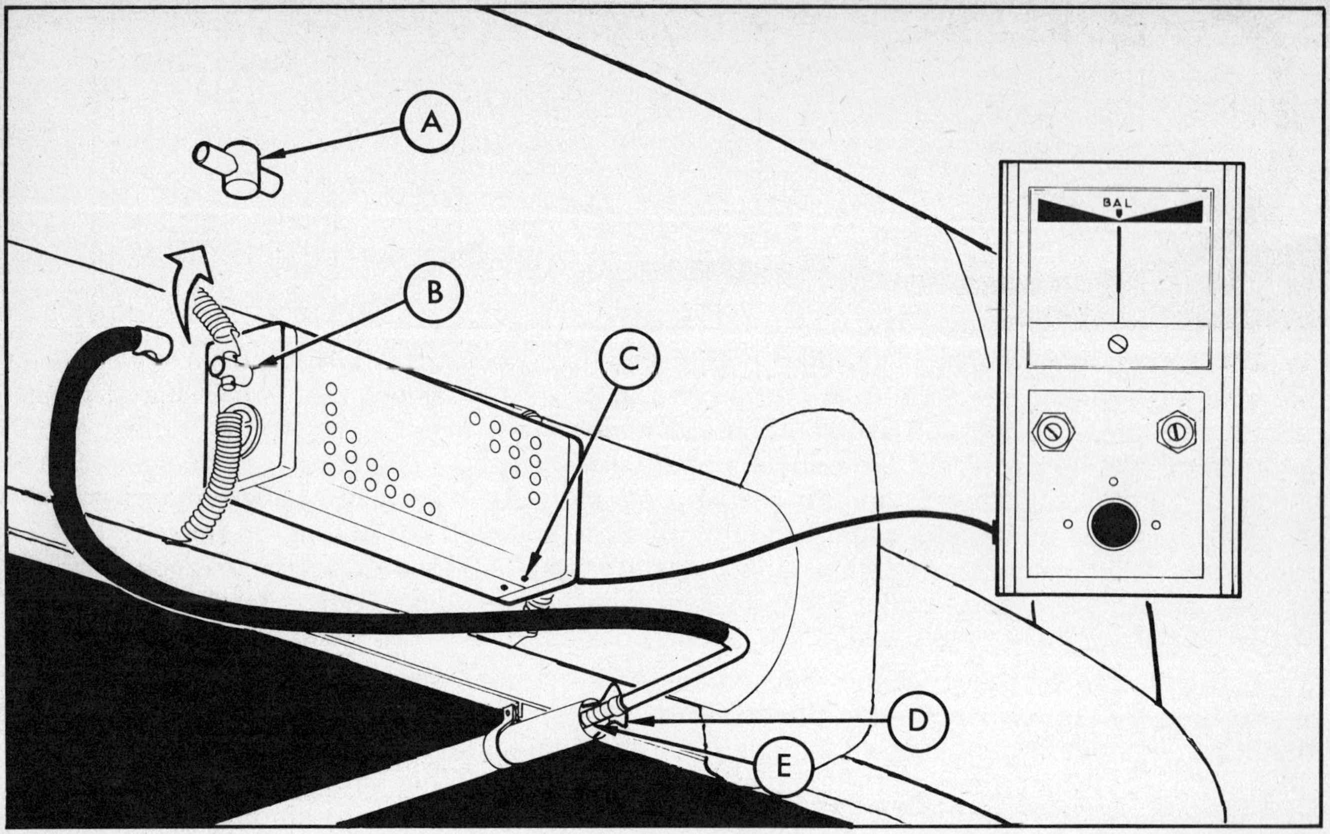

The sensing unit uses either a water trap (A) or drain hole (C) to take care of excessive moisture picked up from the exhaust gas by the pickup tube. The unit is secured to the rear bumper of the car near the exhaust tailpipe by a set of snap-on springs or a flexible strap.

The pickup tube is nothing more than a flexible hose (usually neoprene) equipped with a metal probe or tube (E) at one end and some form of clamp (D) to hold it in the tailpipe. The opposite end of the pickup tube fits into or over a nipple (B) on the sensing unit to complete the connection.

The sensing unit and meter are usually connected by a plug-in electrical cord. The exhaust gas content determination is made by the sensing unit's gas cell and displayed on the meter scale, once the scale has been properly calibrated.

LIMITATIONS IN USE

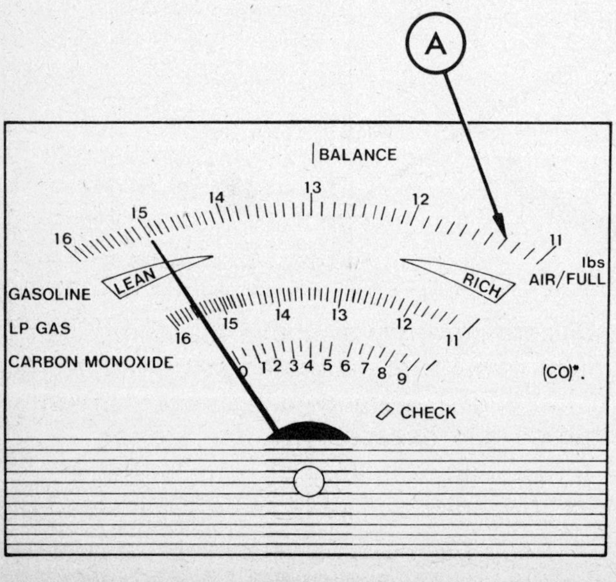

The use of emission controls requires a lean air/fuel ratio of approximately 14:1 at idle speed. Such excessively lean mixtures on late-model cars complicate the work of the analyzer. While many meter scales are calibrated as high as 16:1, the units are unable to actually read that high with accuracy. A ratio greater than 15:1 on the meter air/fuel scale (A) produces exhaust gases rich in carbon dioxide (CO_2) with very little if any hydrogen content. As this particular mixture reduces the thermal conductivity of the sensing unit's gas cell, the analyzer tends to become confused. The meter needle can reverse its reading, going all the way to the rich end of the scale. This peculiarity of all exhaust gas analyzers will trouble you most when checking part-throttle cruising speeds (the cruising mixture test).

METER DISPLAY SCALES

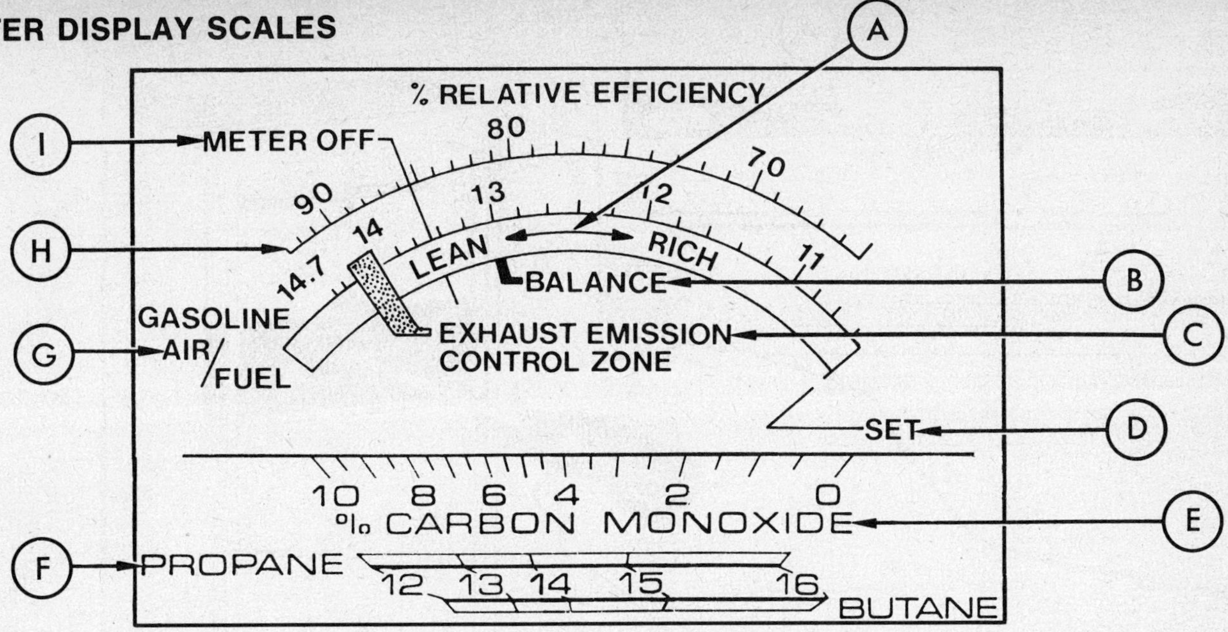

Each manufacturer has designed the appearance of his meter display scale differently, but all contain settings and scales that provide essentially the same basic information to the user. Regardless of the display scale design, you will usually find the following clearly marked: METER OFF (I), SET mark (D), BALANCE mark (B), AIR/FUEL ratio scale (G) and LEAN-RICH scale division (A).

In addition to these primary functions, some meters also contain specifically marked EMISSION CONTROL ZONES (C), CARBON MONOXIDE scale (E), LP (liquid petroleum) gas scale (F) and PERCENT OF RELATIVE EFFICIENCY scale (H). For quick reading, critical zones on some display scales are called out in red or green, to contrast with the black-on-white readout of basic data.

CALIBRATING THE ANALYZER

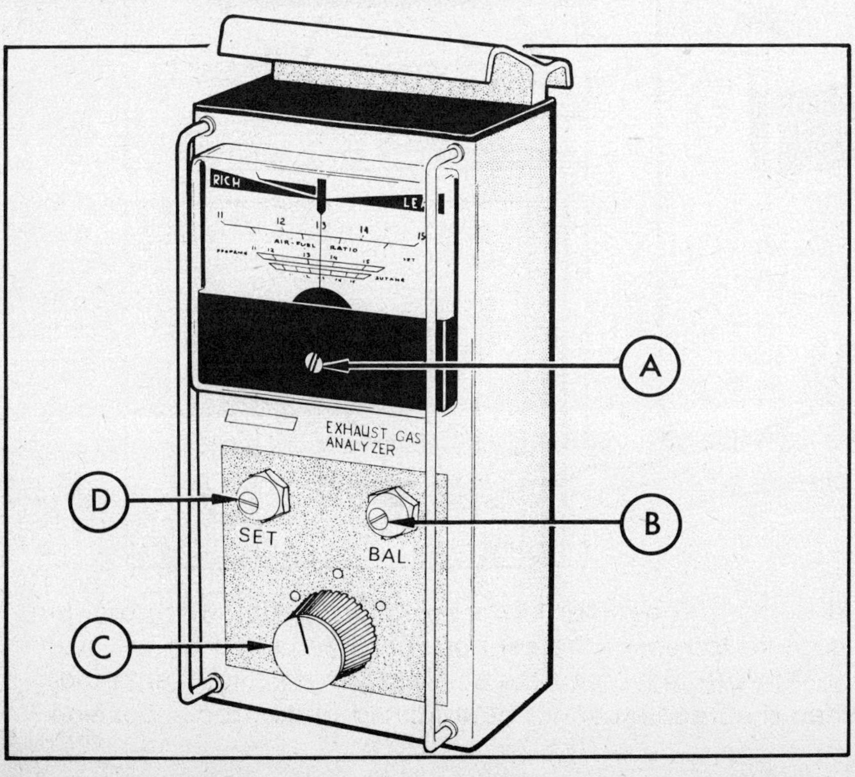

For accurate readings, the analyzer must be calibrated (adjusted) each time it is used. Calibration should be done *before* inserting the pickup probe in the tailpipe. Each unit should have an adjustment screw (A) or a knob or switch marked BALANCE (B) and one marked SET or CHECK (D) located on the analyzer below or to one side of the meter dial. One or more of these functions may also be combined with a selection mode switch (C) that contains other features, such as an off/on switch or automatic timer.

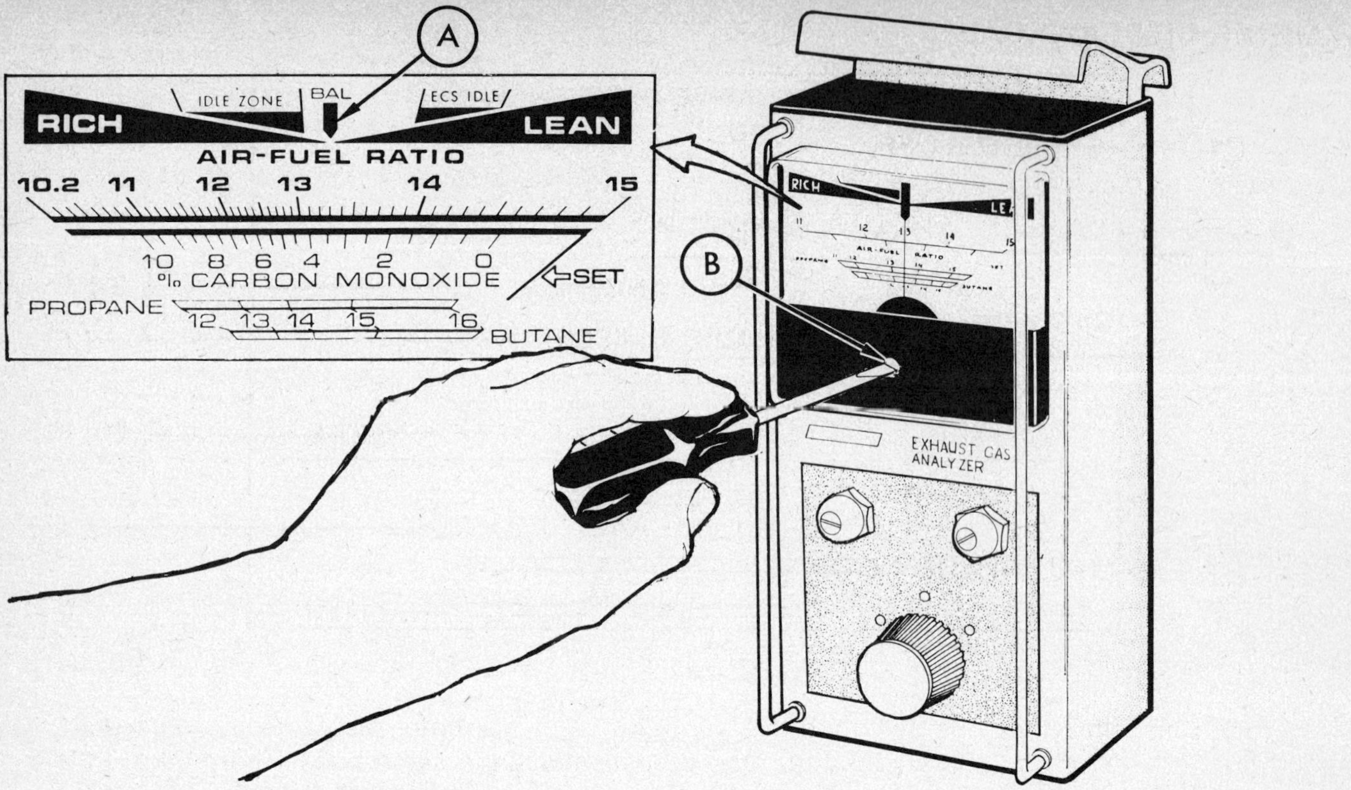

While calibration sequences may differ from one meter to another (check your unit's manual), the sequence usually begins by determining that the meter needle rests in the center of the BALANCE band or mark (A) on the meter scale. If it does not, use a small screwdriver to adjust the needle's position by turning the balance adjustment screw (B).

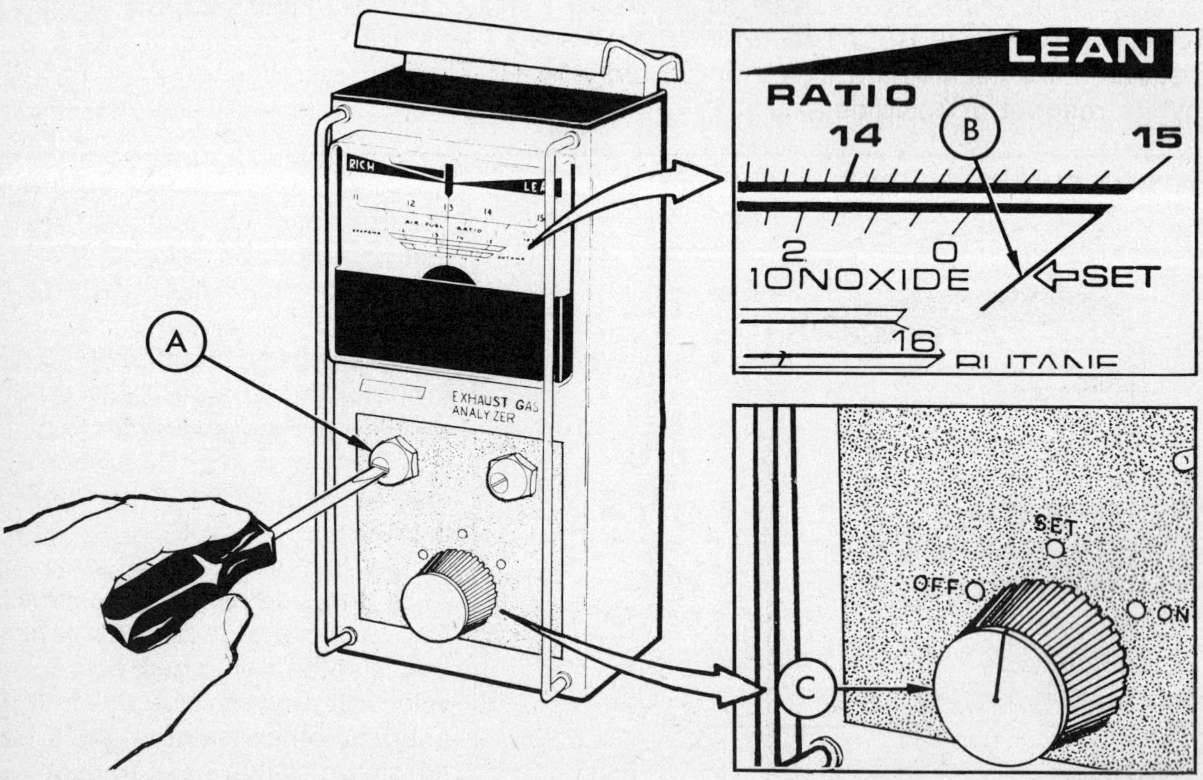

Adjust the SET or CHECK control (A). If the meter has a selection mode switch (C), turn it to SET. The meter needle should come to rest in the set position (B) marked on the scale. If it does not, adjust the set control (A)—you may need the screwdriver here too—until the needle reaches the set line (B). When the needle cannot be adjusted to its proper position, replace the battery.

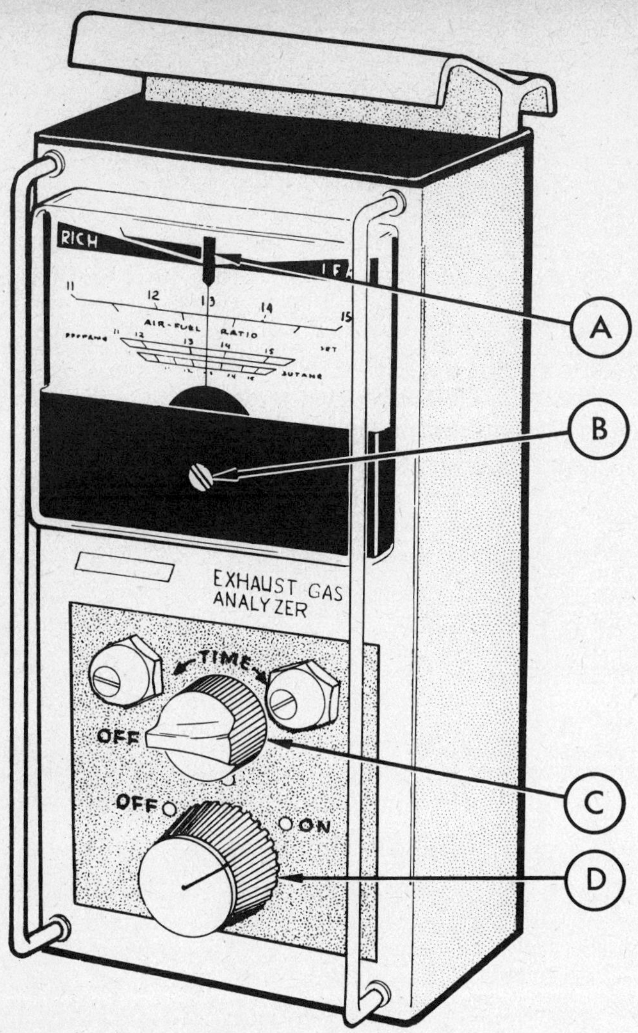

If the analyzer has an on switch (D), it should now be turned to ON. The meter needle will automatically go to the center of the BALANCE band or mark (A). If it does not center perfectly, readjust with the balance screw (B).

When the analyzer is to be used for periods of time longer than eight to 10 minutes, it may be necessary to readjust the SET control because of the drop in battery voltage caused by the constant draw of the unit. Some analyzers have a built-in timer (C) that operates to turn off the unit after approximately eight minutes of use. If so, the timer functions as an "on" switch and must be used by turning the switch or knob as far as it will go in the direction indicated by an arrow. Units equipped with timers can be turned on a second time for extended use without readjustment if you cannot complete the test sequence within the time allotted.

Once the analyzer has been properly calibrated, it is ready for use. It should not be bumped or jarred, however, and the adjustment controls must not be disturbed, or you will have to go through the entire calibration procedure again from the very beginning.

PRE-TEST PREPARATIONS

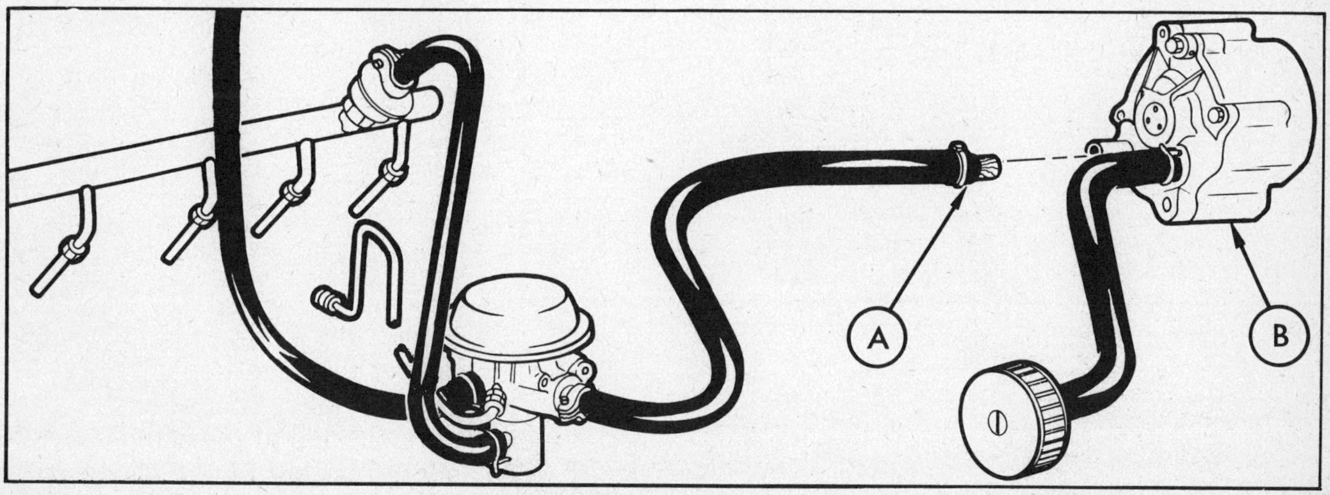

To assure readings that are as accurate as possible when using the exhaust gas analyzer, it is important that the following inspection procedure be carried out carefully before connecting the unit to the car. If any defects are found, they must be corrected before testing with the analyzer. This procedure applies to analyzer use *after* a tune-up as well as before, since there is no uniform agreement on exactly what an engine tune should include.

1. If the engine is equipped with an air pump (B) as part of the emission control system, the manifold connection (A) to the pump must be disconnected and plugged to get a correct air/fuel ratio reading.

2. The PCV valve (C) must be operating correctly on closed-crankcase ventilation systems.

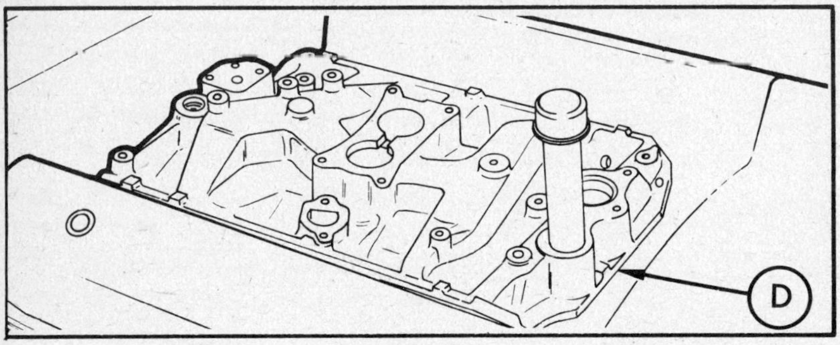

3. The intake manifold (D) should be checked for vacuum leaks, leaks in the manifold gasket, loose manifold bolts and faulty fittings or hoses.

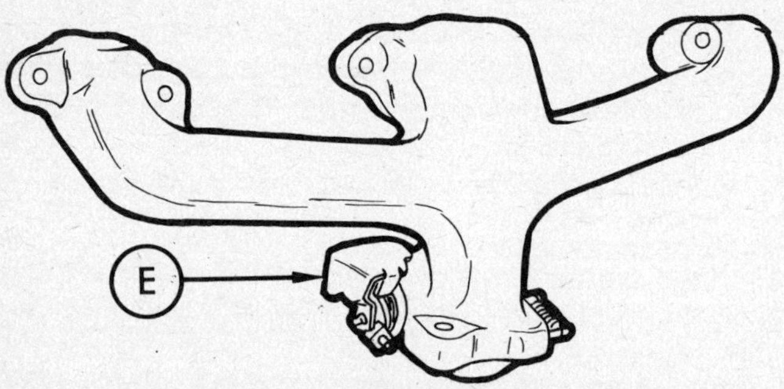

4. The heat control damper or heat riser valve (E) in the exhaust manifold must work freely; check to make sure that its spring is not broken.

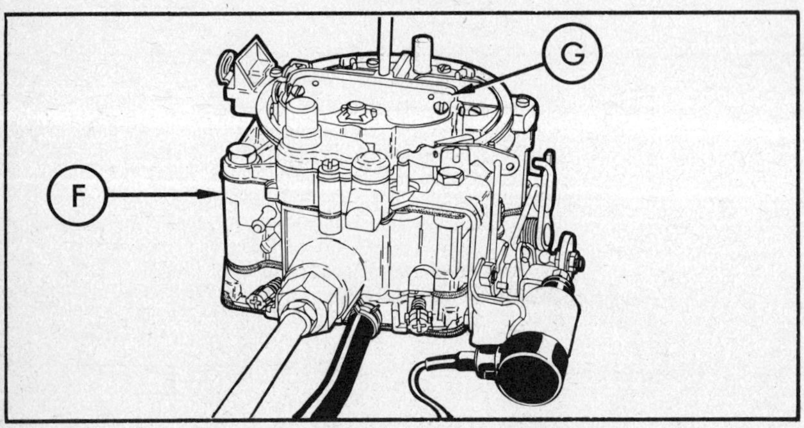

5. The carburetor (F) should be checked for loose bolts, vacuum leaks, fuel leaks and a binding or sticking choke (G).

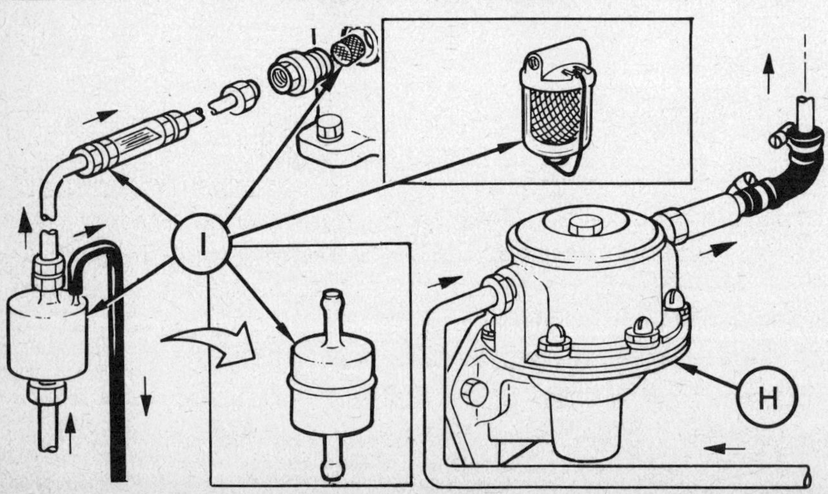

6. Since the fuel pump (H) must deliver its rated capacity or volume, leaks and poor connections at the pump should be corrected. Clean or replace dirty fuel filters (I).

7. Pressure or suction leaks, bad connections or damaged fuel lines must be corrected.

8. The fuel tank must contain at least 25% of its capacity, and the fuel vent system should be free from obstructions.

CONNECTING THE ANALYZER

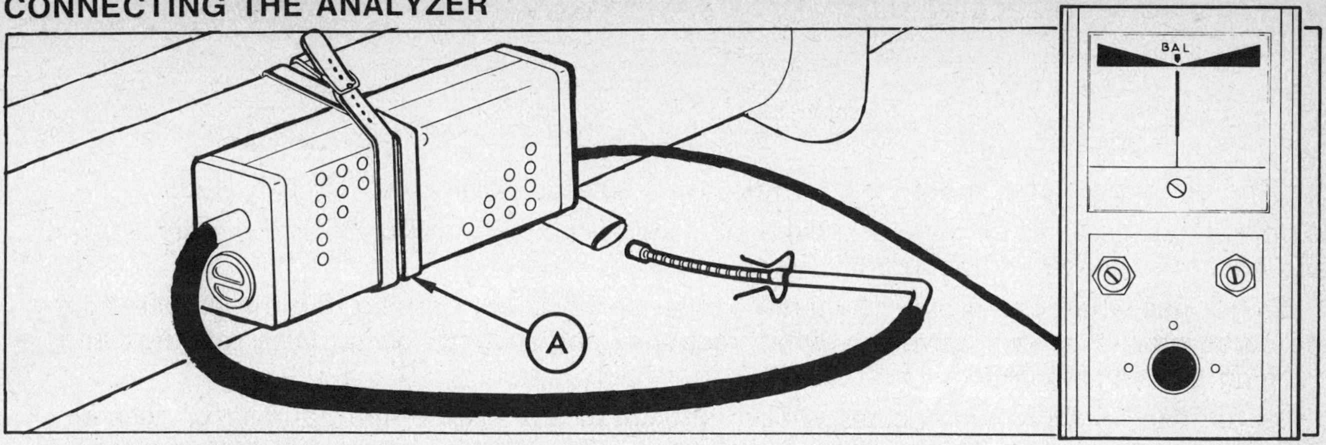

Once the meter has been calibrated and the pre-test preparations completed, the exhaust gas analyzer is ready for use.

WARNING: DO NOT START AND OPERATE AN ENGINE IN AN ENCLOSED AREA. DEATH CAN OCCUR FROM CARBON MONOXIDE POISONING.

Start the engine and let it run until it reaches normal operating temperature. Then position the sensing unit on the bumper near the tailpipe, fastening it in place with the springs or straps (A) provided. When a car is equipped with dual exhaust pipes, one will contain a heat control damper or heat riser valve (usually in the right exhaust pipe just under the exhaust manifold). This valve will prevent sufficient exhaust gas from coming through the pipe to produce a reading on the analyzer meter. Look under the hood of the car and check each exhaust manifold to see which side contains the valve. Then position the sensing unit on the bumper beside the opposite or ''open'' pipe.

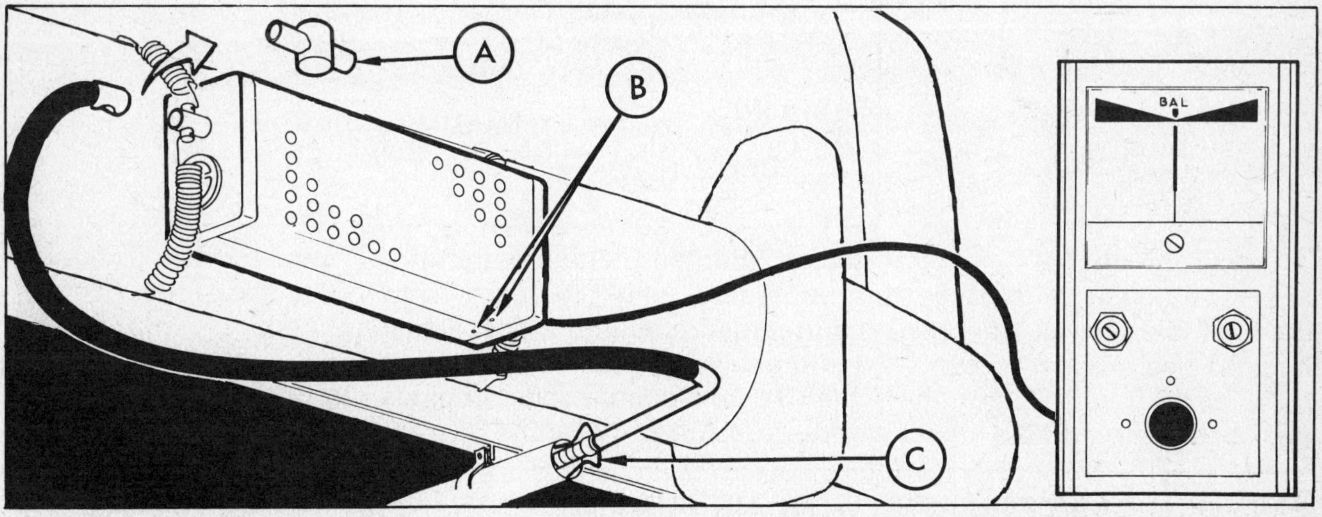

Those sensing units that contain a drain hole (B) should be angled down to allow any water that may collect to drain from the hole. If a water trap (A) is used, it must be inserted between the sensing unit and pickup tube. The water trap must be kept in good condition if the analyzer is to work properly; its desiccant (water-absorbing material) should be reconditioned periodically. This can be done several times by removing and heating the desiccant in an oven for an hour or so at 350-400° F, but it should be replaced when it turns a pink color.

When the sensing unit is fastened securely to the car's bumper, attach the pickup tube with its probe. Make certain that the engine has reached its proper operating temperature before inserting the probe all the way into the exhaust pipe. Be sure the clip or clamp (C) that holds it in place is secure.

EXHAUST GAS ANALYZERS

CAUTION: LIKE ANY OTHER PRECISION UNIT, THE ANALYZER SHOULD BE HANDLED WITH CARE. DO NOT DROP THE UNIT. IF THE SENSING UNIT IS NOT SECURELY FASTENED TO THE BUMPER, IT MAY FALL OFF AND BE DAMAGED.

The analyzer should not be used under the following conditions:

1. Do not use on a cold engine and exhaust system—moisture will drain into the sensing unit and affect the speed and accuracy of readings.

2. Do not use when carburetor cleaners or other solvents are passing into the engine from the carburetor—these will adversely affect response and reading, giving false readings and harming the sensing unit.

3. Do not use on two-cycle engines, diesel engines or any engine that "burns" or uses an excessive amount of oil—the oil in the exhaust gas will affect the sensitivity of the pickup probe and sensing unit.

IDLE TEST PROCEDURES

As a correct exhaust gas analysis can be made only from a properly tuned engine, its ignition timing and valve adjustments should be within factory specifications. If they are not, they will affect mixture and economy, and any carburetor adjustments made as a result of the reading will be incorrect. Test procedures for air/fuel ratio differ according to the type of emission control system (or lack of it) used by the manufacturer.

PRE-EMISSION CARS (BEFORE 1968)

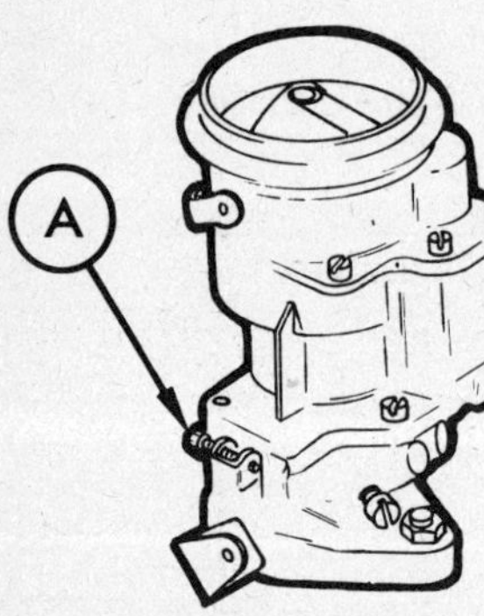

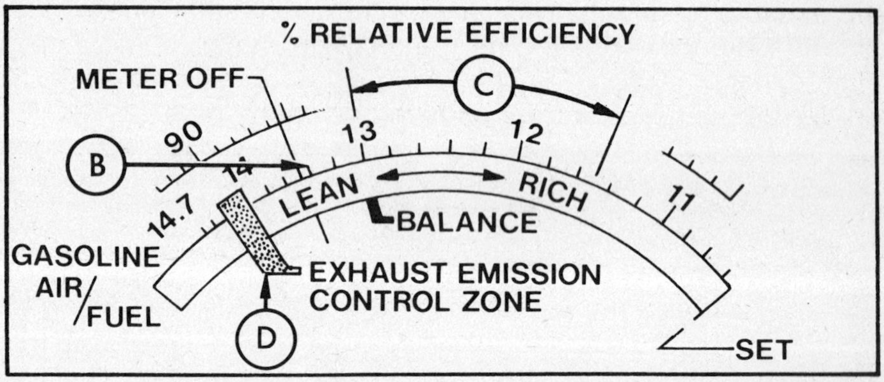

With the engine running at normal operating temperature, attach a tachometer to the engine to check the idle speed. If necessary, adjust the carburetor idle speed screw (A) for the best idle or best lean idle within the manufacturer's specifications. As high-compression engines use a richer fuel mixture, the analyzer meter needle should hold a steady reading between 13.0 and 13.2 on the scale (B). Low-compression engines run on a leaner mixture; the needle should hold steady between 11.5 and 13.0 (C).

EMISSION CARS EQUIPPED WITH AIR PUMP

If an air pump system is used as part of the engine's emission control package, the quarter-turn rich or the lean-drop method of carburetor adjustment (discussed at the beginning of this chapter) will be specified by the manufacturer in order to properly set the car's idle speed for testing. Which method should be used depends entirely upon the specifications for that particular engine and model year. The specs should be followed exactly.

After adjustment, the meter needle will hold a steady reading within specifications for the air/fuel ratio (if given); this will *usually* be in the 14.0 to 14.2 range (D). Wait approximately 15 seconds after making adjustments to let the engine and gas analyzer stabilize before taking a reading. Recheck the idle speed with the tachometer while making the necessary adjustments.

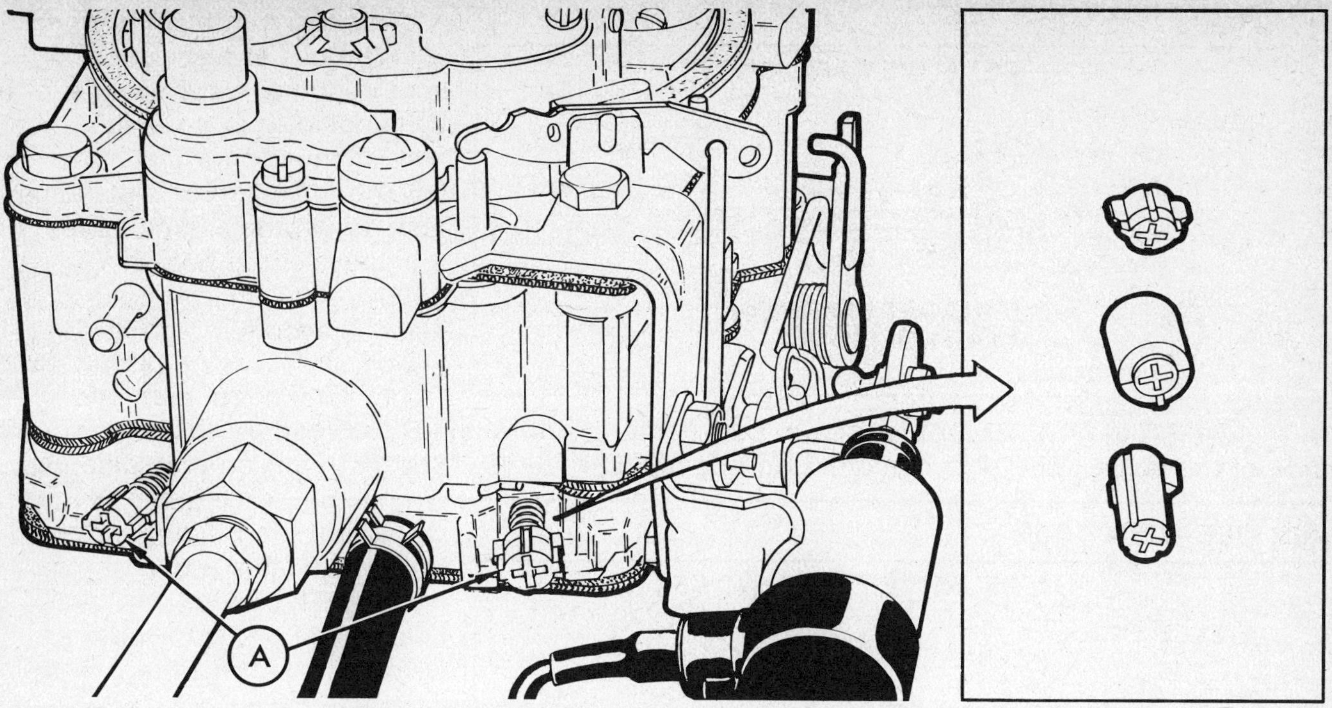

Those carburetors equipped with idle limiter caps (A) on the mixture screw(s) and which do not respond sufficiently to adjustment within the range permitted by the caps need repair. Older carburetors not so equipped which do not respond to adjustment of the idle mixture screw(s) within the range specified by the manufacturer (usually ¼ to 2½ turns open) also need repair.

EMISSION CARS WITHOUT AIR PUMP

On cars whose emission control package does not use an air pump, the idle speed should be adjusted to specifications using either the lean-drop or the quarter-turn rich method. The manufacturer will specify which to use, and his recommendation must be followed carefully. Also check the specifications for any advice concerning air conditioning units and headlights during testing. On some six-cylinder engines, you need to turn the headlights on high beam.

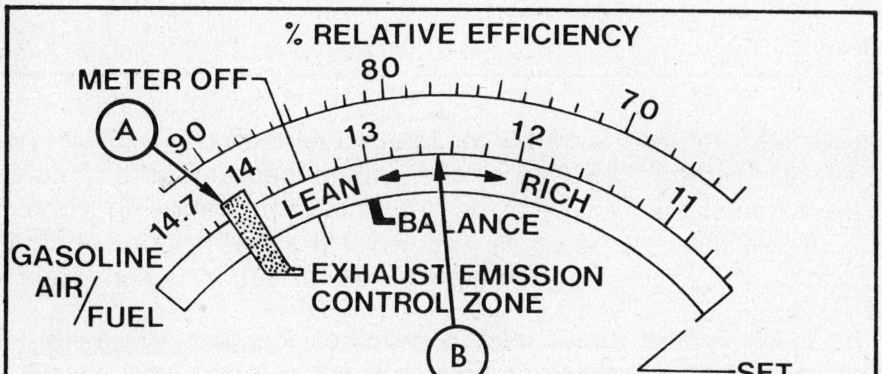

Check the idle speed with a tachometer. If the idle air/fuel ratio does not fall within the manufacturer's specifications (usually the 14.0 to 14.2 range [A] on the meter scale), turn the idle mixture screw(s) out slowly until the meter needle reaches a 12.5 reading (B). Watch the tachometer to maintain the correct idle speed throughout the adjustment procedure.

Now turn the idle mixture screw(s) in, 1/16 turn at a time, waiting for meter stabilization before making further adjustments. Turn screw(s) until a 14.2 reading (A) on the meter scale is obtained. Idle speed should be within factory specifications when the air/fuel ratio reaches 14.2. If not, reset the idle speed and check the air/fuel ratio to be sure it has not changed.

CRUISING MIXTURE TEST

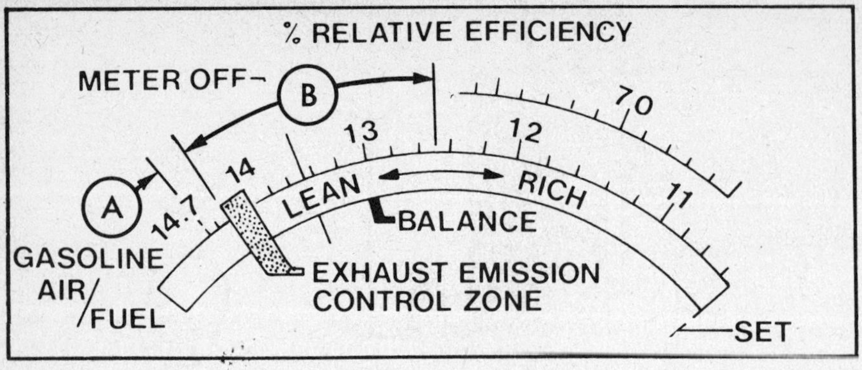

This test can be performed with the car at rest under no-load conditions, but the use of a dynamometer or an actual road test is required for testing under load conditions. For a stationary test, adjust the engine speed to a steady setting between 1,800-2,400 rpm and hold for at least 30 seconds. The meter needle should read 14.4 to 14.7 (A) for late-model cars and between 12.5 to 14.4 (B) for older cars. Low-compression engines will read on the richer (low) side of the scale while high-compression engines tend to read leaner (high) on the scale.

AIR CLEANER TEST

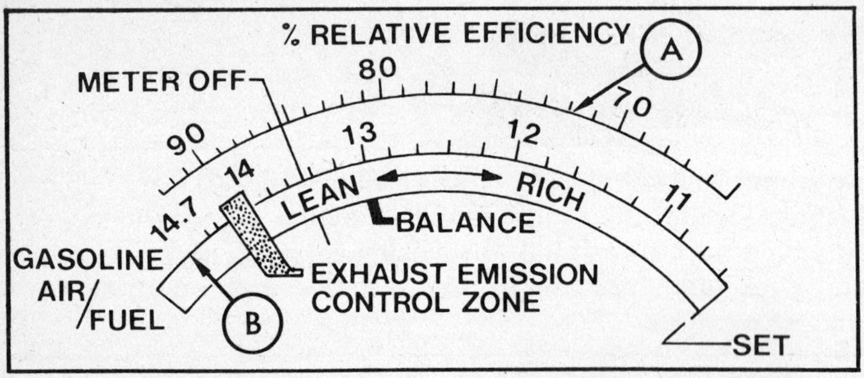

When the reading is too low, it indicates a rich mixture. The cause can be determined by removing the air cleaner. At this point, you must use care in interpretation of the meter needle's reaction. On meters with a PERCENT OF RELATIVE EFFICIENCY scale (A), a change of three or more marks in the reading indicates the need for cleaning or replacement of the air filter. However, excessive movement toward the lean side of the scale may mean that the filter was designed to enrich the mixture and that the restriction it causes is therefore quite normal. Check the factory shop manual if you suspect filter design.

If the meter has no PERCENT OF RELATIVE EFFICIENCY scale, removal of the air cleaner should bring the overly rich reading on the AIR/FUEL RATIO scale (B) back to normal. If it does, clean or replace the air filter to correct the problem. But if removal of the air cleaner has little or no effect on an excessively rich reading, the carburetor requires adjustment or even repair.

REVERSE READINGS

Should the meter needle read richer than 14.4 for late-model cars (or 12.5 for older ones), you may have a "reverse reading," caused by the unbalanced amounts of carbon dioxide and hydrogen in the exhaust gas. This can be determined by removing the air cleaner to lean out the mixture. Wait a few seconds and cover the carburetor opening to purposely enrich the mixture. Keep the opening covered for 15-20 seconds and watch the meter needle carefully.

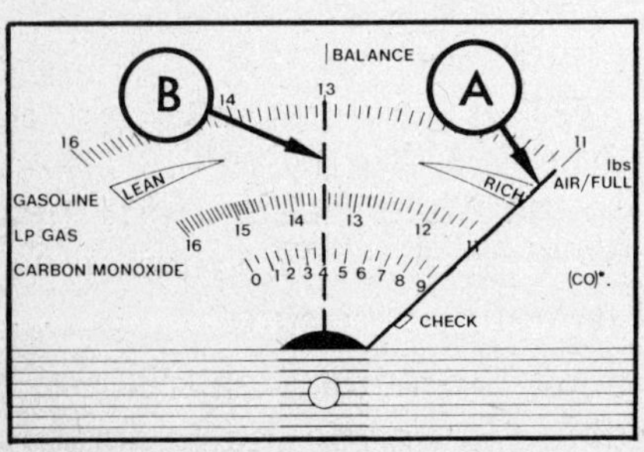

As the richer mixture reaches the gas cell in the sensing unit, the meter needle is forced to move from its excessively rich reading (A) on the scale to a less rich position. Now remove the air cleaner to lean out the mixture once more. Watch the needle. It should continue in the direction of a less rich reading (B). If it swings back to the excessively rich position (A), you are getting a reverse reading. In such cases, the most accurate reading you can obtain is to repeat the covering/uncovering sequence of the carburetor and note exactly how far the needle deflects into the lean area of the scale before returning to the rich side. This is only an approximate indication of what the true reading should be.

ACCELERATOR PUMP TEST

Adjust the engine speed to 1,000 rpm. Allow the meter to stabilize and note the reading. Now rapidly accelerate the engine by opening and closing the throttle quickly. Watch the meter needle—it should move quickly to a low (rich) reading a few seconds later and then return as quickly to its original position. Failure to deflect and return means that the carburetor accelerator pump is not operating correctly and indicates the need for carburetor repair or adjustment. It can also mean that the meter is not working correctly; have an experienced mechanic check the carburetor to be certain.

INTAKE MANIFOLD AIR LEAKS

Air leaks in the intake manifold will cause a leaner-than-normal mixture reading. If your air/fuel ratio is higher than normal, check the manifold carefully. Put a few drops of kerosene on suspected air leak areas and watch the meter needle. If it moves toward the rich side of the scale, you have located a leak which should be corrected before accurate testing can be done.

CARBON MONOXIDE TEST

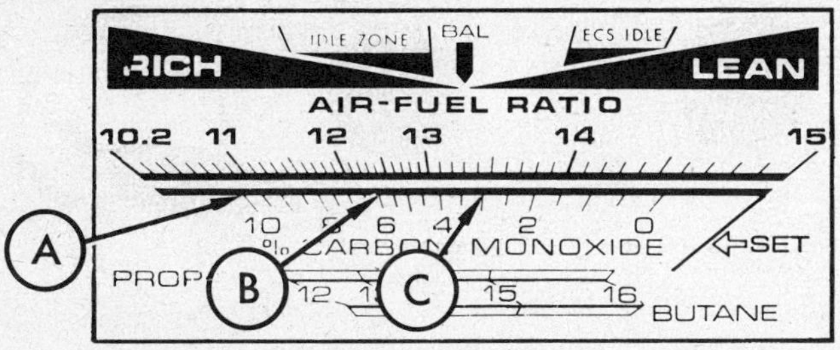

Meters with a PERCENT OF CARBON MONOXIDE or CO scale (A) are designed to test the CO emission level. Those meters using a Wheatstone Bridge in their circuitry will only give an indication of the CO reading and are not accurate. Some states now refuse to accept a test by such an instrument to clear an emission control violation. This test may not be legal in your state. Check with your state highway patrol.

With the engine running at idle speed and a tachometer connected to make certain that the speed is correct, note the reading on the CO scale. It should be 6% or less (B) for non-emission controlled engines and 3% or less (C) if the engine is equipped with smog devices.

Now set the engine speed to 2,500 rpm and note the change in the CO scale reading—it should be lower than the idle speed reading. Exactly how much lower depends upon the engine and year of manufacture. Refer to factory specifications to interpret the reading exactly.

LP GAS ENGINES (LIQUID PETROLEUM)

ENGINE SPEED (rpm)	PROPANE	BUTANE
800—1,500	14.0—14.9	12.8—14.4
1,500—1,800	13.6—14.3	13.3—14.0
2,000—2,200	14.5—16.0	14.2—15.4

Engines using propane or butane gas for ignition are tested in the same manner, but air/fuel ratio readings should be made from the scale marked LP gas, butane or propane, depending upon the meter scale designation. While checking manufacturer's specifications is recommended for this test, the table at left can be used as a general guide to correct air/fuel ratios at specified engine speeds.

METER CARE

After use, the exhaust gas analyzer requires a certain amount of care to keep it in good condition. Disconnect the sensing unit and hang it up vertically to drain out any water that may have condensed in the gas cell. Also drain the pickup tube.

Remove batteries if the unit is not going to be used for a period of several weeks. This will prevent damage to the unit from corroding or leaking batteries.

Keep the plastic meter face clean by wiping with a clean cloth moistened with a 10:1 water-detergent solution. Do not use gasoline, thinner or solvent as a cleaning agent, as these will destroy the anti-static quality of the plastic and cause inaccurate readings as a result. Heavy applications can even dissolve the plastic face.

Store all analyzer components in a dry place.

ALTERNATOR/GENERATOR/REGULATOR TESTER

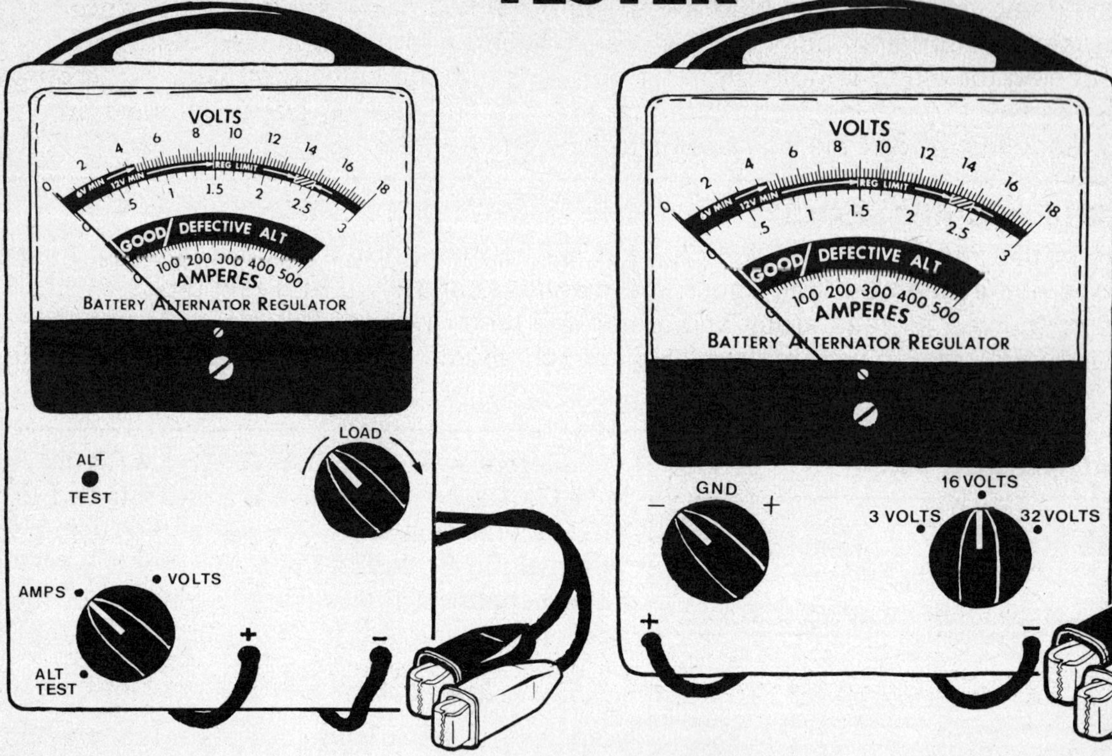

While its name tends to overwhelm the casual user, the alternator-generator-regulator tester is really nothing more than a voltmeter, ammeter and resistance unit combined in a single package. It is the packaging of the three that helps determine the cost of such a unit, hereafter referred to as an AGR tester. In the less expensive units, you must change connections in order to switch the unit function from that of a voltmeter to an ammeter or an ammeter to a voltmeter, and the resistance function is separate. On the more expensive AGR testers, once the connections have been made, you change functions back and forth just by flipping or rotating a selector switch. Also, on such models the resistance function is internal, so it is automatically connected with the tester. AGR testers are used primarily to check the operation of the automotive charging system.

THE CHARGING SYSTEM

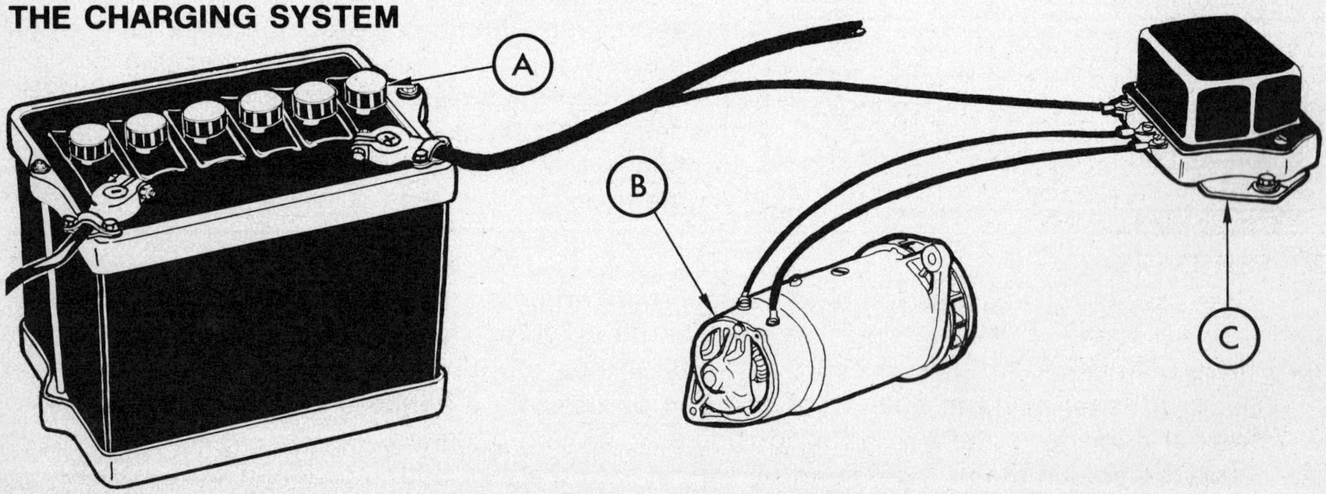

In its simplest form, a charging system consists of a device to generate electricity (B), a storage container (A) to hold the electricity and a regulator (C) to control the voltage and current flow. These basic components appear on the automobile in two slightly different approaches: the generator and the alternator systems.

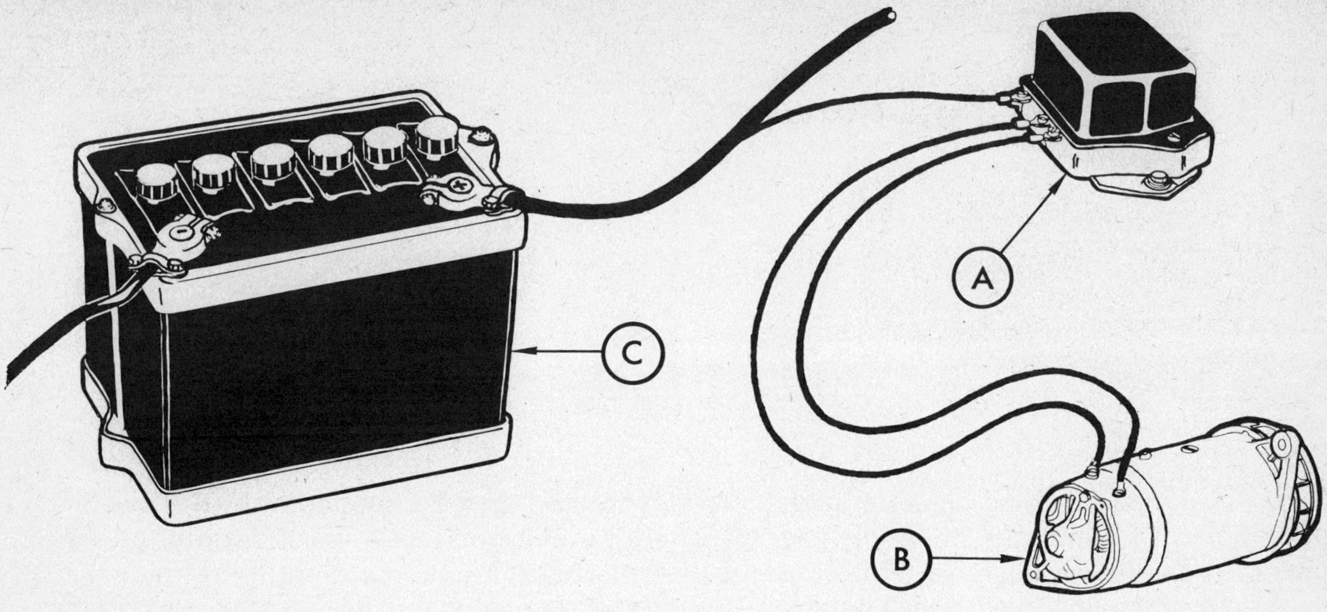

A generator (B) creates direct (DC) current that is sent to the storage container or battery (C). Its output (both current and voltage) is controlled by a generator regulator (A), which usually contains three electromagnetic relays or switches—one to control voltage, one to control current and a cutout relay that prevents the flow of current from reversing itself and returning to the generator. These are mounted on a single unit base with a protective cover over them.

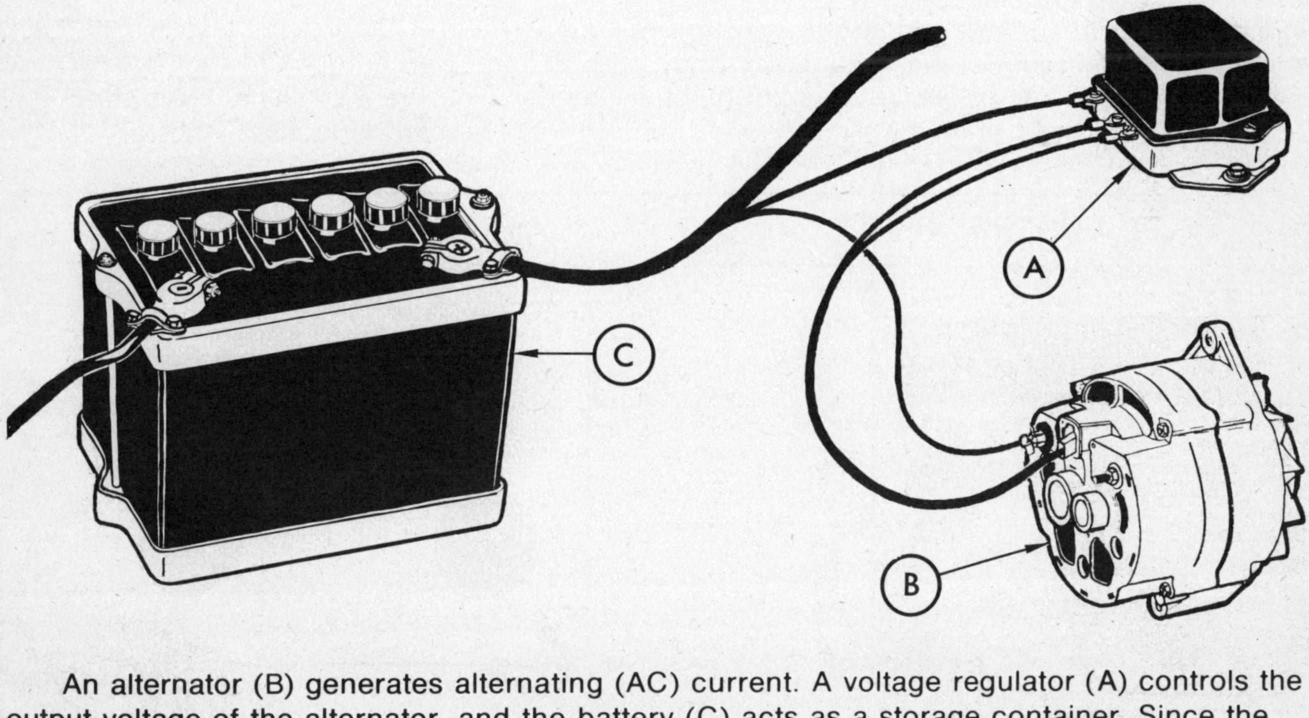

An alternator (B) generates alternating (AC) current. A voltage regulator (A) controls the output voltage of the alternator, and the battery (C) acts as a storage container. Since the automobile uses direct current to operate its electrical functions, diodes (rectifiers) are used to convert AC current generated by the alternator into DC current for storage in the battery. These diodes are housed inside the alternator (B). Since the diodes are actually one-way switches (electricity will pass through them in only one direction), a cutout relay to prevent current flow from the battery back to the alternator is not needed. Nor is a current regulator relay necessary, because an alternator is self-limiting in its current output.

ALTERNATOR/GENERATOR/REGULATOR TESTER

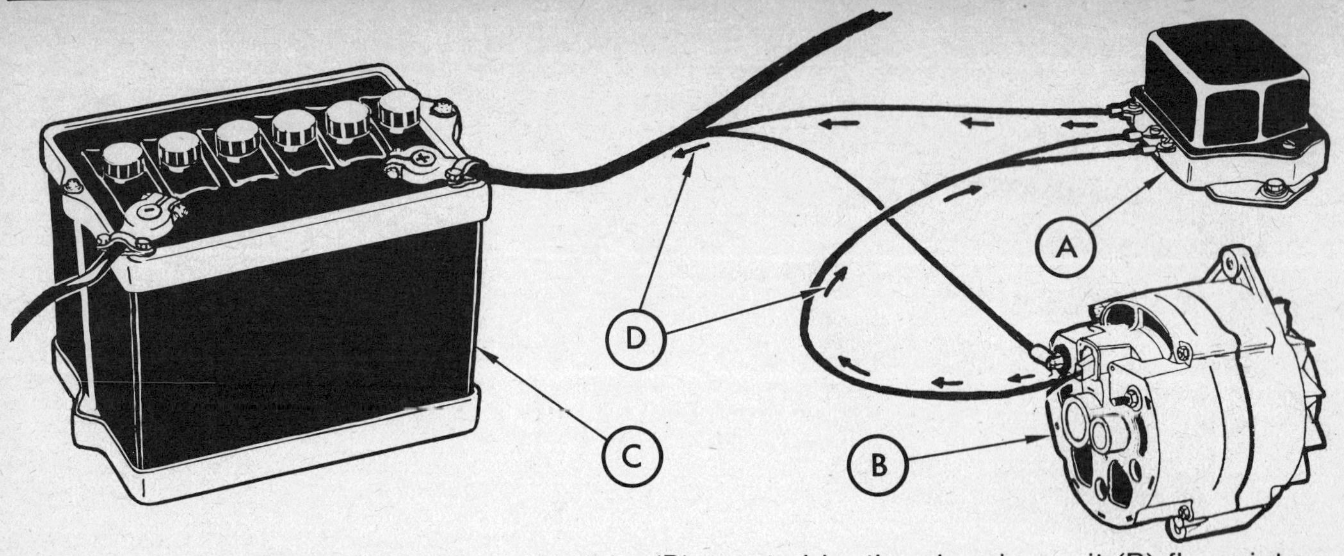

In a normal charging situation, electricity (D) created by the charging unit (B) flows into the battery (C). When the battery is low, the charging unit puts out a lot of current. As the battery voltage rises because of the current being fed into it, the charging unit voltage also increases. If this increase went unchecked, charging unit voltage would rise far beyond the requirements of the system and burn itself out, as well as burning out those units dependent upon it for power. To prevent this, the regulator (A) cuts into the system at a predetermined voltage, holding it to a set maximum.

If the charging unit is faulty and does not deliver the amount (current) and quality (voltage) of electricity that it should, or if the regulator does not allow the electricity to flow into the battery (generator systems only) or keep it from flowing back out, the battery will be undercharged. It may not be able to deliver consistent power In an amount sufficient to operate the starter motor.

However, if the charging unit delivers *more* electricity than it should (because the regulator does not control the voltage correctly), the battery overcharges, damaging the ignition breaker points as well as all light bulbs in the car. The breaker points burn and the life of the bulbs is prematurely shortened. Such problems can be located with the help of the AGR tester.

AGR TESTER CONTROLS

While the design of AGR testers differs considerably in outward appearance from manufacturer to manufacturer, all have a control switch for choosing the function of the unit. This switch may be marked according to function (A) or it may be combined with the voltage selector (B). The voltage selector is used to choose the meter range to be read, according to the voltage of the system being tested.

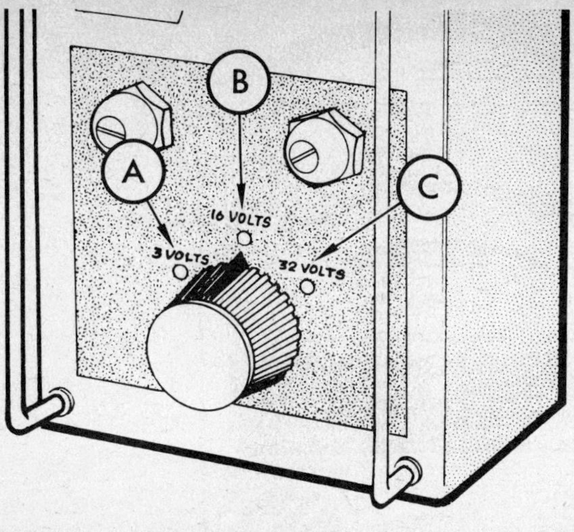

Voltage markings on the voltage selector differ from tester to tester, but you will usually find a 3-volt position (A) used for measuring voltage losses, a 16, 17 or 18-volt position (B) used for testing six- and 12-volt electrical systems and a 32-volt or higher position (C) used for testing 24-volt electrical systems. The average user will find little use for this upper range.

CAUTION: NEVER ATTEMPT TO TEST VOLTAGES HIGHER THAN THAT SPECIFIED BY THE SWITCH'S UPPER LIMIT. CONNECTING THE TESTER LEADS ACROSS A CURRENT FLOW THAT IS GREATER THAN THE UNIT'S CAPABILITY AS STATED BY THE UPPER RANGE POSITION WILL BURN OUT THE TESTER AND MIGHT EVEN DAMAGE THE SYSTEM BEING TESTED.

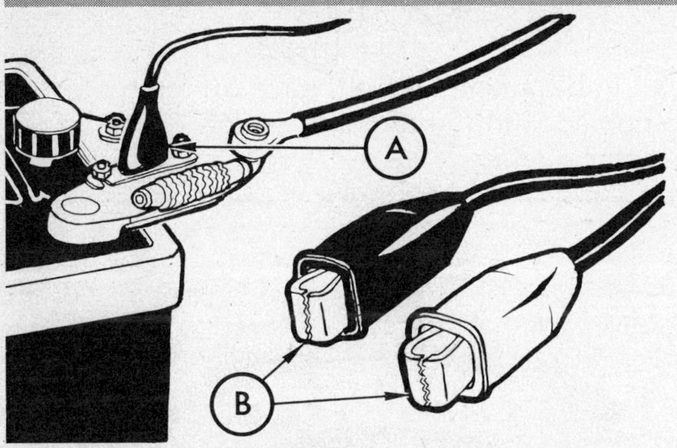

So if you're tempted to find out what's wrong with the family TV set by pressing your AGR tester into double duty, forget it.

AGR testers in which the resistance function is separate are equipped with an amperage shunt lead (A) in addition to the twin leads (B) used to connect the tester to the charging system. This shunt lead is connected to a battery post adapter, which is supplied with the tester for convenience in testing resistance.

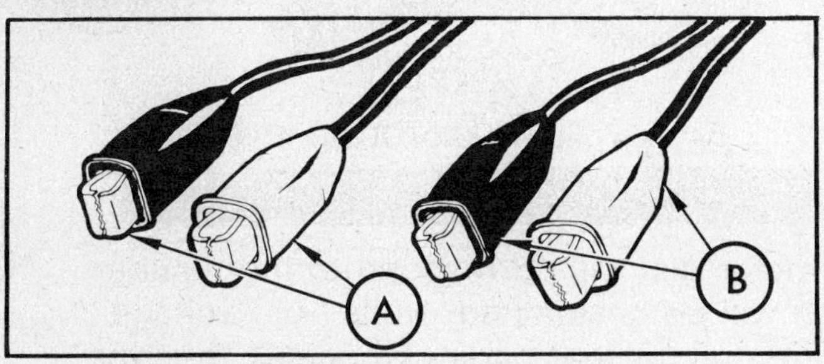

Some AGR testers may be equipped with more than one set of leads. If so, one set is for using the voltmeter function (A) and the other for the ammeter function (B). The proper use for each set should be marked clearly on the leads. If not, check your instruction manual or contact the manufacturer.

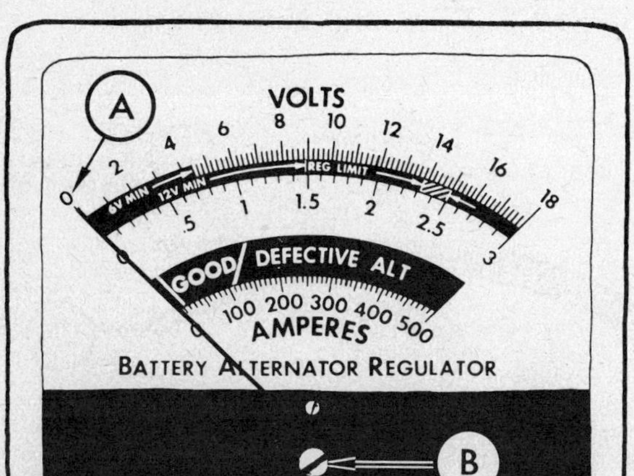

You should find a zero adjusting screw (B) located at the base of the meter needle on the dial face. The needle must rest on the zero mark (A) before any tests are made. If it does not, turn the adjusting screw with a screwdriver until the needle is at zero.

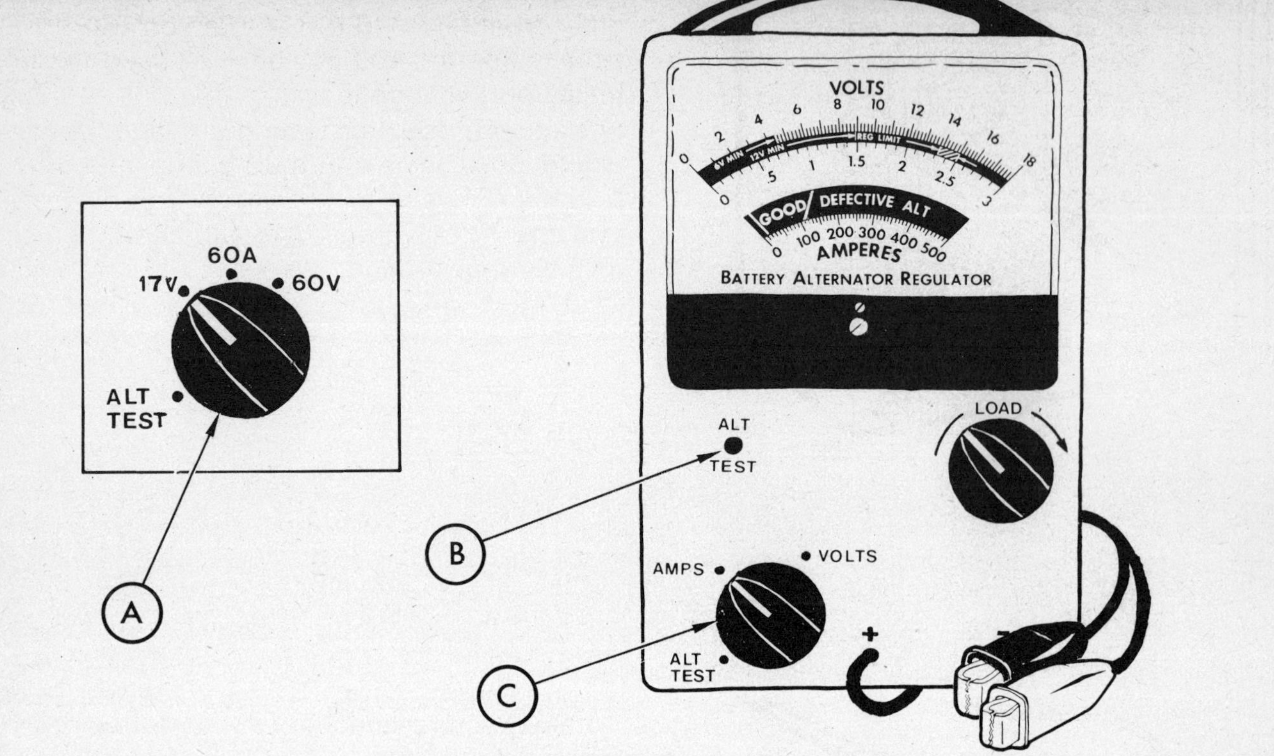

A few AGR testers, especially those combined in a multimeter unit, have an alternator test button (B) and an alternator test position combined with the function switch (C) or voltage selector (A). These are used to test alternator diodes.

AGR TESTER SCALES

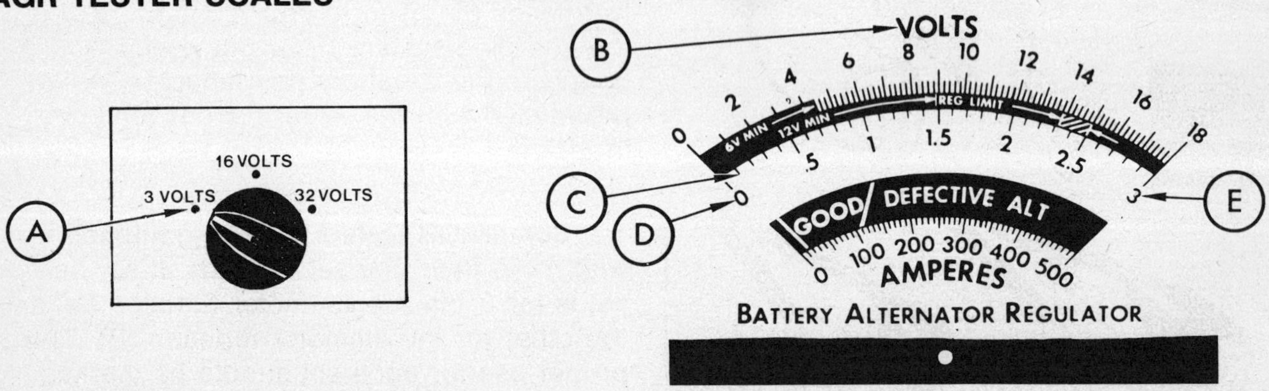

The number of voltage scales found on the meter dial will correspond to the number of positions marked on the voltage selector. Each will be designated "volts" or "voltage" (B) and marked from 0 to the limit specified on the dial for that particular scale. Thus the scale to be read when using the 3-volt position (A) is marked 0 (D) at one side and 3 at the other (E), usually in 0.1-volt divisions (C).

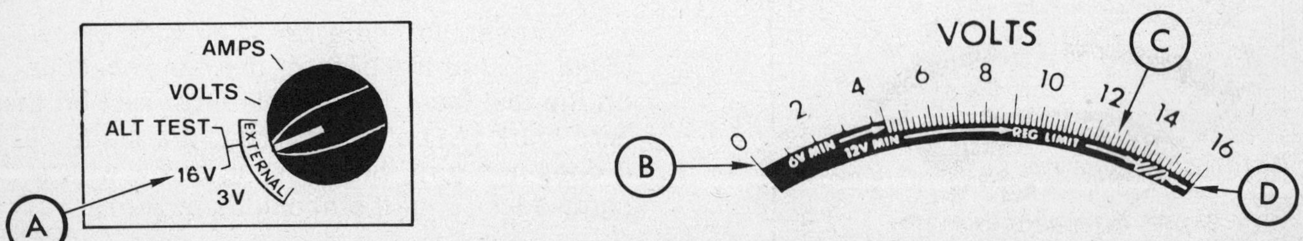

The scale to be read when using the 16-volt position (A) is marked 0 (B) at one side and 16 at the other (D), usually in 0.2-volt divisions (C).

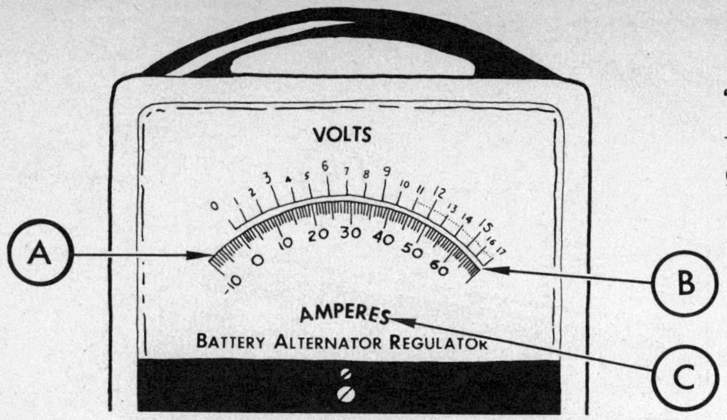

In addition, there will be a scale marked "amps" or "amperes" (C), reading from 0 or −10 at one side (B) to an upper limit of 60 or 90 at the other side (A). The amp scale may be graduated in either 1- or 5-amp divisions.

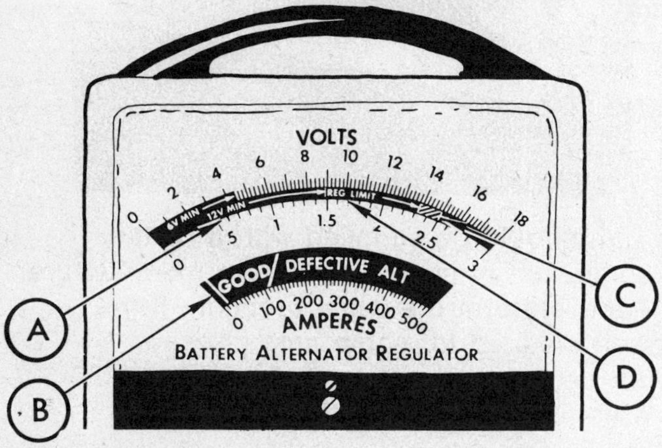

Those meter dials found on testers with an alternator test function simplify your reading by including either a color-coded GOOD/DEFECTIVE scale (B) for the alternator or a coded voltage scale marked with test points for six- and 12-volt batteries (A), a regulator limit (D) and a "replace regulator" zone (C).

CONNECTING THE AGR TESTER

AGR tester connections depend upon the test being performed and the system to be checked. Connection patterns for each individual test are given under that section and are for negative ground systems. If the car being tested uses positive ground, reverse the leads from the pattern specified.

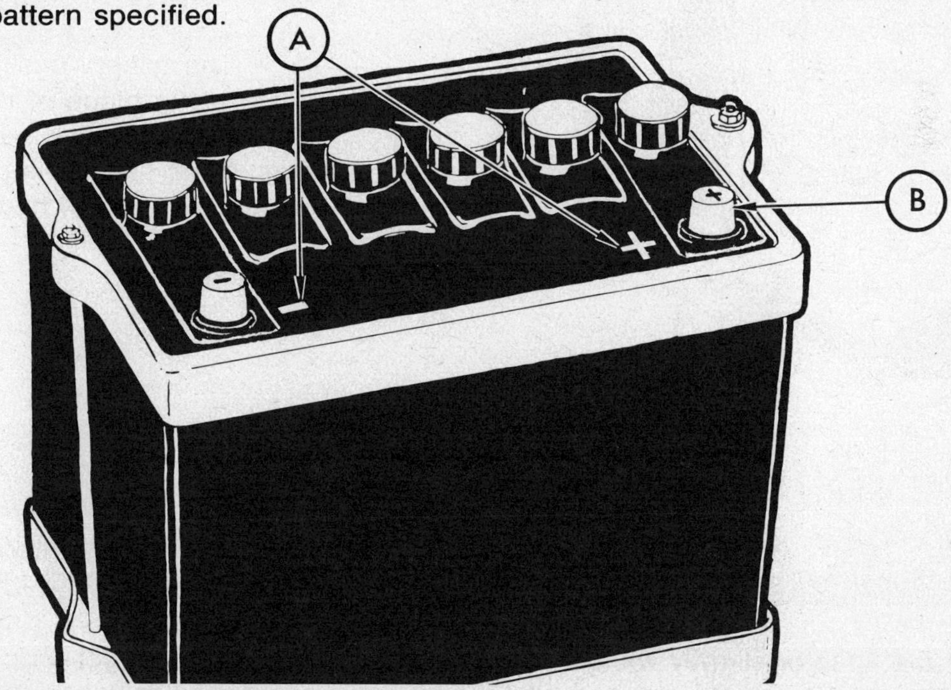

Your car's ground can be determined by locating the battery cable which is connected to the engine or frame. The battery terminal post to which the other end of this cable connects should be marked with a + (positive) or − (negative) sign. The marking may appear on top of the battery post (B) or on the battery case beside the post (A). Generally speaking, all American cars are negative ground.

BATTERY POST ADAPTER—EXTERNAL RESISTANCE FUNCTION

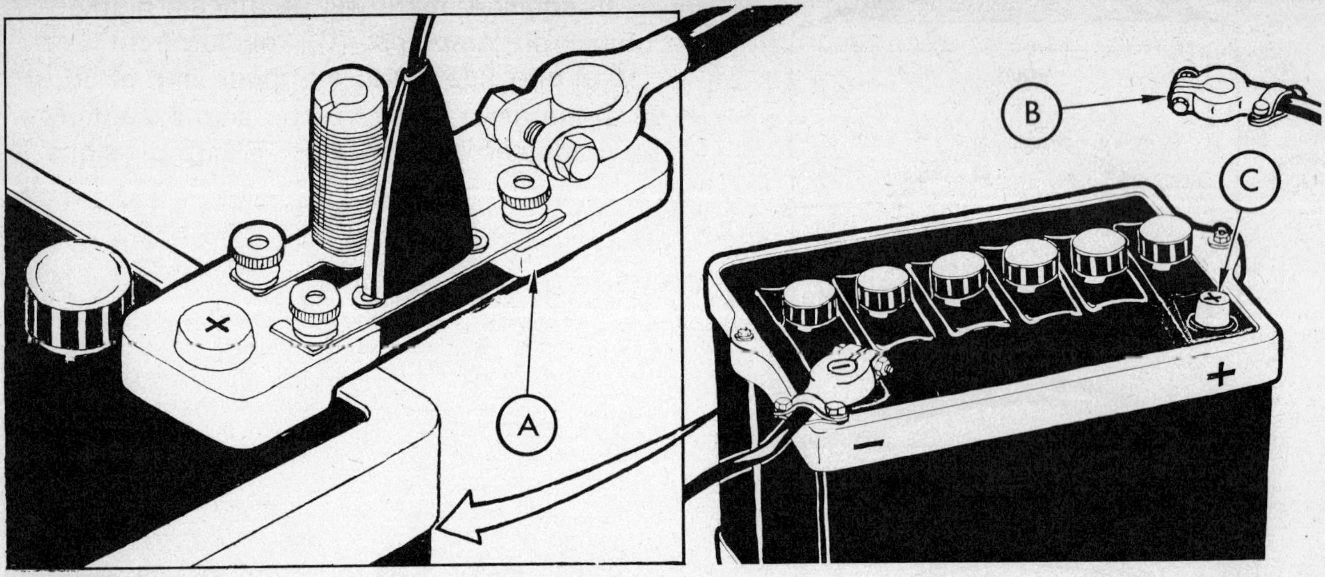

AGR testers which use an external resistance unit should be supplied with a battery post adapter (A). While these adapters may differ somewhat in appearance from manufacturer to manufacturer, all contain the same components and perform the same function. If the battery cable (B) and terminal post (C) have not been cleaned for some time prior to installing the adapter, it is a good idea to clean both before proceeding. This prevents the possibility of loose or dirty connections, either of which will interfere with test results.

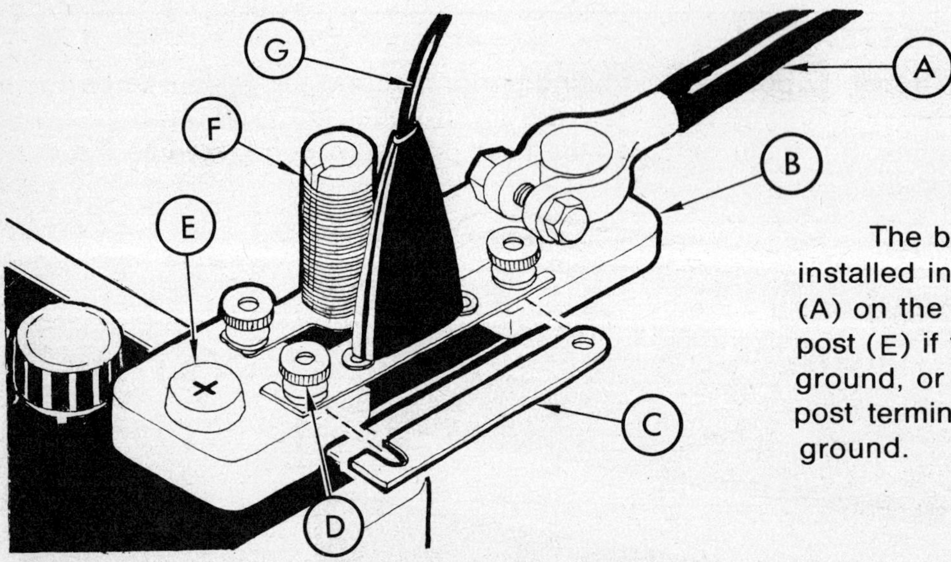

The battery post adapter (B) is installed in place of the battery cable (A) on the positive battery terminal post (E) if the car is negative (−) ground, or on the negative battery post terminal if the car is positive (+) ground.

CAUTION: ALWAYS DISCONNECT THE GROUND CABLE FIRST, THEN THE "HOT" CABLE. A BATTERY IS CAPABLE OF PRODUCING VERY HIGH CURRENT THAT CAN CAUSE SEVERE SHOCK AND PERSONAL INJURY IF THIS PRECAUTION IS DISREGARDED.

Rotate the adapter slightly while applying downward pressure to get a good connection when installing. A ¼-ohm resistance unit (F) simulates a charged battery condition to allow accurate voltage regulator measurements to be made. An ammeter shunt lead (G) from the AGR tester connects to a shunt lever (D), which is used to start the engine and take ammeter readings. When the ammeter shunt is removed, a linkage switch (C) is used to start the engine. The switch should be in an OPEN position or *removed completely* whenever the shunt lead (G) and shunt lever (D) are in place.

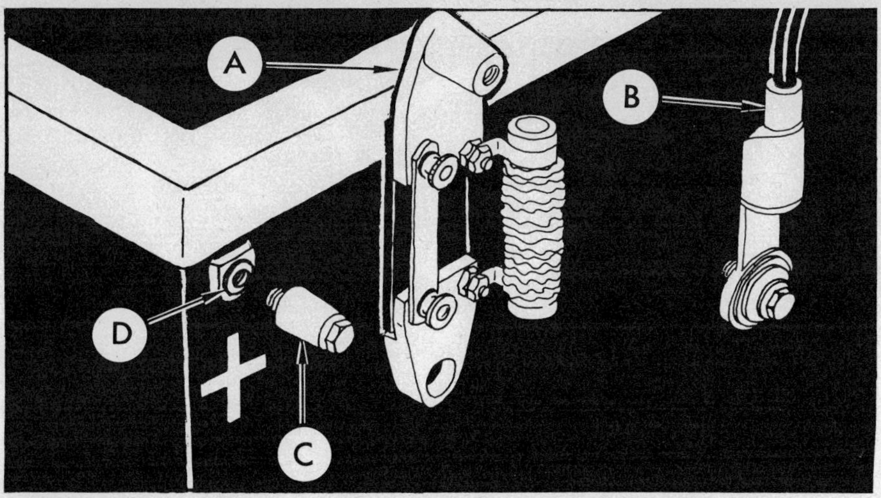

A slightly different battery post adapter is furnished for use with some new batteries, whose terminals are side-mounted. Unscrew the positive battery cable (B) from the battery and screw it to the adapter (A). A post (C) must be screwed into the battery terminal (D) before the adapter can be installed.

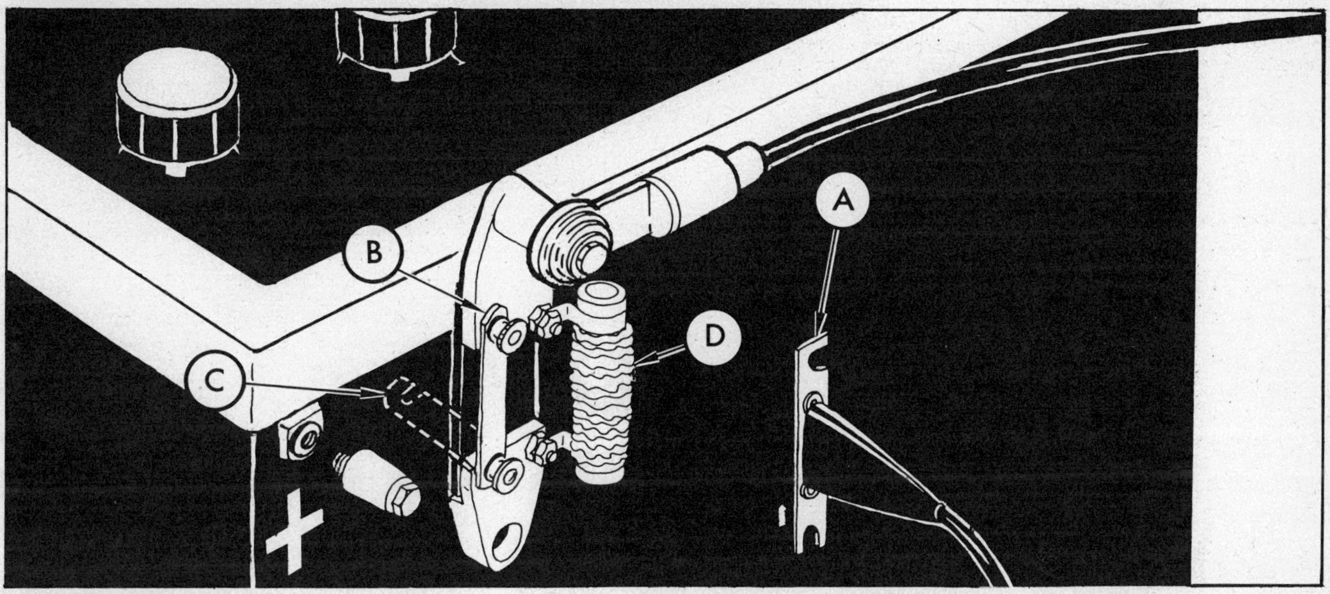

CAUTION: TO START THE ENGINE, THE SHUNT LEVER (A) OR LINKAGE SWITCH (B) MUST BE IN A *CLOSED* POSITION. ONCE THE ENGINE IS RUNNING, THE SHUNT LEVER (A) OR LINKAGE SWITCH (B) CAN BE OPENED (C) FOR TEST PURPOSES WITHOUT HARM TO THE ¼-OHM RESISTANCE UNIT (D) FAILURE TO OBSERVE THIS PRECAUTION WILL RESULT IN BURNING OUT THE ¼-OHM RESISTANCE UNIT.

PRECAUTIONS IN USE

To prevent damage to the tester or the system being tested, the following precautions must be observed:

1. Do not polarize an alternator (see DC generator polarity); it is unnecessary.

2. Do not reverse polarity of an alternator by connecting the battery backwards, as this will burn out the alternator diodes.

3. Do not operate the alternator on open circuit (with the output [BAT] terminal open).

4. Do not ground or short out any alternator or regulator terminals.

5. Do not close regulator cutout relay contacts by hand when the battery wire is connected to the regulator. This will create a harmful high-current arc across the contacts.

6. Do not ground the generator field while the field lead is connected to a Delco-Remy double-contact regulator, because it will burn up a set of voltage regulator contacts.

BATTERY CAPACITY TEST

When the charging system is not working correctly—a condition indicated by the car's ammeter (A) or discharge light (B) on the instrument panel—the first part of the system to check is the battery.

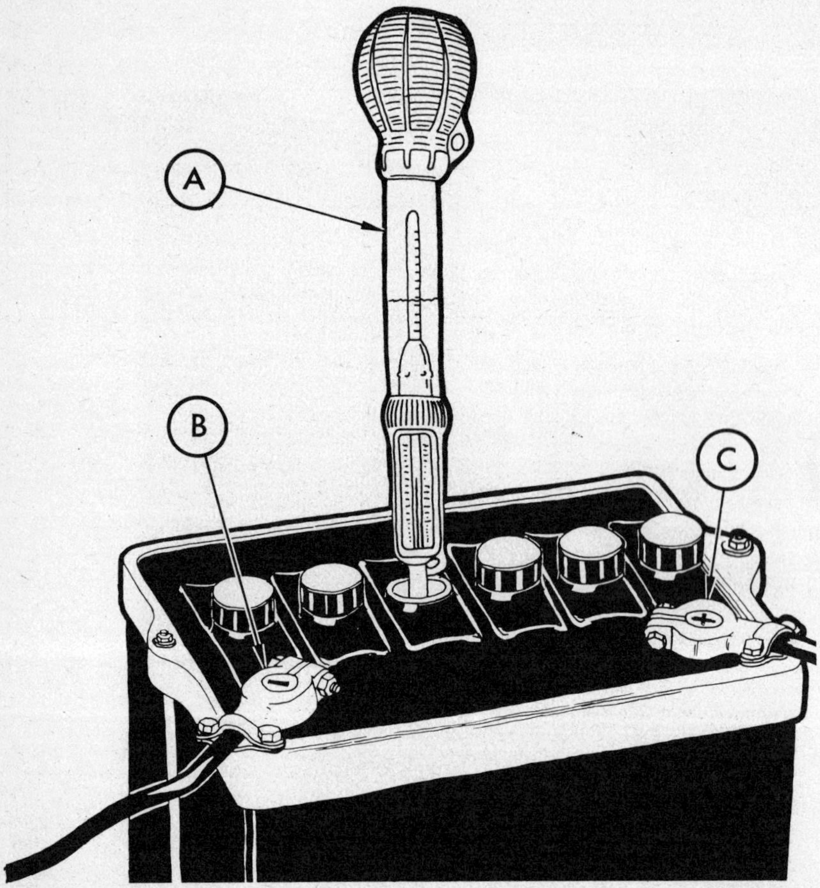

To perform the capacity test, the battery must be fully charged. Since much of a battery's capacity is lost at low temperatures, do not test for capacity at extremely low temperatures. Use a hydrometer (A) to check the specific gravity as the first step. Each battery cell must read a minimum of 1.240 on the hydrometer scale.

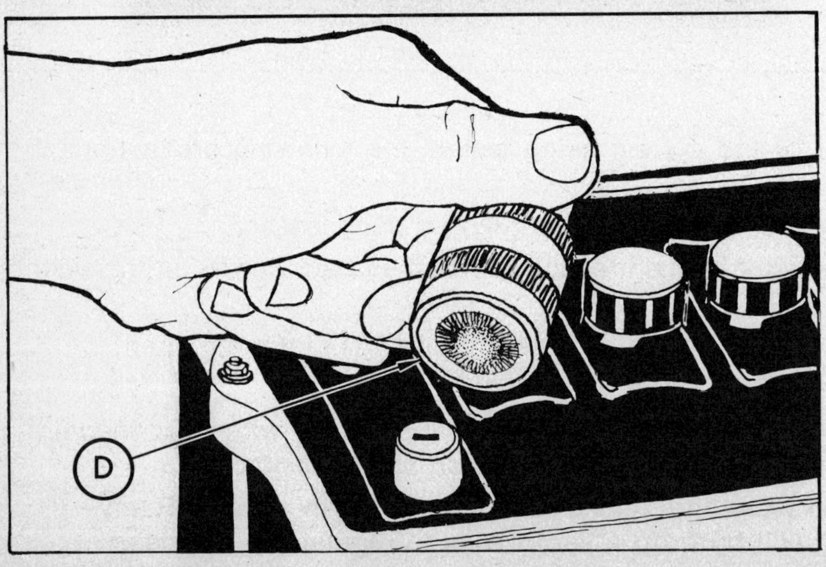

Remove the battery ground cable from its terminal post (B) and clean with a battery terminal cleaning tool (D). Now remove the positive or "hot" cable from its terminal post (C) and clean. Apply a thin coating of light grease or Vaseline to each post. Then replace the positive cable and tighten the cable clamp securely. Replace the ground cable last and tighten its clamp.

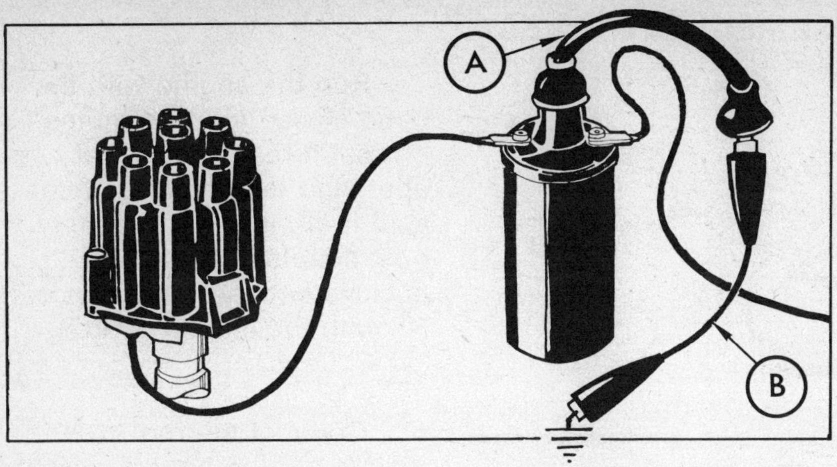

Run the engine until it reaches normal operating temperature, then shut it off. Ground the secondary coil wire (A) with a jumper lead (B) to prevent the engine from starting.

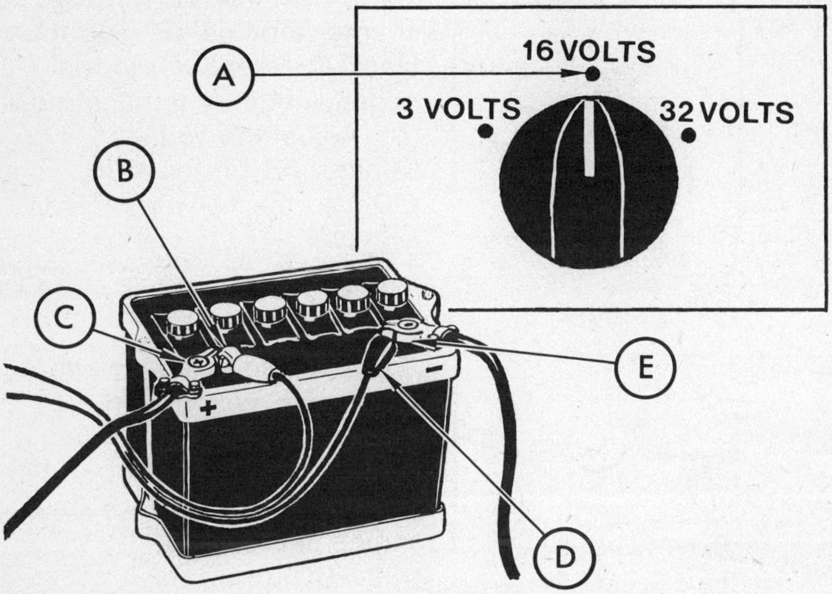

Adjust the voltage selector of the test unit to the 16-volt position (A) or equivalent, then connect the red AGR lead (B) to the positive battery terminal post (C) and the black AGR lead (D) to the negative battery terminal post (E).

Connect a remote start switch (see that chapter) and watch the meter scale as you crank the engine for 10 to 15 seconds. The remote start switch lets you crank the engine from under the hood without the help of another person sitting in the car. The starter should crank normally.

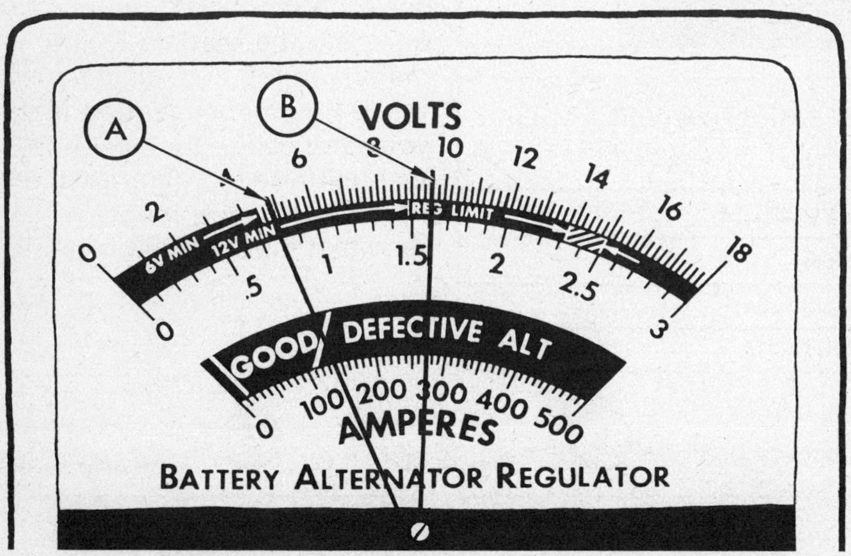

When the battery and starter systems are functioning normally, the 16-volt meter scale will read 9.6 volts or more (B) for a 12-volt system. A six-volt system should read 4.8 volts or more (A).

Readings below 9.6 or 4.8 volts (depending upon the battery being tested) indicate a defective battery, battery cables or starter motor. In the majority of cases, the battery is at fault and must be replaced.

SYSTEM TEST PROCEDURES

To simplify test procedures for alternator- and generator-equipped systems, the two are described separately. This will prevent reader confusion and possible damage to the AGR tester or to the charging system due to incorrect connections. Alternator systems are described below, and they are followed by generator system tests.

ALTERNATOR/GENERATOR/REGULATOR TESTER

ALTERNATOR SYSTEM—VOLTAGE REGULATOR TEST

Run the engine with the hood closed for 15 minutes or until it reaches normal operating temperature, then shut it off. Install the battery post adapter (B) on the positive battery terminal post (C) with the linkage switch (A) CLOSED.

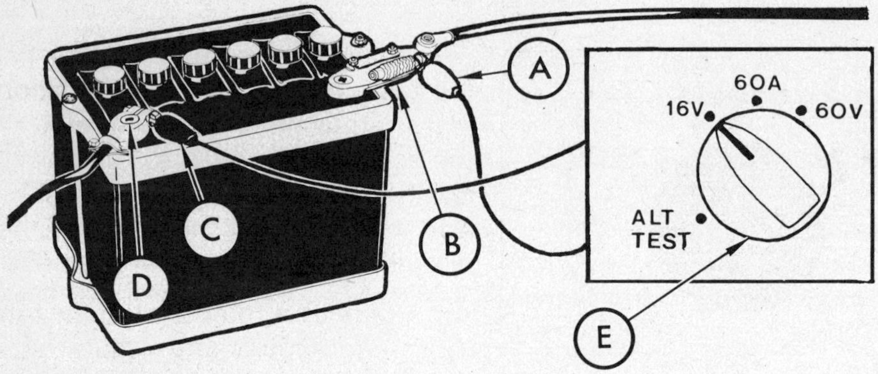

Connect the red AGR lead (A) to the battery post adapter terminal (B) and the black AGR lead (C) to the negative battery terminal post (D). Adjust the voltage selector (E) on the AGR tester to the 16-volt position.

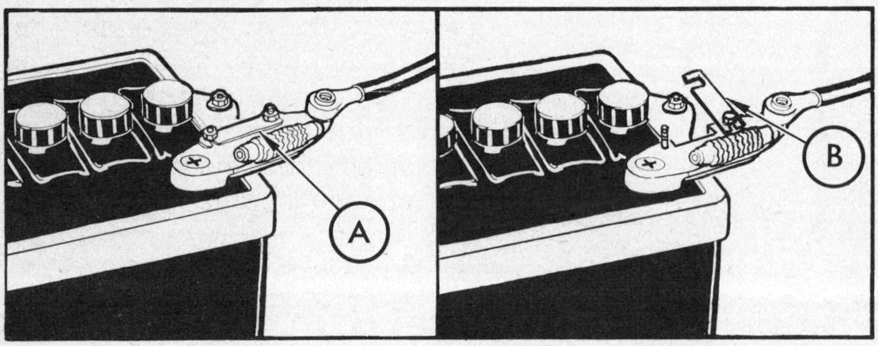

Check to make sure that the linkage switch is in a CLOSED position (A). Then start the engine and open the linkage switch (B). Increase engine speed from idle to 1,200-1,500 rpm (revolutions per minute), read the 16-volt scale on the meter dial and compare the reading to the manufacturer's specifications. While it is recommended that you refer to the factory shop manual for exact figures, the regulator voltage range provided here can be used.

ALTERNATOR TYPE	REGULATOR VOLTAGE RANGE—12-VOLT SYSTEM
Chrysler	13.3—14.4
General Motors	13.5—14.4
Ford	13.8—14.5
Motorola	13.7—14.5

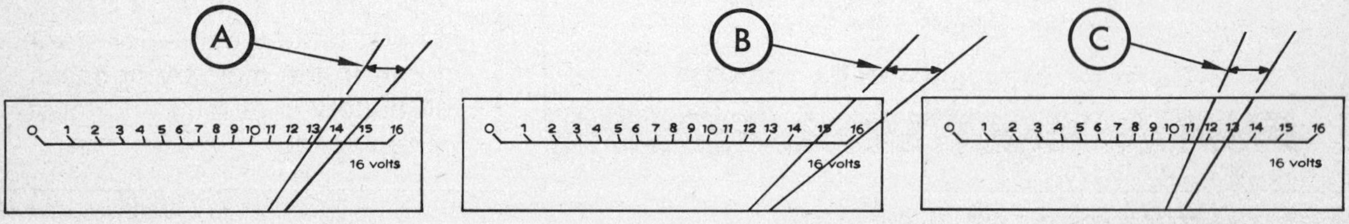

A voltage reading within the specified range (A) indicates that the regulator is good. A higher-than-normal reading (B) means that the regulator is faulty and must be adjusted or replaced. If the voltage reading is lower than normal (C), the problem can be in either the regulator or alternator. The current output test is required to determine which unit is not working properly.

ALTERNATOR SYSTEM—CURRENT OUTPUT TEST

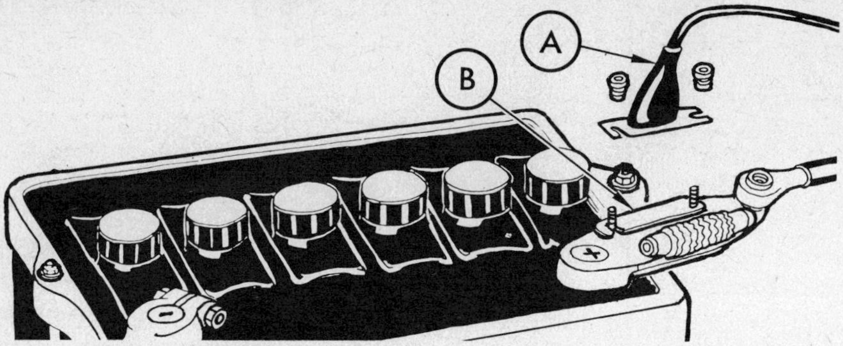

Maximum output from the alternator is necessary to run this test. With the engine shut off and not running, remove the linkage switch (A) from the battery post adapter and replace with the shunt lever and shunt lead (B). The shunt lever must be in the CLOSED position.

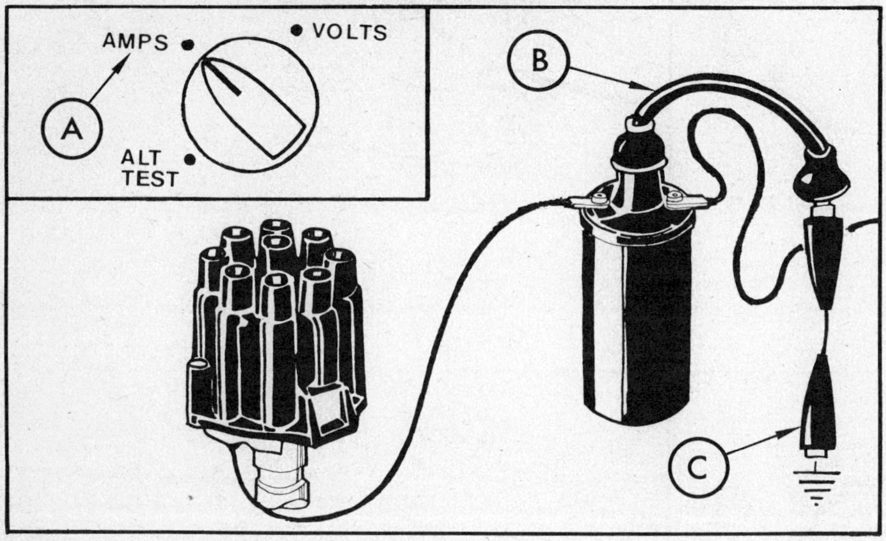

Switch the meter to read amps (A). Remove the secondary coil wire (B) and ground to the engine with a jumper lead (C). Turn on the headlights for two or three minutes, then crank the engine for 15 to 30 seconds to reduce battery voltage. Turn off the headlights and reconnect the secondary coil wire (B) securely.

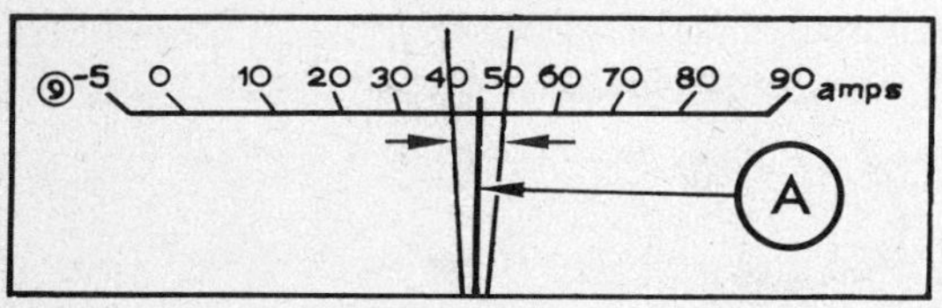

Start the engine and bring its speed up to approximately 2,000 rpm while reading the AMP scale. Add 5 amps to the current reading to compensate for ignition system draw (9 amps for transistorized ignitions) and compare the reading to the manufacturer's rated output. A plus or minus (±) tolerance is provided to allow for temperature variation, so the reading should fall within 3° of the specified amperage (A) on either side. If it does, the alternator is good.

A professional test unit eliminates the need to run the battery down by turning on the headlights and/or cranking the engine. It draws off current from the alternator to make it put out the maximum. The trouble with the battery rundown method described above is that you are never quite sure you have run the battery down sufficiently. Furthermore, if you do not make your current test very quickly, the battery starts to built up almost immediately and maximum current does not flow.

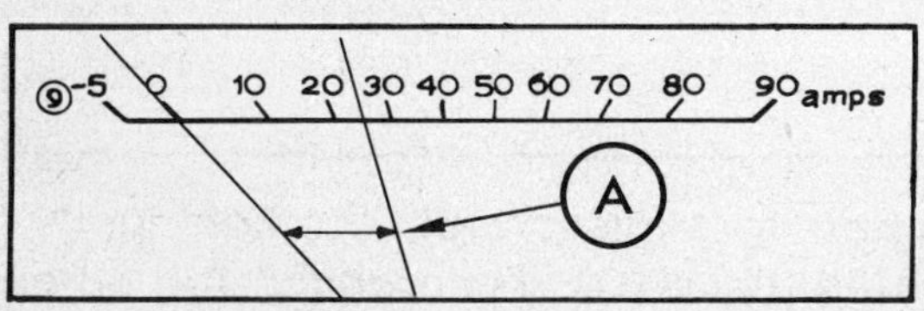

A low reading (A) on the AMP scale means that the voltage regulator must be eliminated from the circuit and the test repeated to determine whether it is the alternator or the regulator that is not working properly.

VOLTAGE REGULATOR ELIMINATION

Eliminating the voltage regulator can be done either at the alternator or at the regulator. The method you use will depend upon the amount of difficulty encountered in getting to the alternator, as access may be limited on some cars. Both methods are covered here. As the first step, determine whether the alternator is an "A" or "B" circuit.

ALTERNATOR/GENERATOR/REGULATOR TESTER

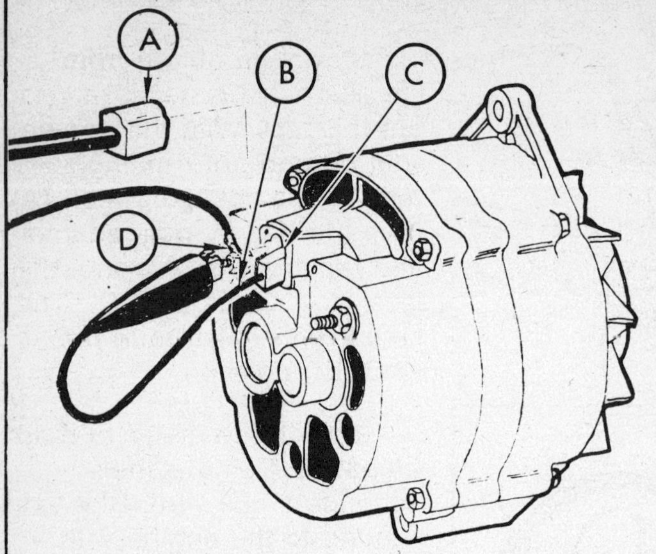

DELCO-REMY

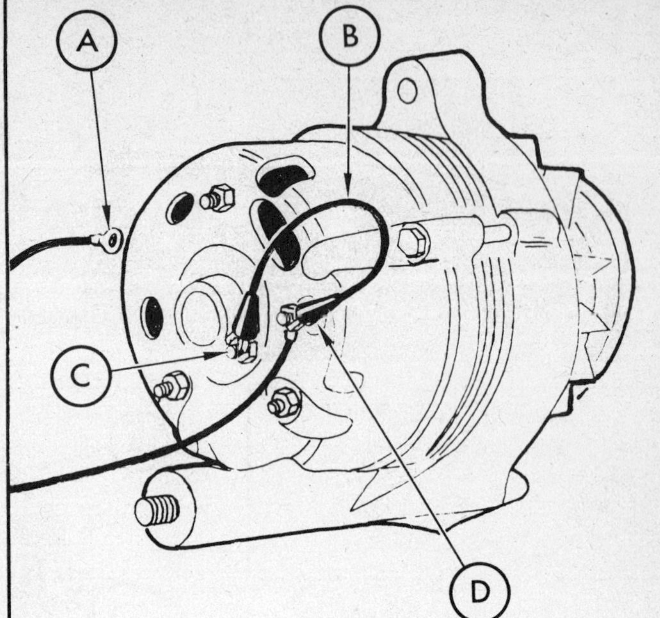

FORD

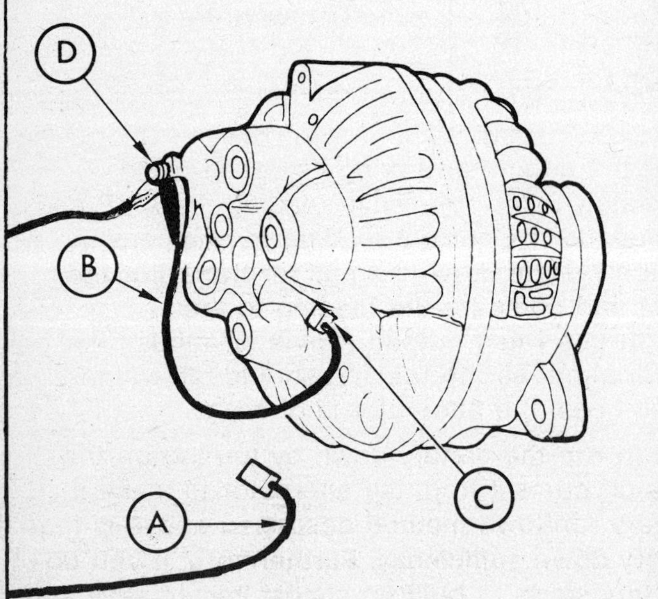

CHRYSLER

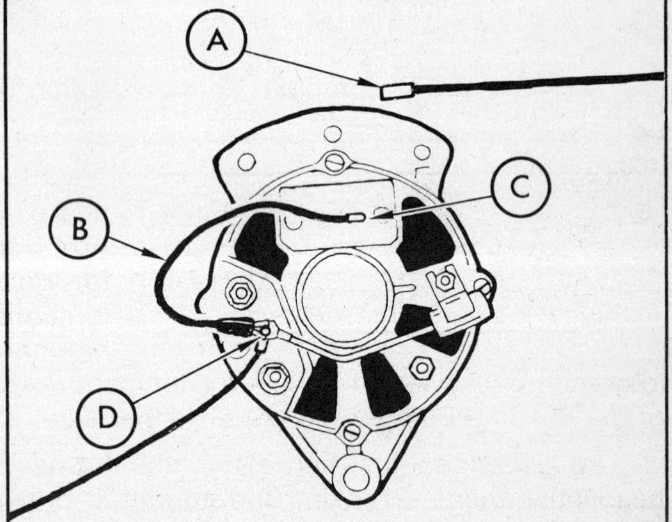

MOTOROLA

To eliminate the voltage regulator with "B" circuit alternators, disconnect the field lead (A) at the alternator and plug or clip a jumper lead (B) in its place (C). Then attach the other end of the jumper lead to the battery terminal (D) on the alternator.

CAUTION: WHEN DOING THIS, WORK SLOWLY AND CAREFULLY. THERE IS MORE THAN ONE TERMINAL AT THE REAR OF MOST ALTERNATORS, AND A WRONG CONNECTION WILL RESULT IN AN INCORRECT TEST AND POSSIBLY EVEN DAMAGE TO THE SYSTEM.

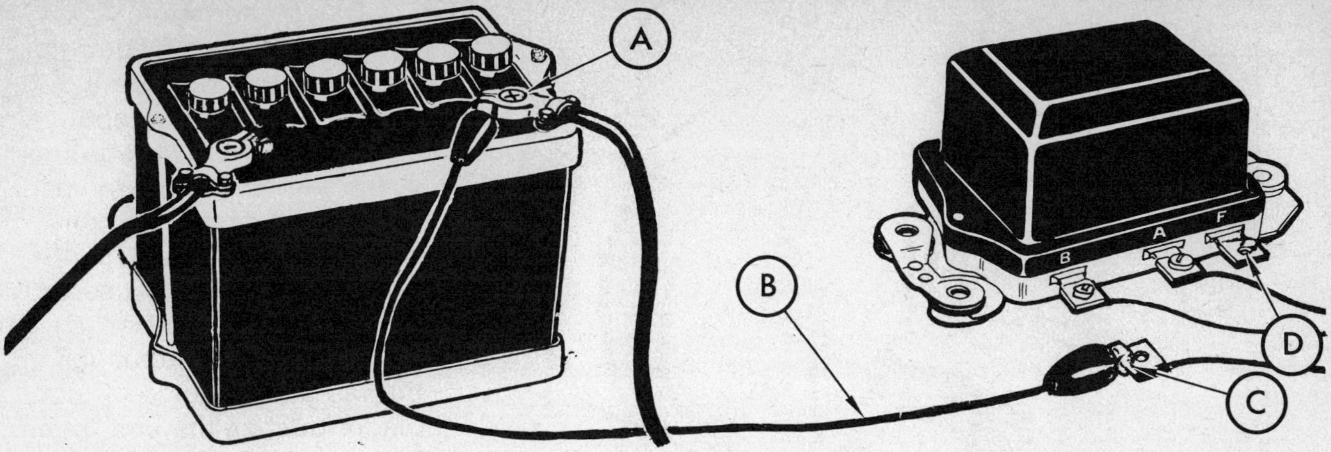

When placement of the alternator or other accessory units makes access difficult, the voltage regulator can be eliminated at the regulator connection by removing the field lead (C) from the field terminal (D). Connect a jumper lead (B) to the field lead (C) and run it directly to the positive battery terminal post (A).

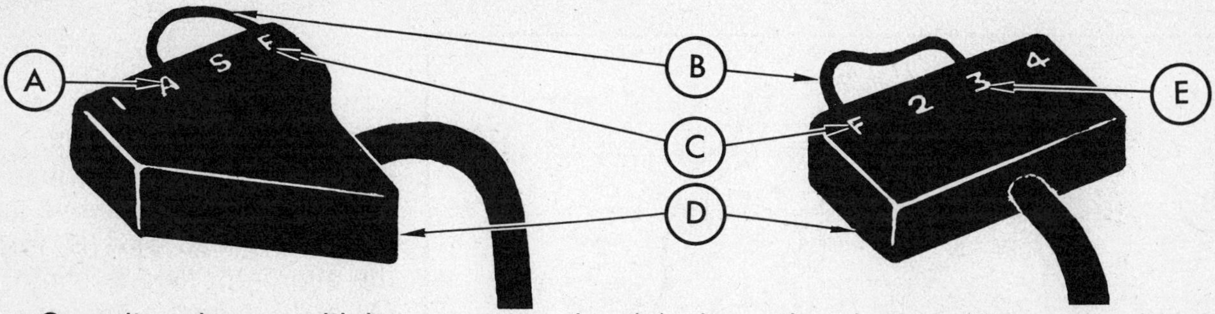

On units using a multiple connector plug (D), the entire plug must be removed and a jumper lead (B) connected between the field lead (C) and the battery lead (A, E).

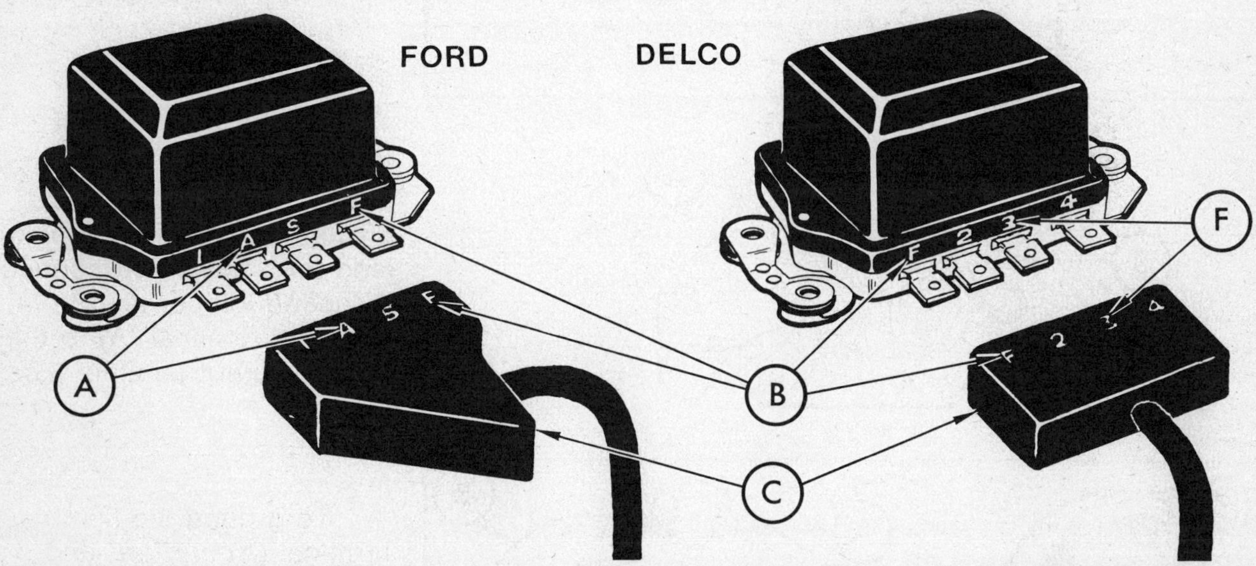

The position of the wires in the connector plug (C) and their markings differ among regulators, but generally speaking, Ford and Autolite units are marked "F" for field (B) and "A" for battery (A), while Delco regulators are marked "F" for field (B) and "3" for battery (D). If you have any doubts or if the markings are different from those we describe, refer to the factory shop manual before proceeding.

Electronic voltage regulators are of "A" circuit design and located inside the alternator, except for Chrysler products, whose regulators are firewall-mounted. The field lead of electronic or solid-state regulators must be disconnected and a jumper lead connected between the field terminal and a ground on the engine. How this is done depends upon the design of the regulator.

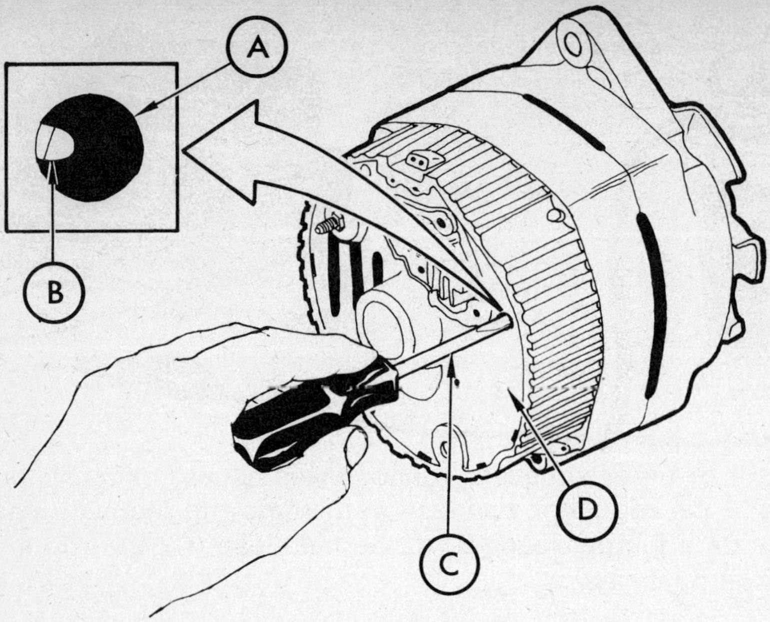

Delco alternators equipped with an electronic voltage regulator have a small hole (A) in the end frame (D). If you were to look inside the hole, you would see a small metal tab (B) about 1 inch back. Ground the tab to the end frame by inserting a screwdriver (C) until the blade tip touches the tab and the side of the blade touches the end frame. This completes the ground.

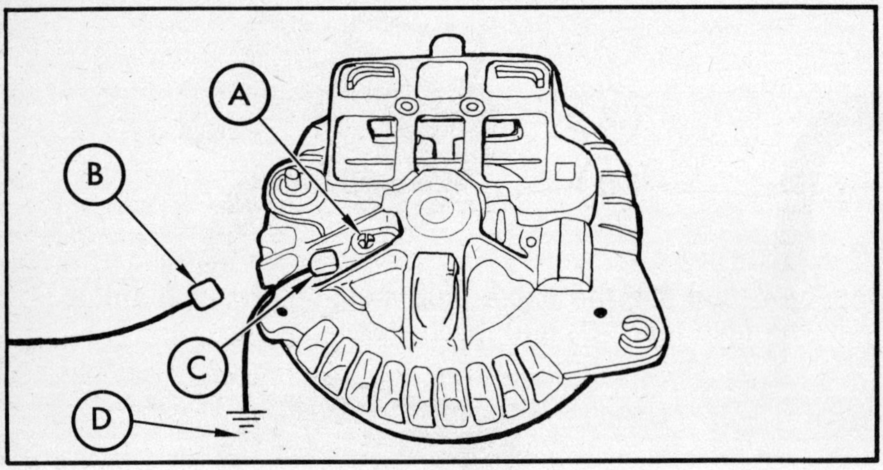

On Chrysler alternators equipped with an electronic voltage regulator, remove the green field lead wire (B) from the alternator field terminal (A) and connect a jumper lead (C) from the field terminal (A) to a ground (D).

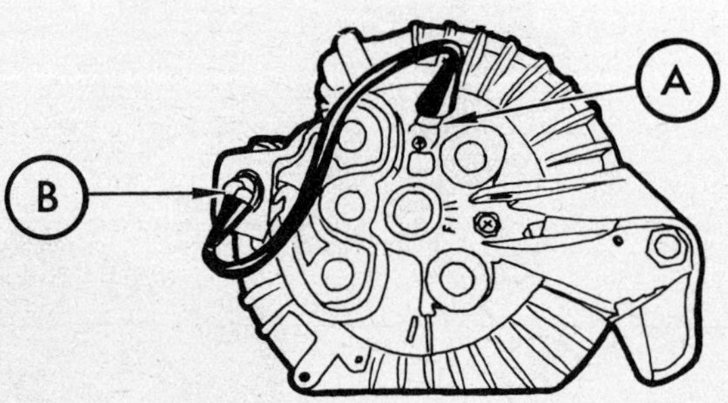

If the Chrysler alternator uses a dual field terminal, disconnect the second field lead (blue wire) and connect a second jumper lead from that field terminal (A) to the battery output terminal (B).

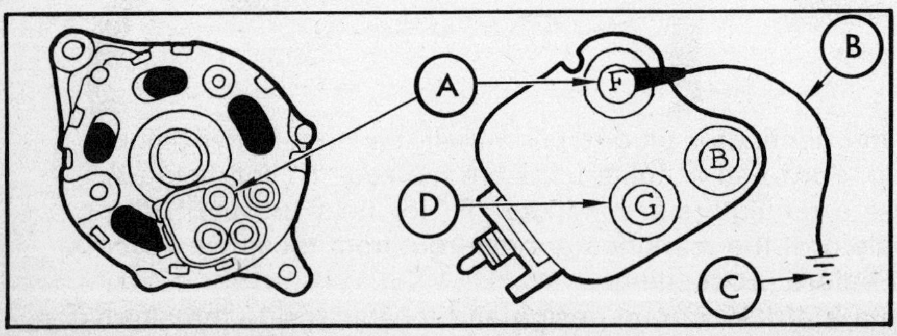

To ground the field terminal (A) on Ford and Autolite electronic voltage regulators, remove its cap. Connect the jumper lead (B) to the field terminal (A) and to a nearby ground (C).

CAUTION: DO NOT CONNECT THE JUMPER WIRE TO THE GROUND TERMINAL (D) ON THE REGULATOR.

ALTERNATOR CURRENT OUTPUT TEST—VOLTAGE REGULATOR ELIMINATED

Once the voltage regulator has been eliminated from the circuit, perform the current output test as before. Start the engine and increase the speed slowly until you reach 2,000 rpm *or* the rated output of the alternator as indicated on the meter's AMP scale.

CAUTION: DO NOT EXCEED THE RATED OUTPUT AS SPECIFIED.

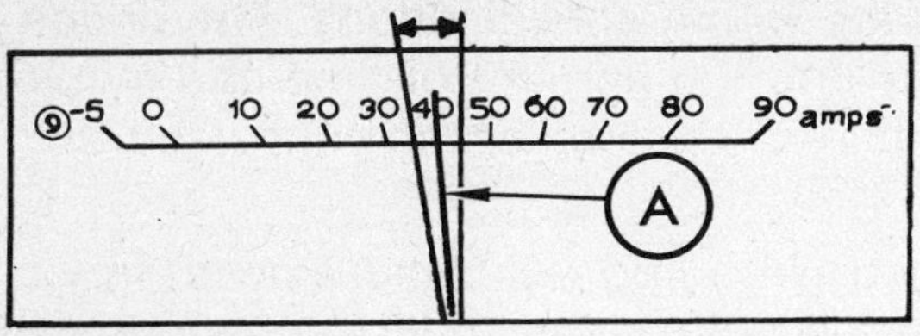

If the meter needle reads within 5 amps of the rated output (A) at 2,000 rpm, the alternator is working correctly. This means that if a lower-than-normal voltage reading was obtained during the voltage regulator test, the regulator is defective or improperly adjusted.

When the rated output is not reached with the voltage regulator eliminated, either the alternator is defective or charging circuit resistance is excessive. Before replacing the alternator, perform the circuit resistance check.

CIRCUIT RESISTANCE—ALTERNATOR

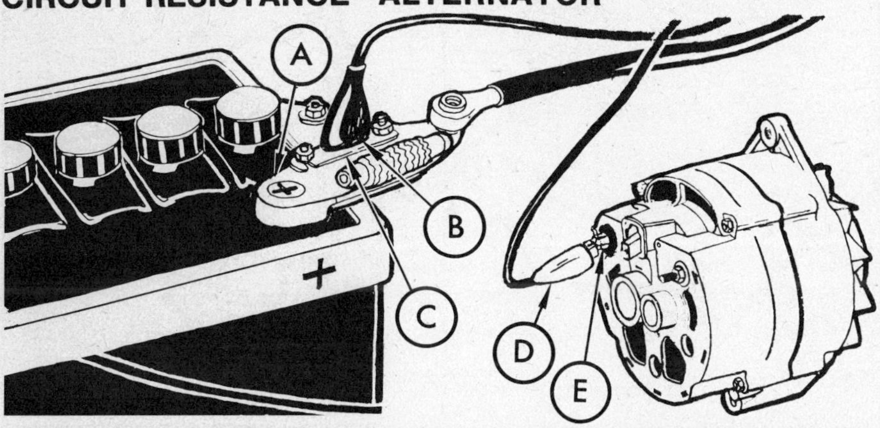

Under some conditions, the alternator can be working properly while the battery's charging rate continues to be insufficient—due to excessive voltage loss in the circuit. The circuit resistance test is used to determine whether or not excessive resistance is present.

To perform this test, the battery post adapter (A) must be in place with the shunt lever (C) *closed.* Set the meter to read the lowest voltage scale possible (usually 3 volts) and turn on the headlights. Start the engine and bring its speed up to about 1,200 rpm.

To test the insulated circuit, connect the black AGR lead (B) to the battery post adapter (A) and the red AGR lead (D) to the alternator terminal marked BAT (E). A voltage drop of more than 0.3 volts indicates excessive resistance.

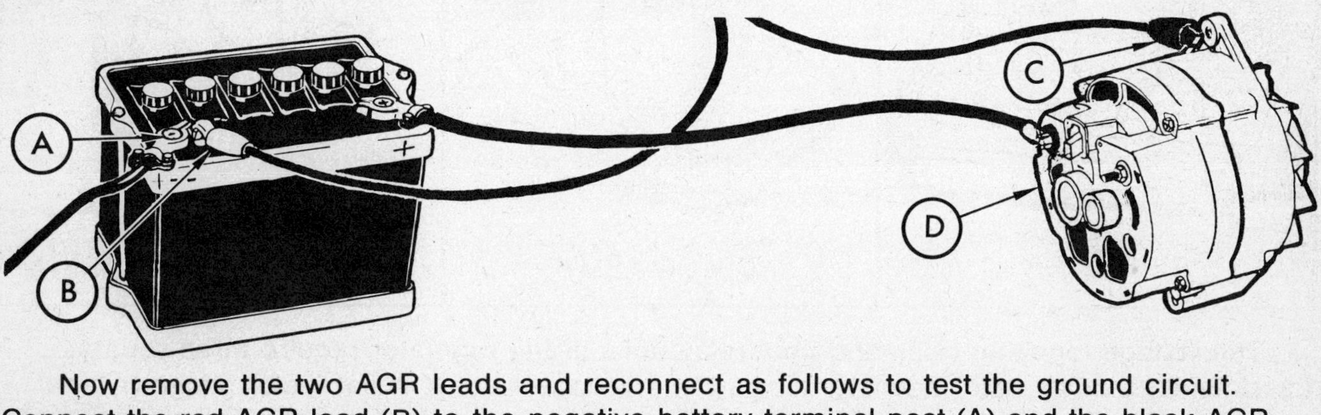

Now remove the two AGR leads and reconnect as follows to test the ground circuit. Connect the red AGR lead (B) to the negative battery terminal post (A) and the black AGR lead (C) to the alternator case (D). A voltage drop of more than 0.2 volts on the meter scale indicates excessive resistance in the ground circuit. Where excessive resistance is found, the cables must be replaced.

VOLTAGE REGULATOR ADJUSTMENT—ALTERNATOR SYSTEMS

If testing sequences indicate faulty operation of the voltage regulator in a "B" circuit design, there are two ways in which the problem can be corrected—by replacement or by adjustment of the unit. This option is not available with "A" circuit designs, because no adjustment is possible with electronic voltage regulators. When these no longer function properly, they must be replaced.

Replacing a faulty regulator with a new or rebuilt one in a "B" circuit design is the quicker but more costly approach. Less expensive but more time-consuming is adjusting the voltage regulator relay and/or field relay to operate within specifications. The option you choose will depend upon your mechanical ability and the amount of time and money at your disposal, as well as the availability of the required replacement in your geographic area.

For those who wish to adjust the defective relay or who find adjustment necessary because no replacement is immediately at hand, the following adjustment sequences can be performed on the car by removing the regulator cover. After making the required adjustments, the cover must be replaced and the regulator retested to make certain that the relay adjustment was correct and that the relay is now operating correctly.

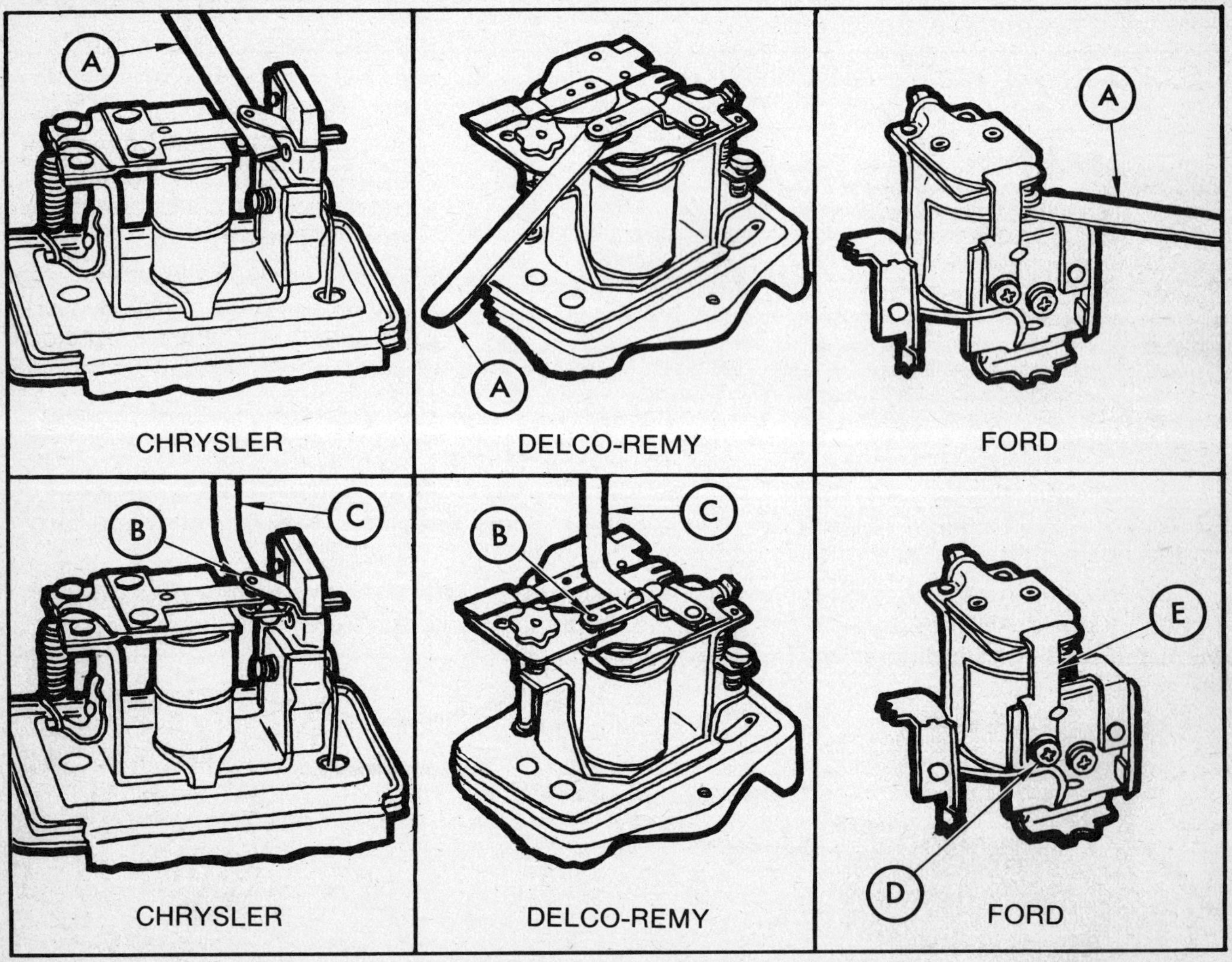

The voltage regulator relay and field relay units of the regulator require three separate adjustments—contact point opening, air gap and voltage setting. A flat feeler gauge (A) that corresponds to the specified point opening is used to measure the gap between regulator points. Adjustment is made according to unit design. On some, you bend the contact arm (B) with tool (C), while on others, you loosen an adjustment screw (D) and move the upper contact arm (E) up or down accordingly to set the specified opening.

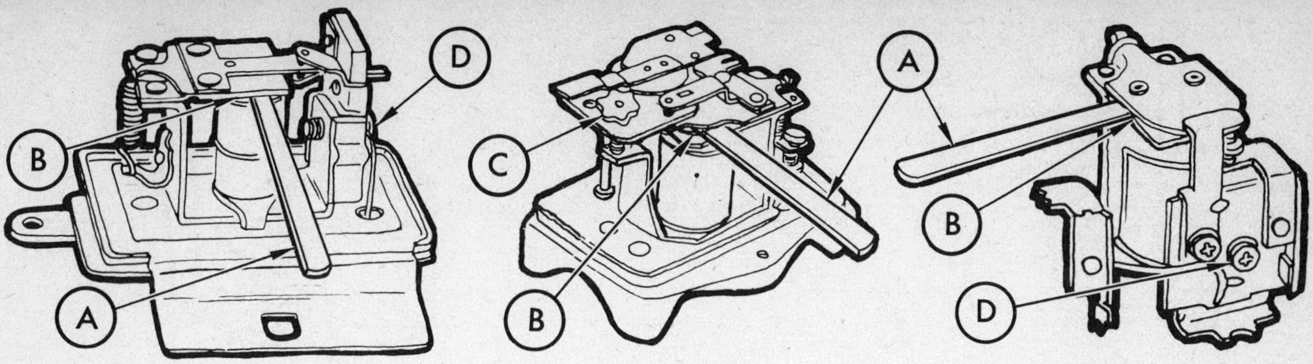

CHRYSLER DELCO-REMY FORD

A feeler gauge (A) is also used to measure air gap (B) between the core and armature. Adjustment of the air gap is made according to design: by turning the air gap adjustment nut (C) or by loosening the contact bracket screw (D) and adjusting the distance between core and armature to set the specified air gap.

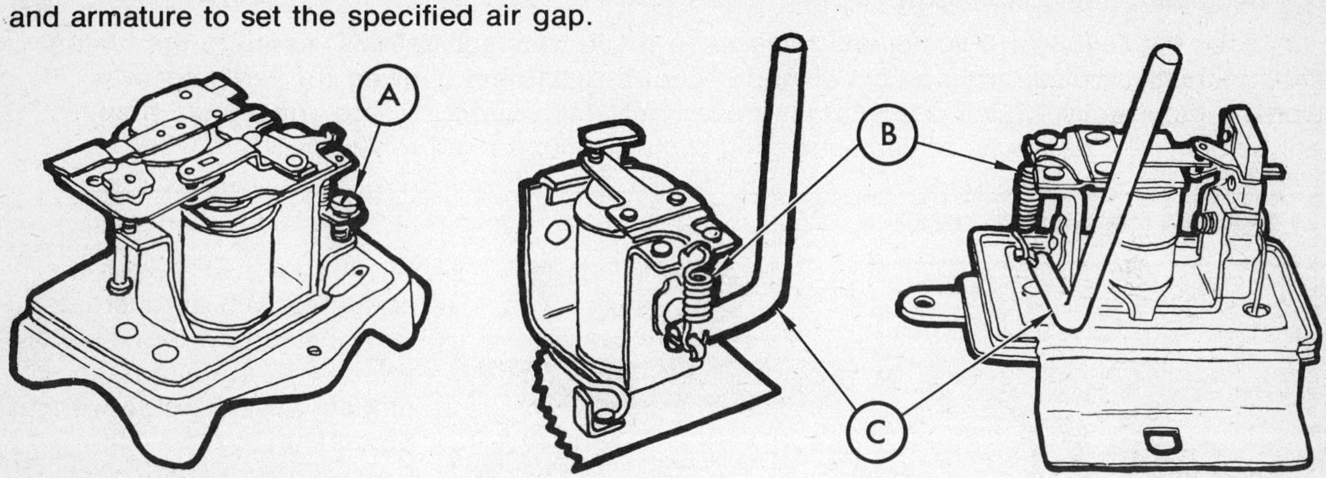

DELCO-REMY FORD CHRYSLER

The voltage setting is adjusted with a screw (A) or by bending the spring hanger (B) down to increase and up to decrease the setting with a bending tool (C).

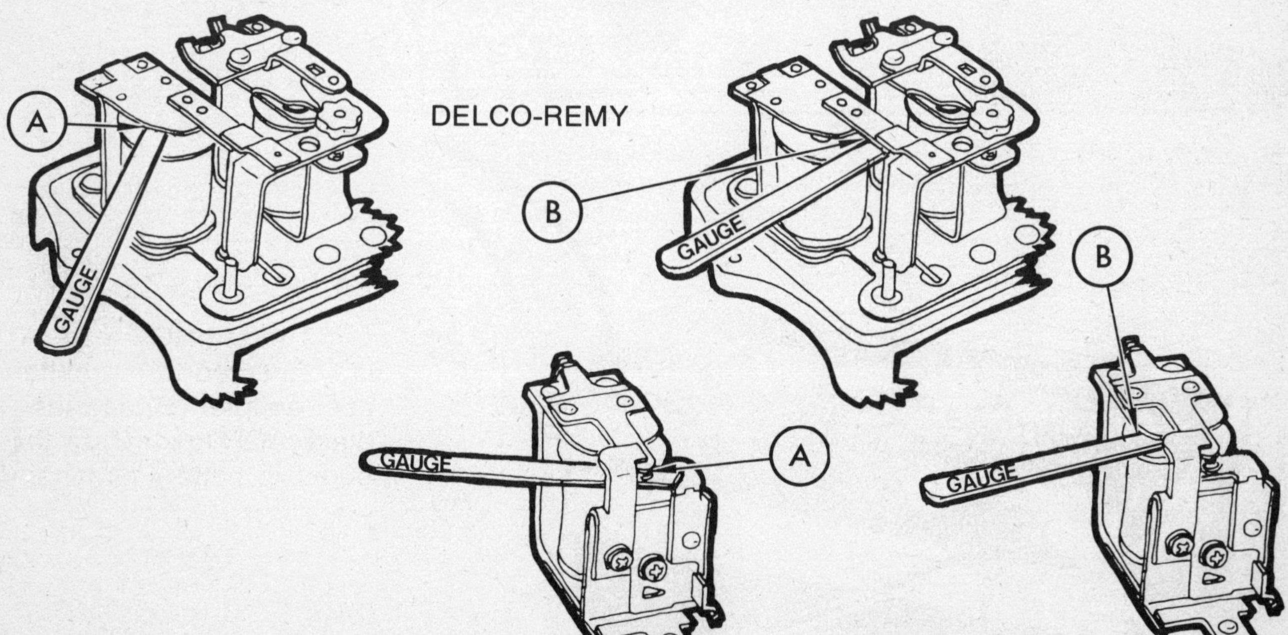

DELCO-REMY

FORD

The field relay point opening (A) is measured and adjusted in the same manner as the voltage regulator relay, as is the field relay air gap (B).

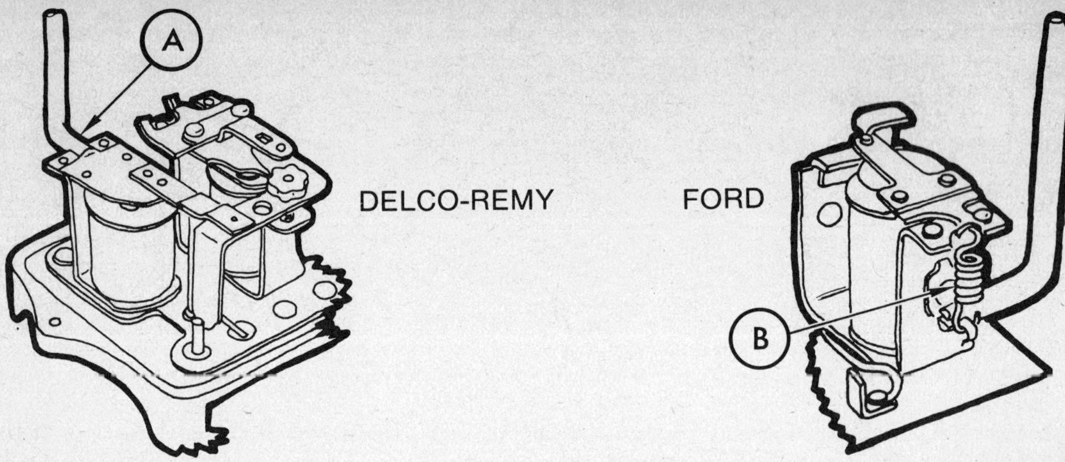

To adjust field relay closing voltage, bend the relay's heel iron (A) or its spring hanger (B), depending upon its design.

Once the regulator adjustments have been made to specifications (found in the factory shop manual), replace the unit cover and repeat the test that showed the regulator was working incorrectly. If your adjustments have been made properly, the unit should now test out as correct. The cause of your charging system disorder will have been removed.

ALTERNATOR DIODE TEST

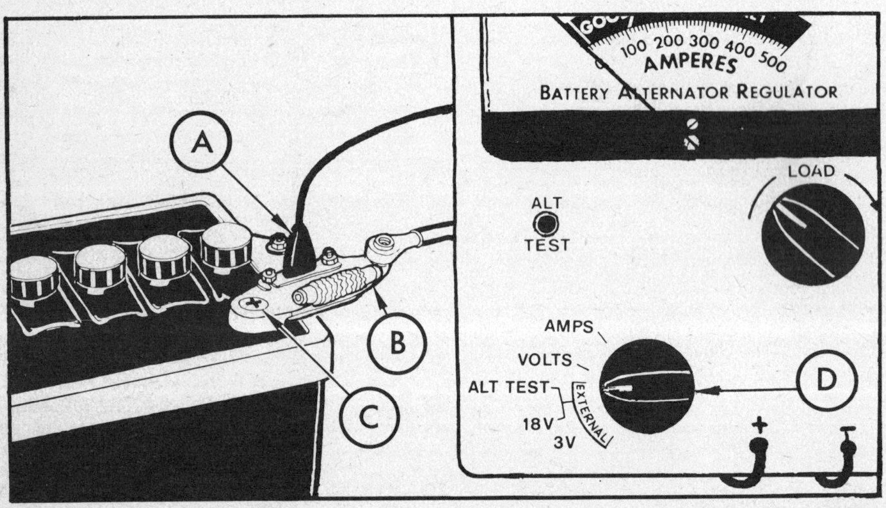

If your AGR tester is equipped with an alternator test position on the selector switch (D), the alternator diodes can be tested without disassembling the unit. Connect the battery post adapter (B) to the positive battery terminal (C) with the shunt lever and lead (A) in a *closed* position.

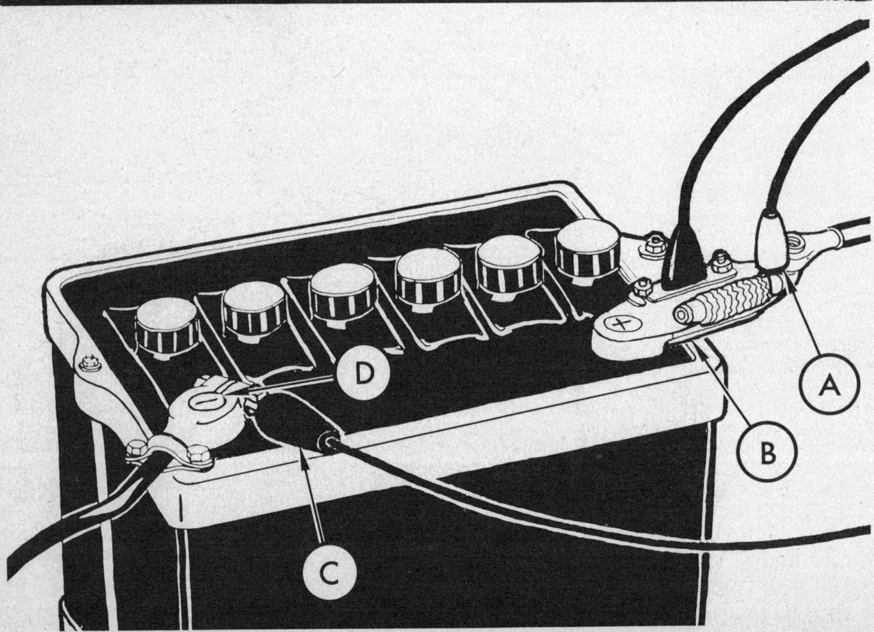

The red AGR lead (A) is connected to the battery post adapter (B) and the black AGR lead (C) to the negative battery terminal (D).

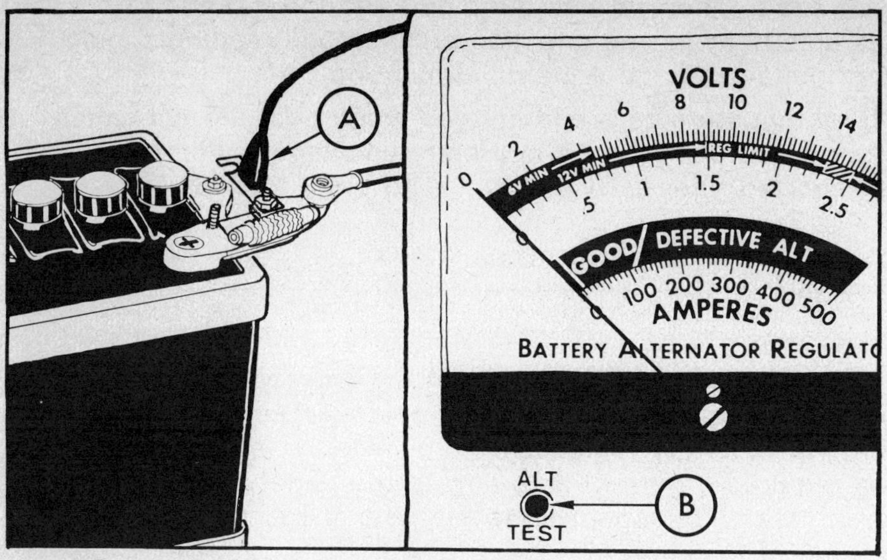

Start the engine and open the shunt lever (A). Turn on the headlights and increase engine speed to about 1,200 rpm. Press the alternator test button (B) and read the scale marked GOOD/BAD ALTERNATOR or GOOD/DEFECTIVE ALTERNATOR.

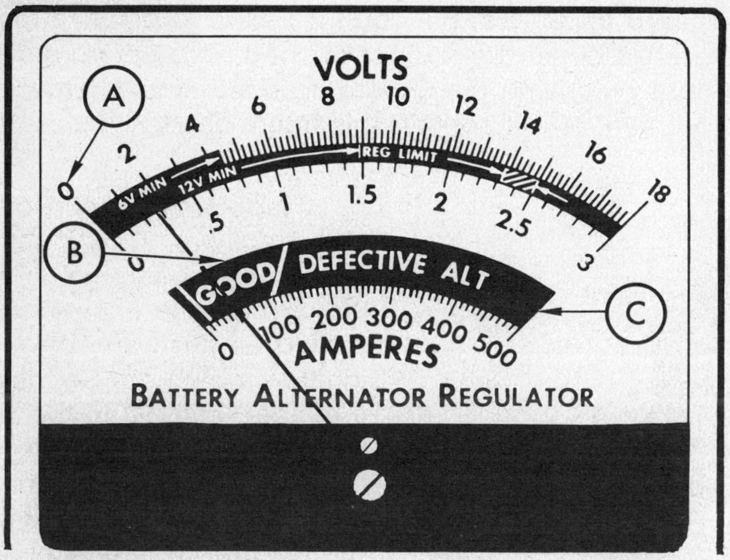

If the alternator diodes are good, the needle will rest in the GOOD band of the scale (B). A zero reading (A) on some scales (check your instruction manual) also means that the alternator is working properly.

Should the needle rest in the BAD or DEFECTIVE band of the scale (C), the alternator contains one or more shorted/open diodes or an open/shorted stator winding. In this case it must be removed from the car for further testing.

DIODE BENCH TEST

Individual replacement of diodes is not possible with most late-model alternators, since the diodes are permanently mounted in one or two heat sinks. If the diodes test out bad in the alternator diode test, simply replace the entire heat sink with another containing new diodes. Often, however, the diodes used in older alternators can be carefully unsoldered and pressed out of the alternator case for replacement. They can also be tested individually with the AGR tester.

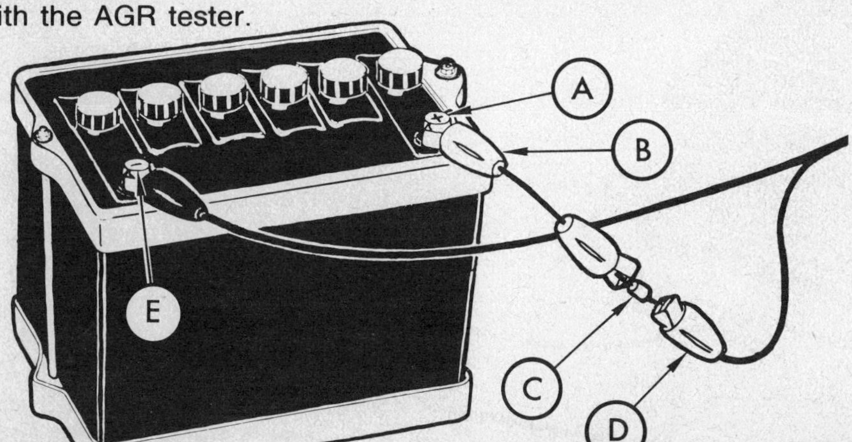

Adjust the AGR voltage selector to the 16-volt position. Connect the black AGR lead to the negative terminal post (E) of a 12-volt battery and one end of a jumper lead (B) to the positive battery terminal post (A). Now clip the other end of the jumper lead to the diode case (C) and then connect the red AGR lead (D) to the diode wire.

Note the reading on the 16-volt scale. Then reverse the diode end-for-end and take a second reading. The diode passes electricity in one direction only, so one reading should be zero and the other should be at battery voltage if the diode is good.

The diode is *open* when both readings are zero and *shorted* when both are full battery voltage. A reading above zero on one end combined with a battery voltage reading on the other indicates a *leaking* (partially shorted) diode. Diodes that read other than GOOD must be replaced.

This ends the testing sequence for alternator systems.

DC GENERATOR POLARITY

Whenever a new generator or regulator is installed or when the old generator is reconnected after testing and/or repair, the generator must be polarized before starting the engine. If this is not done, the generator (or regulator) will burn out in a short time, since correct generator polarity is required if current is to flow to the battery. Polarizing the generator is necessary to restore the residual magnetism to the pole shoes. It amounts to electrifying the fields with a battery.

The use of the AGR tester is not required to do this, but you may have to polarize the generator before or after testing. Therefore we cover the correct method here before moving into the sequence of generator tests. Before polarizing the generator, determine whether it is an "A" or "B" circuit. The factory shop manual will contain this information if you are not certain.

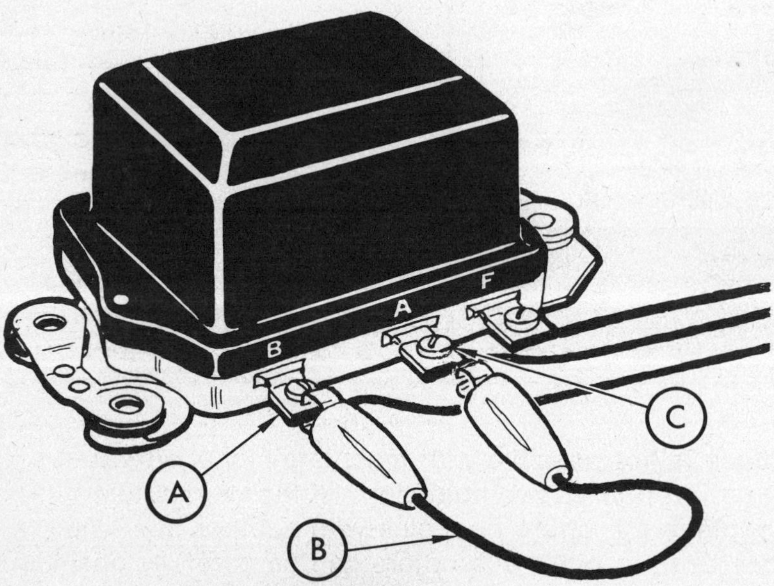

To polarize Delco-Remy and Autolite external-field ground generators using an "A" circuit, the engine should be dead and the ignition switch off. Connect a jumper lead (B) to the battery terminal (marked BAT) of the regulator (A) and touch the other end of the lead to the regulator armature terminal (C)—marked ARM or GEN—for a fraction of a second. If there is a fuse in the regulator-to-battery circuit, bypass it or it may be blown.

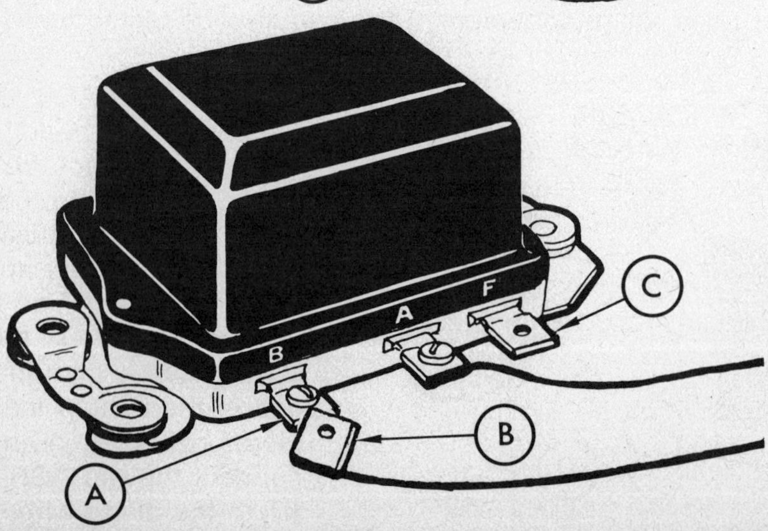

Internal field ground generators using a "B" circuit should also be polarized with the engine dead and the ignition off. Remove the field lead (B) from the regulator field terminal (C), which is marked "F," and touch it for a fraction of a second to the regulator's battery terminal (A), marked BAT.

GENERATOR SYSTEM—VOLTAGE REGULATOR TEST

Run the engine with the hood closed for 15 minutes or until it reaches normal operating temperature, then shut it off. Install the battery post adapter (B) on the positive battery terminal (C) with the shunt switch and shunt lead (A) CLOSED.

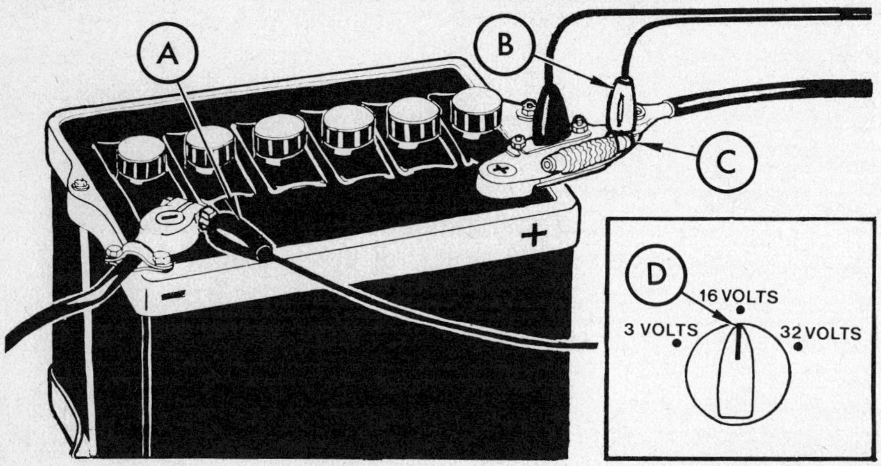

Connect the red AGR lead (B) to the battery post adapter terminal (C) and the black AGR lead to the negative battery post (A). Adjust the voltage selector (D) to the 16-volt position.

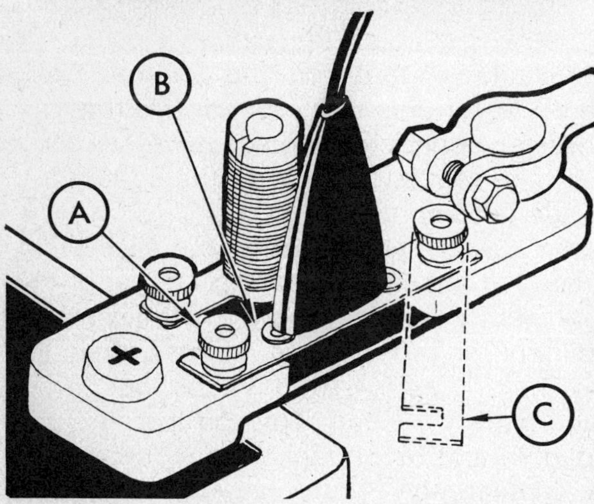

GENERATOR TYPE	REGULATOR VOLTAGE RANGE
All six-volt systems	7.0— 7.6
Chrysler	14.3—14.6
General Motors	13.8—14.8
Ford	14.8—15.4

Check to make sure that the shunt lever (B) is in a CLOSED position (A). Then start the engine and open the lever (C). Increase engine speed from idle to approximately 1,500 rpm, read the 16-volt scale on the meter dial and compare it to the manufacturer's specifications. While we recommend that you check the factory shop manual for exact figures, you can use the regulator voltage range provided here.

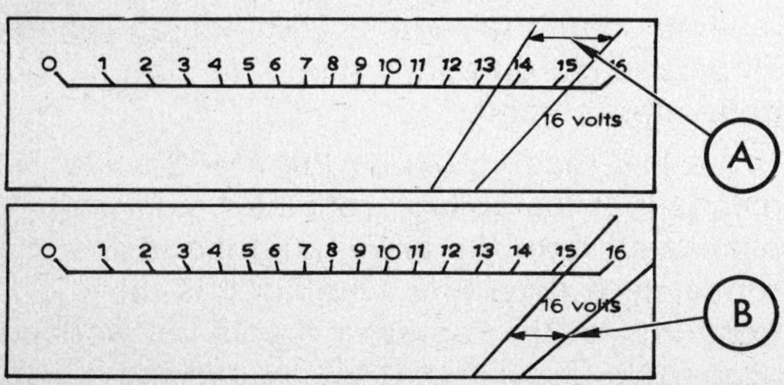

A voltage reading within the specified range (A) indicates that the regulator is good. A reading higher than normal (B) indicates:

- The regulator is not operating correctly and must be adjusted or replaced.
- The generator field is shorted to ground inside the generator unit.

ALTERNATOR/GENERATOR/REGULATOR TESTER

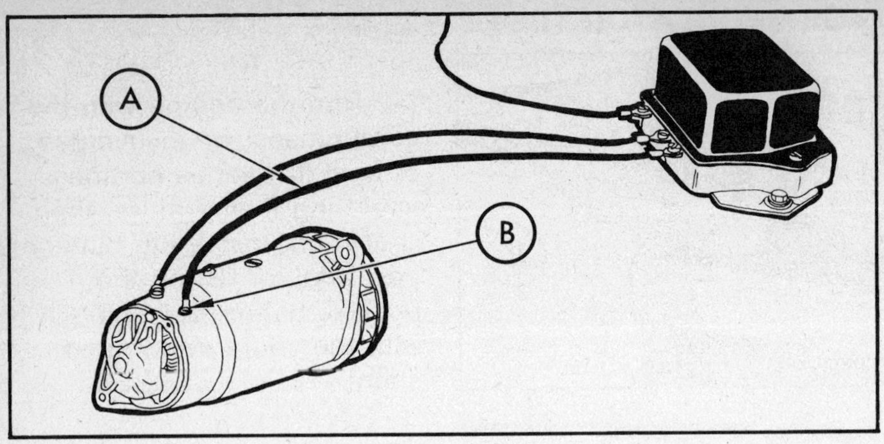

To pinpoint the problem area exactly, remove the field wire (A) at the generator (B) with the engine running at 1,500 rpm. If the voltage reading goes down to zero, adjust or replace the regulator; if the reading remains high, the generator is faulty.

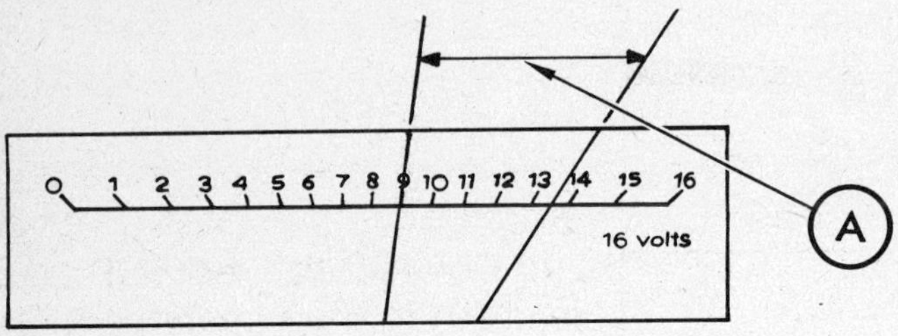

A lower-than-normal reading (A) also indicates that the malfunction is in either the regulator or generator. The current output test is necessary, though, to determine which of the two is at fault.

GENERATOR SYSTEM—CURRENT OUTPUT TEST

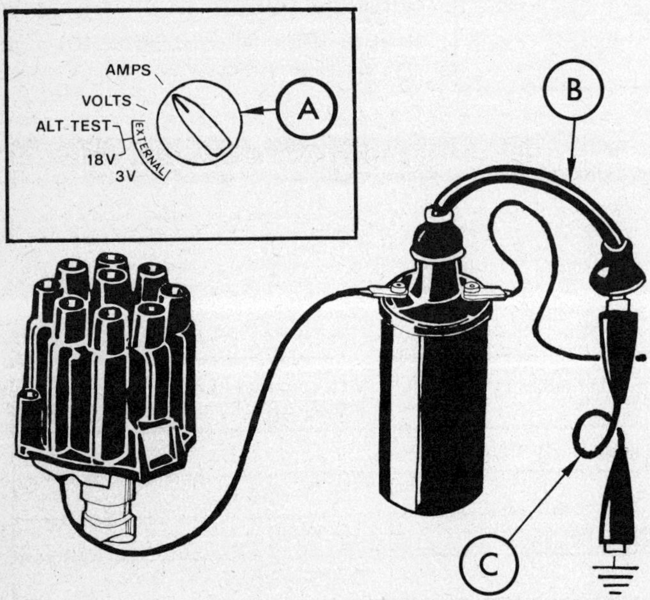

Maximum output from the generator is required to run this test properly. With the AGR tester connected the same way as for the voltage regulator test and the battery post adapter shunt lever in the CLOSED position, switch the meter to read AMPS (A). Remove the secondary coil wire (B) and ground with a jumper lead (C). Turn on the headlights for two or three minutes, then crank the engine for 15 to 30 seconds to reduce battery voltage. Turn off the headlights and reconnect the secondary coil wire (B) securely.

Start the engine and bring its speed up to about 1,600 rpm. Read the AMPS scale and compare to the manufacturer's rated output. A plus or minus tolerance is provided to allow for temperature variations, so the reading should fall within 3° of the specified amperage (A) on either side. If so, the generator is good.

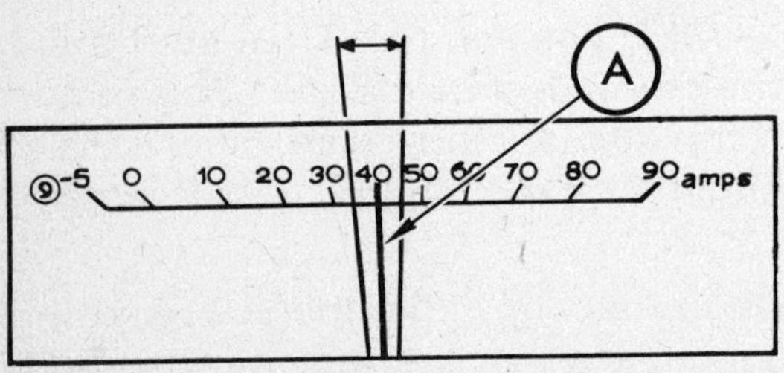

A low reading (A) on the AMPS scale means that the voltage regulator must be eliminated from the circuit and the test repeated to determine whether it is the generator or the regulator that is not working properly.

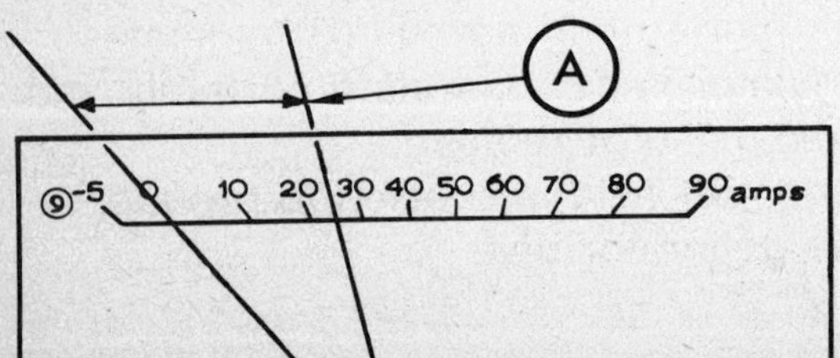

VOLTAGE REGULATOR ELIMINATION

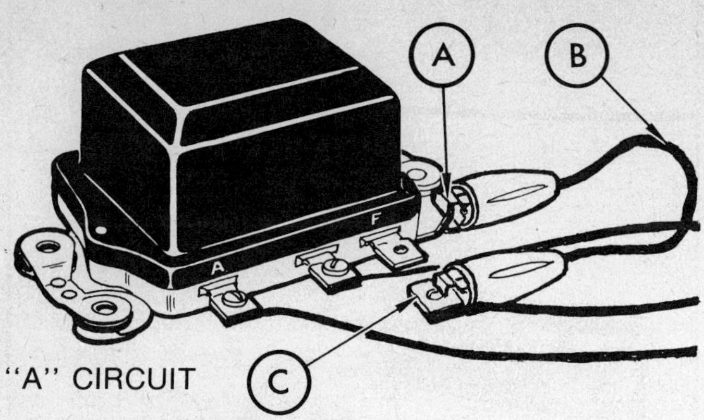

"A" CIRCUIT

To eliminate an "A" circuit voltage regulator at the regulator, disconnect the field lead (C). Connect one end of a jumper lead (B) to the field lead (C) and the other end to a ground (A).

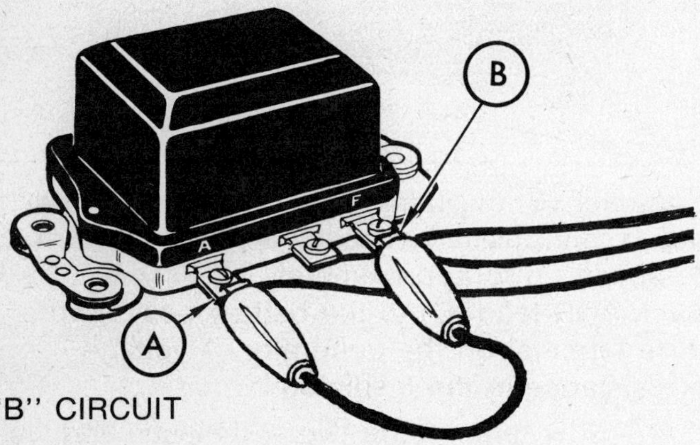

"B" CIRCUIT

To eliminate a "B" circuit regulator, connect a jumper lead from the field terminal (B) to the armature terminal (A). As both terminal placement and markings can vary among different regulators, make certain that your connection is correct (or consult the factory shop manual before continuing).

GENERATOR CURRENT OUTPUT TEST—VOLTAGE REGULATOR ELIMINATED

Once the voltage regulator has been eliminated from the circuit, perform the current output test as before. Start the engine and increase its speed slowly until you reach 2,000 rpm *or* the rated output of the generator as indicated on the meter's AMPS scale.

CAUTION: DO NOT EXCEED THE RATED OUTPUT AS SPECIFIED.

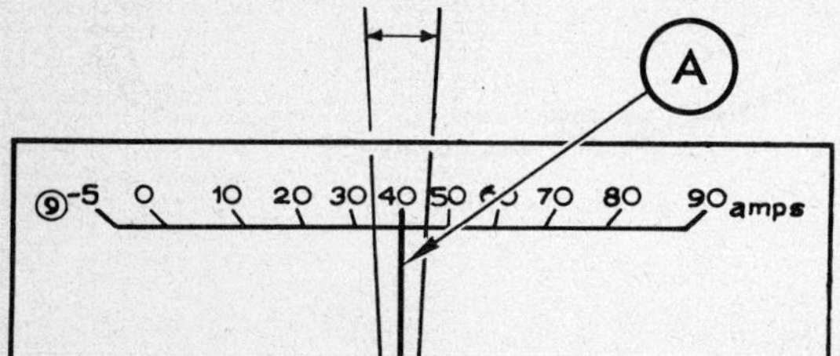

If the current shown by the meter scale approaches to within 5 amps of the rated output (A) at 2,000 rpm, the generator is working correctly. Thus a lower-than-normal regulator reading obtained during the voltage regulator test means that it is the regulator which is not operating correctly and that it must be adjusted or replaced.

When the rated output is not reached with the voltage regulator eliminated, look for a defective generator or excessive charging circuit resistance. Perform the circuit resistance test before replacing the generator.

CIRCUIT RESISTANCE—GENERATOR

It is possible for the generator to be working properly under certain conditions and yet for the battery's charging rate to remain insufficient because of excessive voltage loss in the circuit. The circuit resistance test is used to determine if such excessive resistance is present.

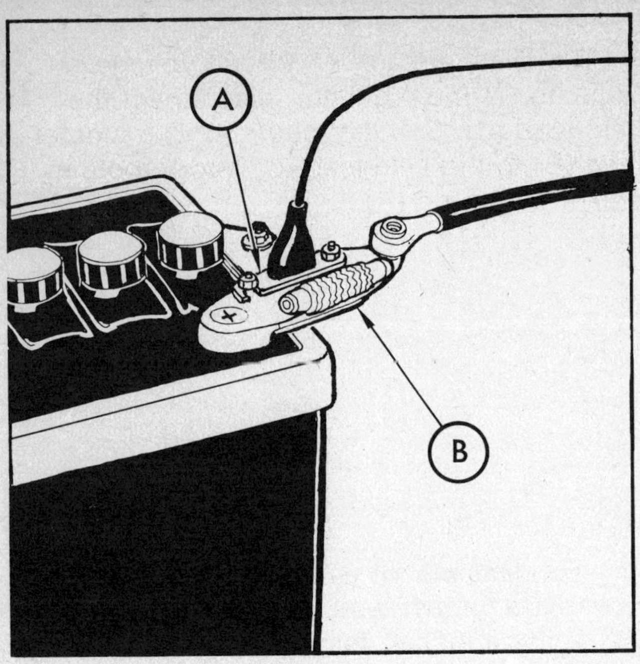

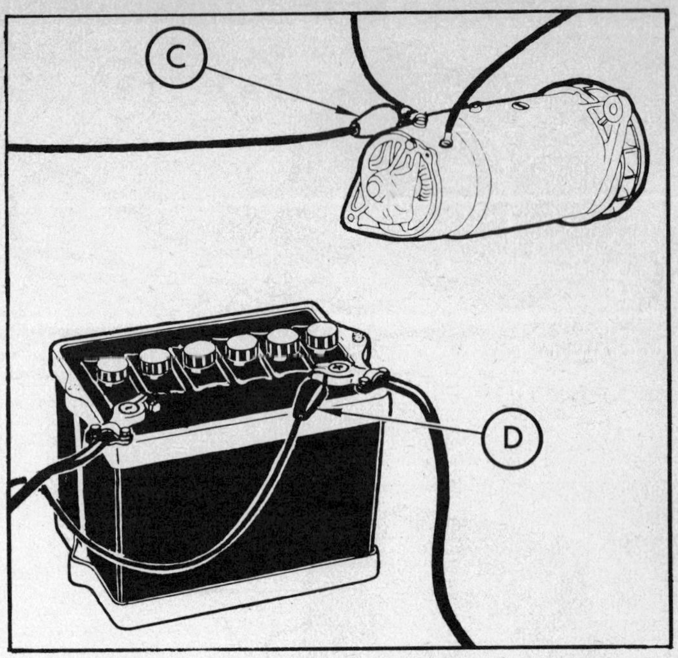

The test is performed with the battery post adapter (B) in place and the shunt lever (A) CLOSED. Set the meter to read the lowest voltage scale possible (usually 3 volts) and turn on the headlights. Start the engine and bring its speed up to approximately 1,200 rpm.

To test the insulated circuit, connect the black AGR lead (D) to the battery post adapter terminal and the red AGR lead (C) to the armature terminal on the generator. A voltage drop of more than 0.8 volts indicates excessive resistance in the insulated circuit.

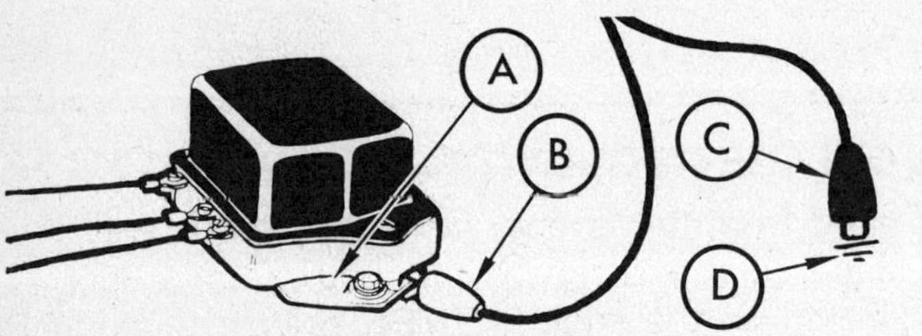

Now remove the two AGR leads and reconnect as follows to test the ground circuits: Connect the red AGR lead (B) to the regulator case (A) and the black AGR lead (C) to a ground on the engine block (D). Read the voltage drop and remove the AGR leads.

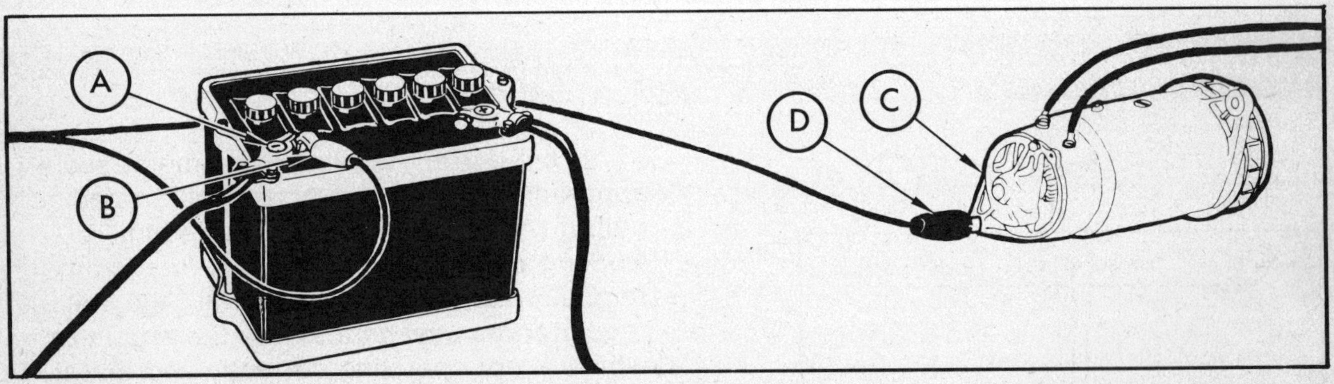

Reconnect the leads a second time but in this manner: The red AGR lead (B) goes to the negative battery terminal post (A) and the black AGR lead (D) to the generator case (C). Read the voltage drop and remove the AGR leads.

A voltage drop greater than 0.1 volt in either of these hookups indicates excessive resistance in the ground circuits.

GENERATOR SYSTEM—CUTOUT RELAY TEST

If the cutout relay breaker points do not open when the engine is shut off, current from the battery will flow back through the circuit to the generator. Also, if the points are not closed when the engine is running, current from the generator will not flow to the battery.

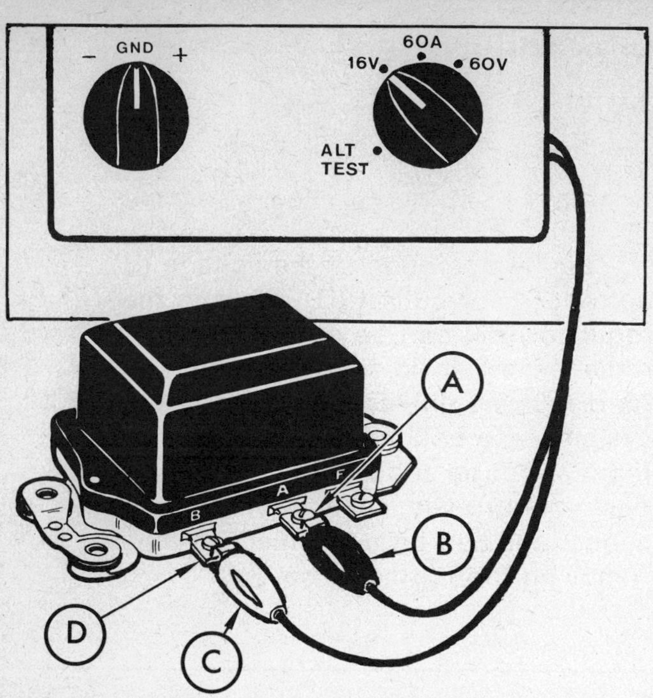

To check the on/off operation of the cutout relay, connect the red AGR lead (C) to the regulator battery terminal (D), marked BAT and the black AGR lead (B) to the regulator armature terminal (A), marked ARM or GEN. With the engine off, the 16-volt scale should read battery voltage. A lower reading indicates that the cutout relay breaker points are stuck closed. They will allow the battery to discharge unless the malfunction is corrected.

With the engine running at approximately 1,200 rpm, the 16-volt scale should read zero volts. If it does not, the cutout relay is not working correctly and must be adjusted. Now turn the engine off. If the meter needle read zero with the engine running at 1,200 rpm, it should return to a battery voltage reading as the breaker points open.

GENERATOR SYSTEM—CURRENT RELAY TEST

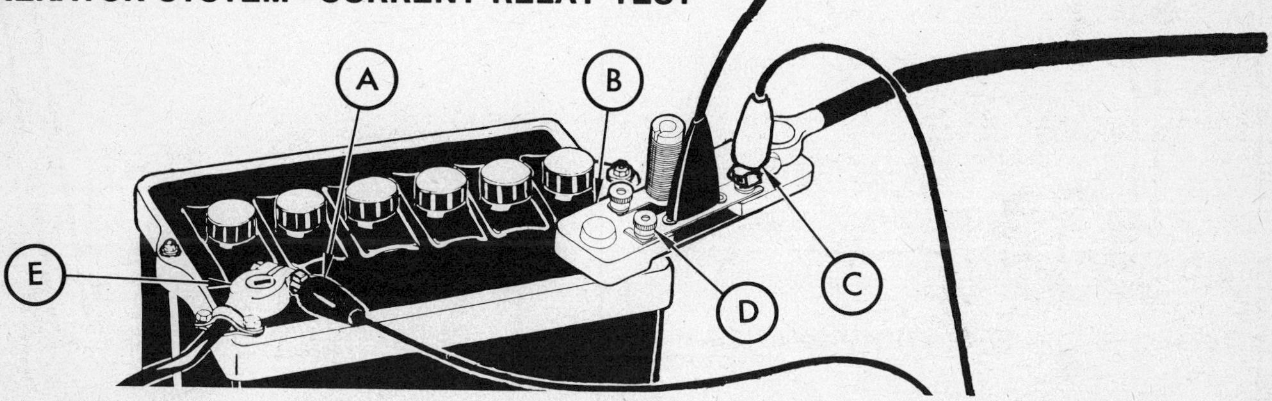

Use the battery post adapter (B) with the red AGR lead (C) connected to it and the black AGR lead (A) connected to the negative battery terminal post (E). The shunt lever (D) must be CLOSED.

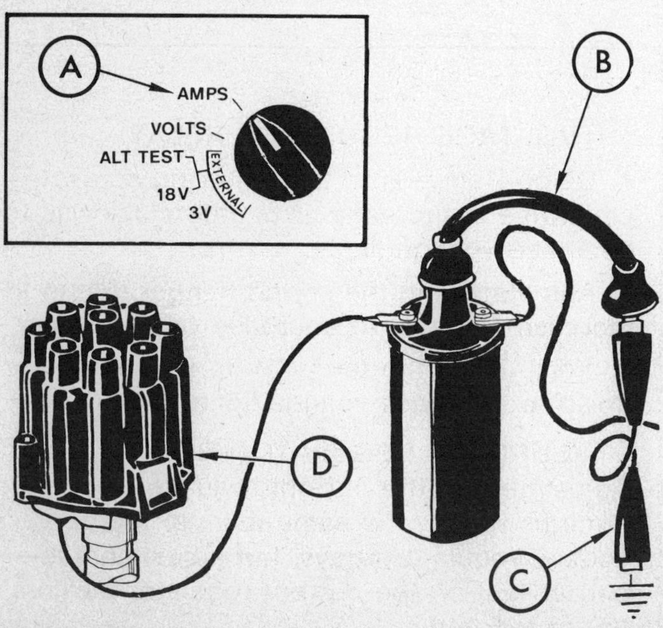

Turn the selector switch to the AMPS position (A). Disconnect the secondary coil wire (B) at the distributor (D) and ground it with a jumper lead (C) connected to the engine block. Crank the engine for 30 seconds, then reconnect the secondary coil wire (B) to the distributor (D).

Now start the engine and increase its speed to about 2,000 rpm while watching the AMPS scale on the meter dial. As the engine speed increases, the current output reading on the AMPS scale should also increase, up to the specified current relay setting provided by the manufacturer, holding at that level until engine speed is decreased. A reading above or below the specified cutout relay setting means that the relay is incorrectly adjusted or defective.

VOLTAGE REGULATOR ADJUSTMENT—GENERATOR SYSTEMS

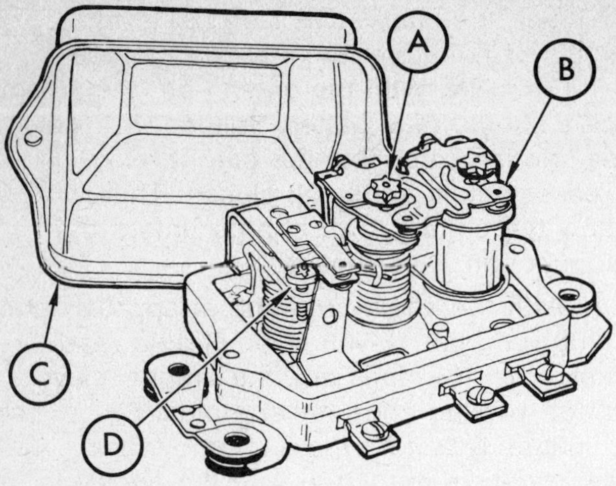

Faulty operation of the voltage (B), current (A) or cutout (D) relays in the regulator unit can be corrected by replacement or adjustment of the unit. As discussed earlier under alternator systems, the option depends upon several factors, but for those who find adjustment desirable or even necessary, the following sequences can be performed on the car by removing the regulator cover (C).

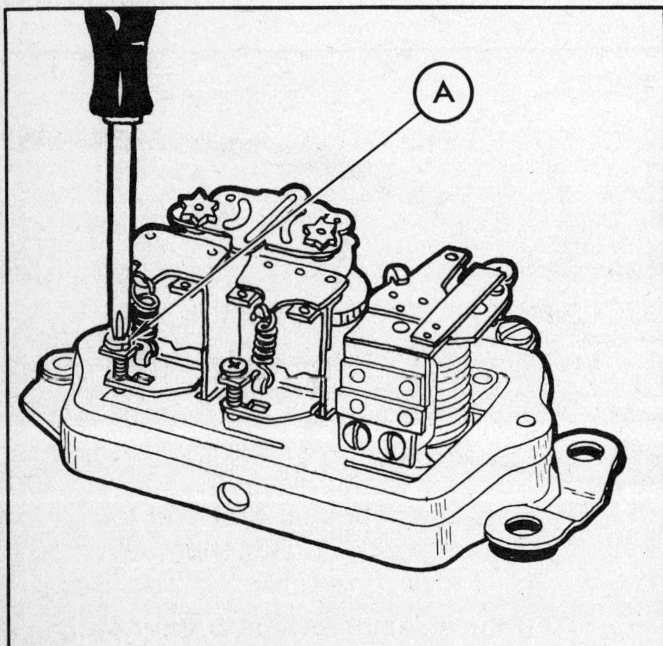

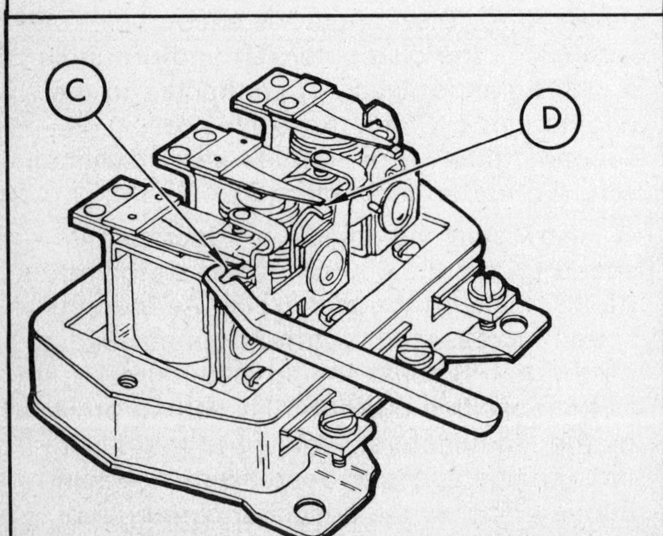

VOLTAGE REGULATOR RELAY:

1. Delco-Remy: Turn the adjusting screw (A) clockwise to increase or counterclockwise to decrease voltage limits.

2. Autolite: Bend the spring hanger (B) up to increase or down to decrease voltage limits.

3. Ford: Bend the tang (C) up to increase or down to decrease voltage limits.

CURRENT REGULATOR RELAY: Adjustment of the current regulator relay is done in exactly the same manner as for voltage regulator relays, but adjustment is carried out on the current regulator portion (D) of the unit.

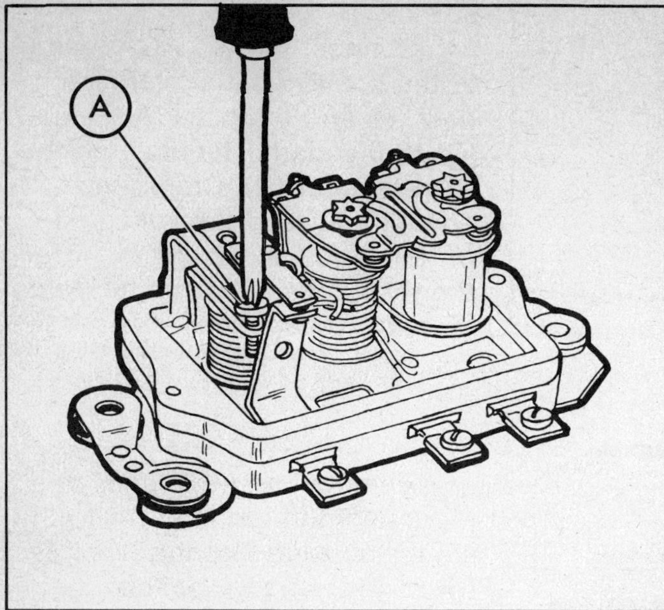

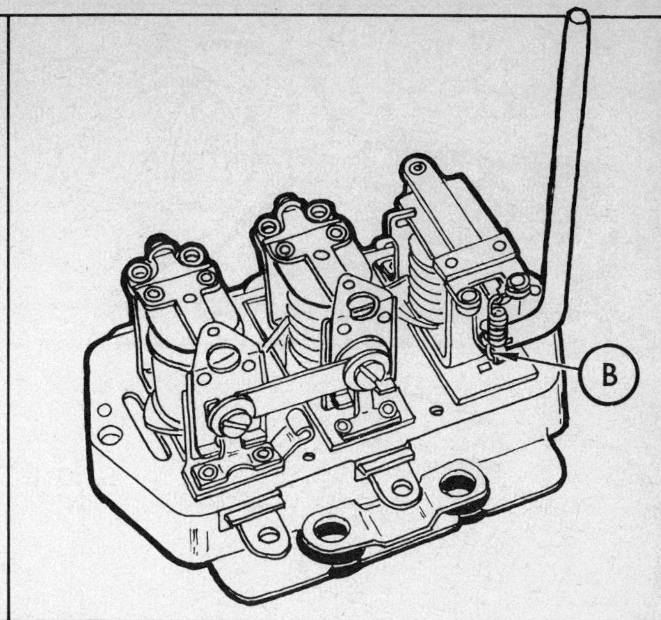

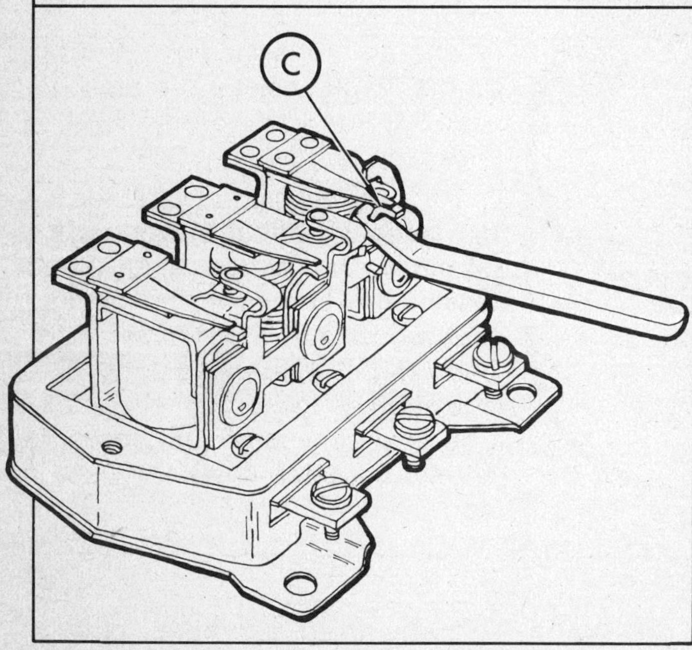

CUTOUT RELAY:

1. Delco-Remy: Turn the adjusting screw (A) clockwise to increase or counterclockwise to decrease closing voltage.

2. Autolite: Bend the spring hanger (B) down to increase or up to decrease closing voltage.

3. Ford: Bend the tang (C) up to increase or down to decrease closing voltage.

After making the required adjustments, replace the regulator cover and retest to make certain that the adjustment was correct and that the faulty relay is now operating correctly. This completes the AGR testing sequence on the automotive charging system, but other uses for the unit are covered below.

STARTER CIRCUIT VOLTAGE LOSSES

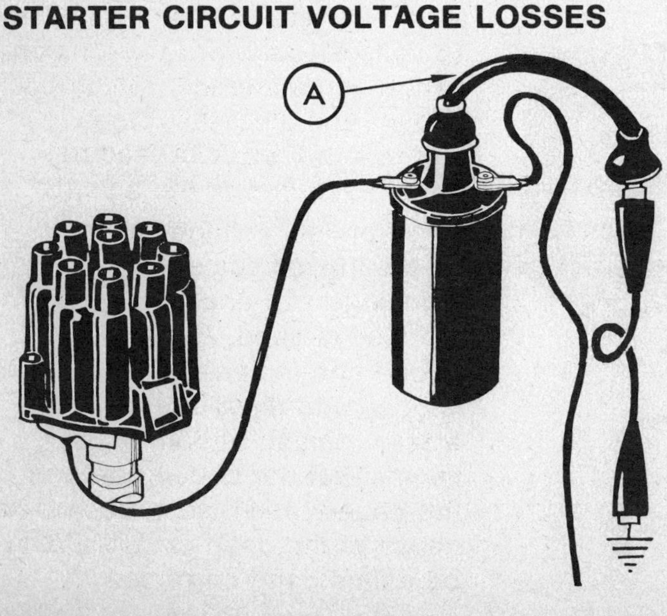

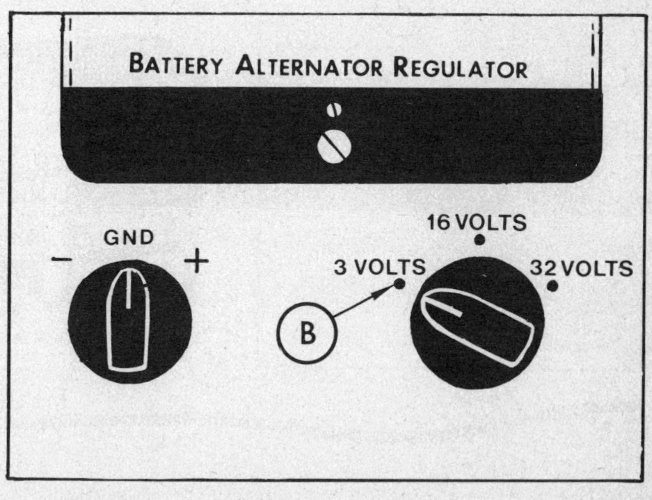

With the AGR voltage selector set at the 3-volt position (B), ground the secondary coil wire (A) to prevent the engine from starting.

ALTERNATOR/GENERATOR/REGULATOR TESTER

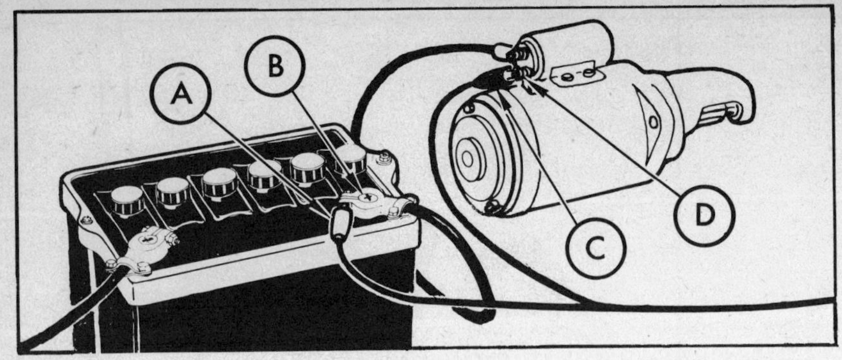

Connect the red AGR lead (A) to the positive battery terminal post (B) and the black AGR lead (C) to the starter terminal on the starter motor (D). Crank the engine and read the voltage drop on the 3-volt scale. If it is more than 0.5 volts (0.3 for a six-volt system), voltage loss is excessive.

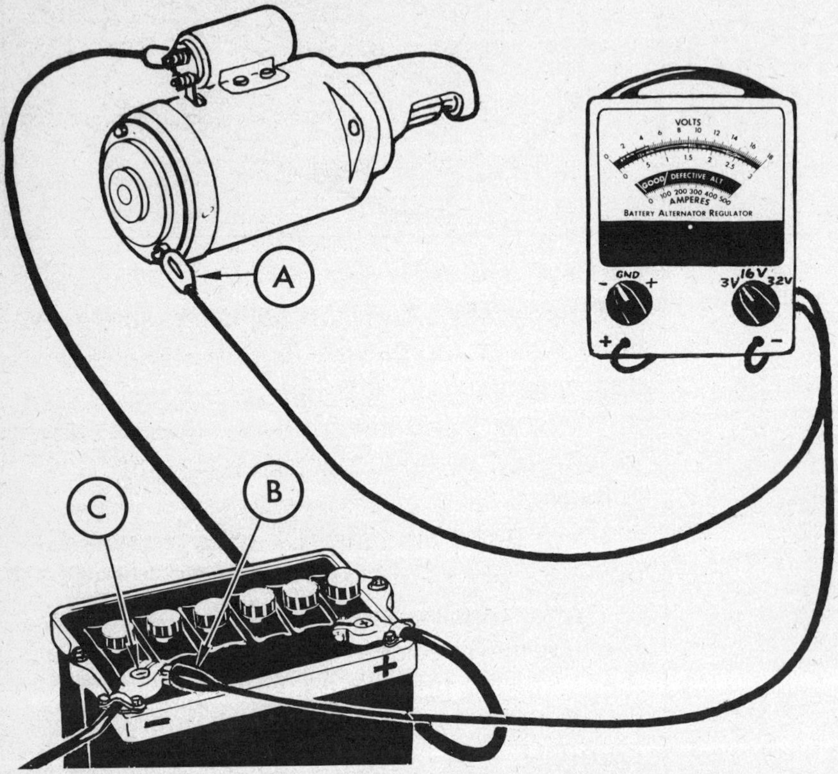

Remove the AGR leads and reconnect the red lead (A) to the starter housing and the black lead (B) to the negative battery terminal post (C). Crank the engine and read the voltage drop on the 3-volt scale; more than 0.2 volts (0.1 for a six-volt system) is excessive.

Excessive high-voltage losses can seriously affect the starter's operation and performance. Such losses may be caused by cables that are defective or too small to carry the current or by dirty, corroded or loose connections. The exact cause of such losses should be determined and corrected.

FUSE TEST

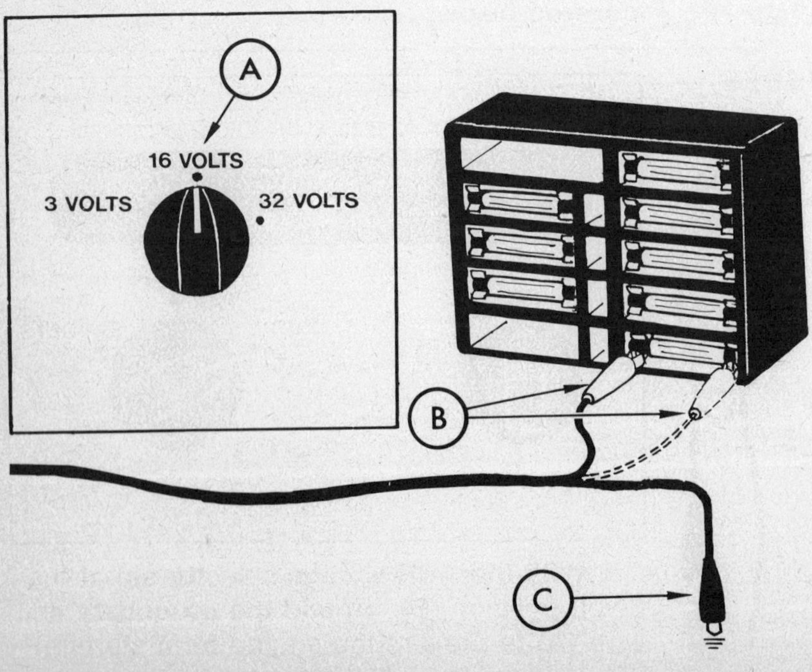

Any fused automotive circuit can be checked with the AGR tester. Set the voltage selector to 16 volts (A) and turn the ignition switch and accessory switch on. Ground the black AGR lead (C) and use the red AGR lead to touch each end of the fuse (B).

If battery voltage can be read on the 16-volt scale while touching each end of the fuse, the fuse is good. A zero reading at one end indicates that the fuse is *open* and must be replaced; a zero reading at both ends means that the circuit between the battery and fuse is *open*. The cause of the open circuit should be located and corrected.

PRIMARY IGNITION SYSTEM—VOLTAGE

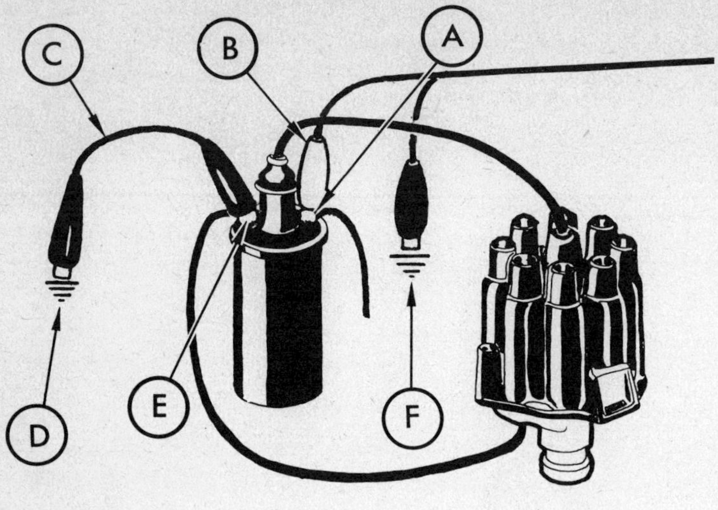

The primary ignition voltage of conventional ignition systems can be checked with the AGR tester. This test will not work on transistorized ignition systems, because the coil is connected differently in the circuit. Move the voltage selector to the 16-volt position (G). Connect a jumper lead (C) from the distributor terminal on the coil (E) to ground (D). The red AGR lead (B) connects to the battery terminal of the coil (A) and the black AGR lead to ground (F).

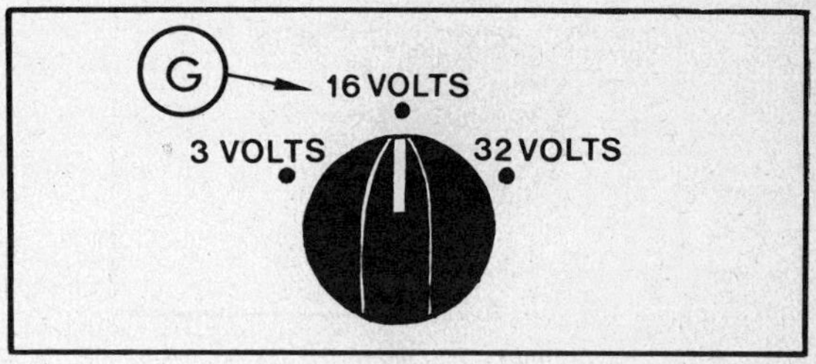

Turn the ignition switch to the ON position, crank the engine and read the 16-volt scale on the meter dial. With a 12-volt system, the meter needle should read a minimum of 9.6 volts (4.8 volts for a six-volt system). A meter needle reading of zero indicates an open circuit between the battery and coil, while a reading above zero but below the 9.6- (or 4.8-) volt minimum indicates excessive resistance between the battery and coil or a weak battery.

Some 12-volt systems are equipped with an external resistor. These should read between 4.5 and 7.5 volts. If the needle reads 9.6 volts or higher on such a system, the resistor has been bypassed or the bypass section of the ignition switch is defective.

PRIMARY IGNITION SYSTEM—CIRCUIT RESISTANCE

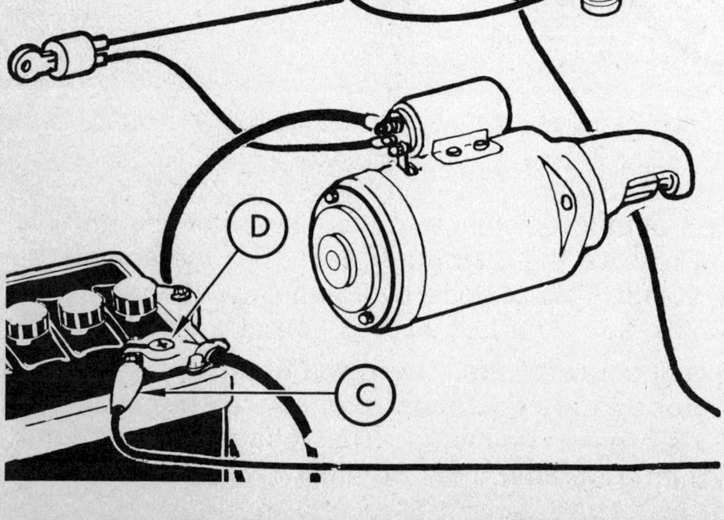

To test primary ignition circuit resistance with the AGR tester, set the voltage selector to the 16-volt position (B). Connect a jumper lead from the distributor terminal on the coil (E) to ground (F). The red AGR lead (C) connects to the positive battery terminal post (D) and the black AGR lead to the battery terminal on the coil (A).

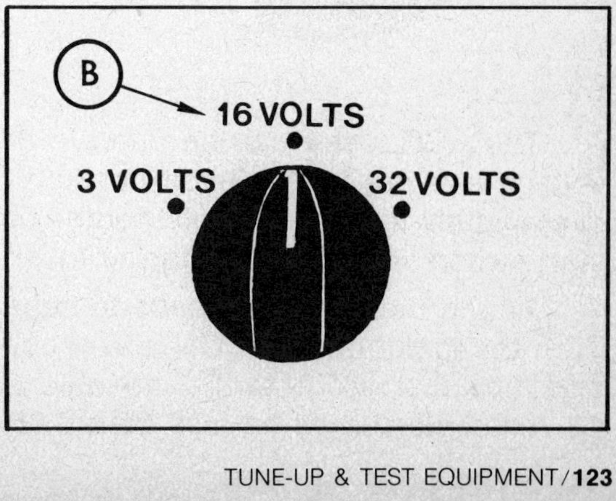

Turn the ignition switch to the ON position and read the 16-volt scale on the meter dial as you crank the engine. If the meter needle reads more than 0.6 volts (0.3 on six-volt systems) while the engine is cranking, high circuit resistance is indicated.

CAUTION: MANY PRE-1963 AUTOMOBILES DID NOT USE A BALLAST RESISTOR BYPASS CIRCUIT. THE METER READING ON SUCH CARS CAN BE AS HIGH AS 5 VOLTS. CHECK THE FACTORY SHOP MANUAL ON OLDER CARS FOR EXACT SPECIFICATIONS IF YOU ARE IN DOUBT.

DISTRIBUTOR CIRCUIT RESISTANCE TEST

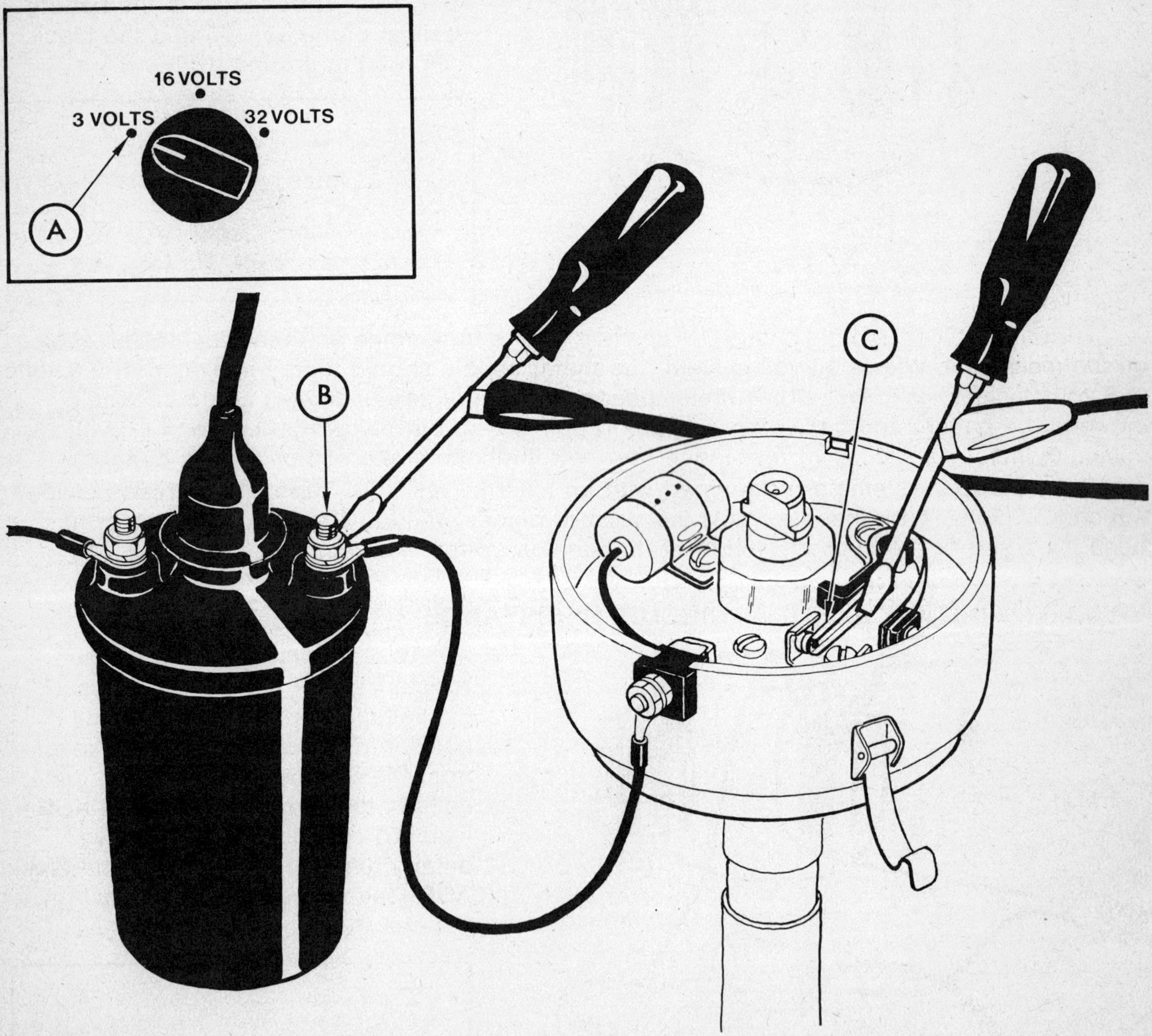

The AGR tester can be used to determine circuit resistance in the distributor. Set the voltage selector to the 3-volt position (A) and remove the distributor cap. Turn the engine over slightly to close the distributor breaker points. This is done by "bumping" the remote start switch so briefly that the crankshaft moves only a part of a revolution at a time.

Attach the two AGR leads to separate small screwdrivers. These will be used as probes. Turn the ignition switch ON and touch one probe to the distributor primary coil terminal (B) and the other probe to the movable arm of the breaker points (C). The voltage drop on the 13-volt scale should be less than 0.05 volts (half a division mark on most 3-volt scales).

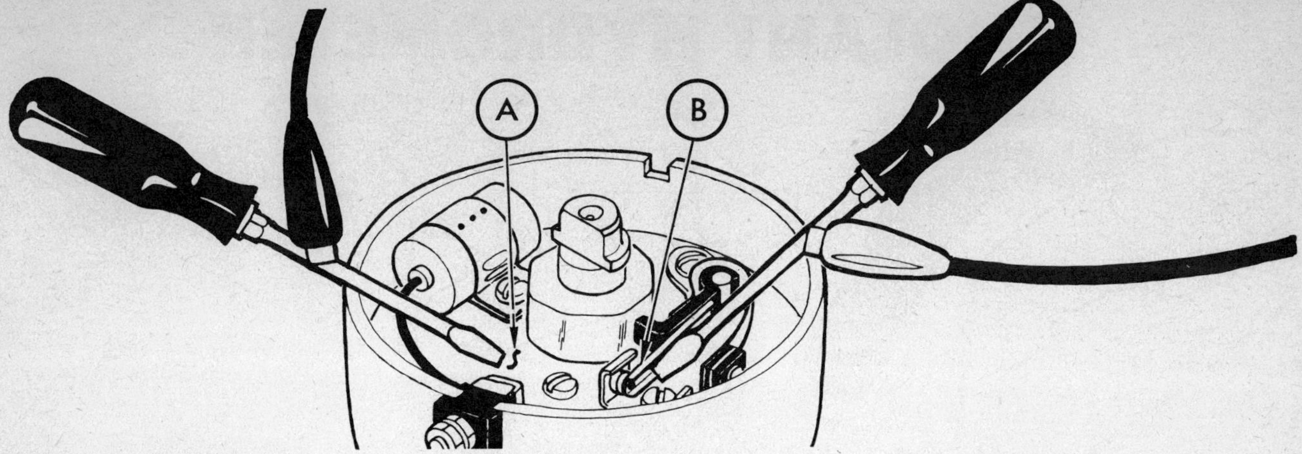

Touch one probe to the breaker plate (A) while holding the other probe on the movable arm of the breaker points (B). The voltage drop on the 3-volt scale should be 0.2 volts or less.

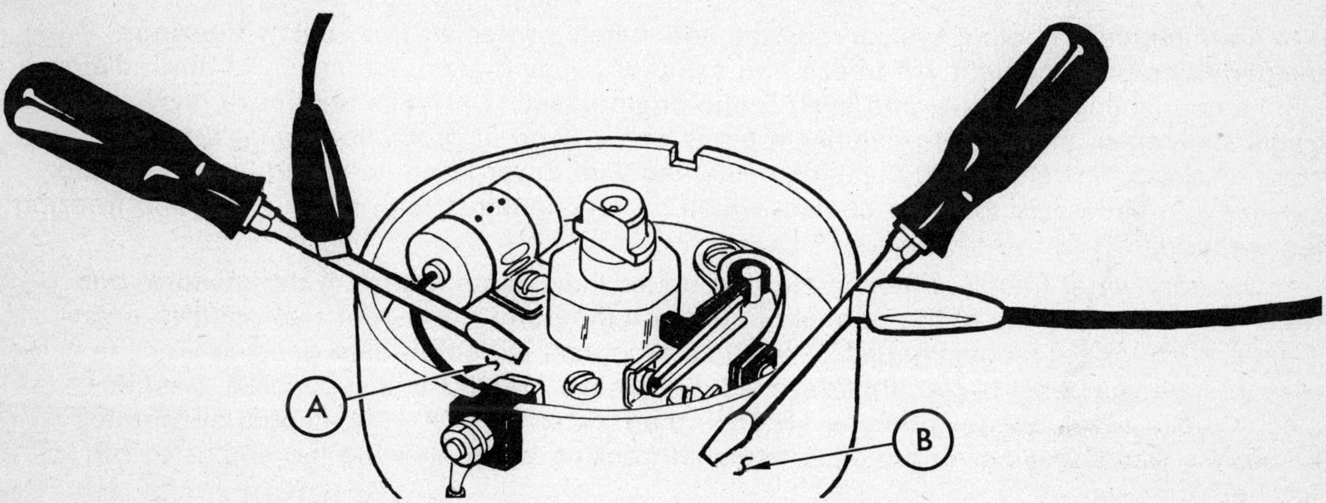

Touch one probe to the distributor housing (B) while holding the other probe on the breaker plate (A). Voltage drop here should be less than 0.05 volts.

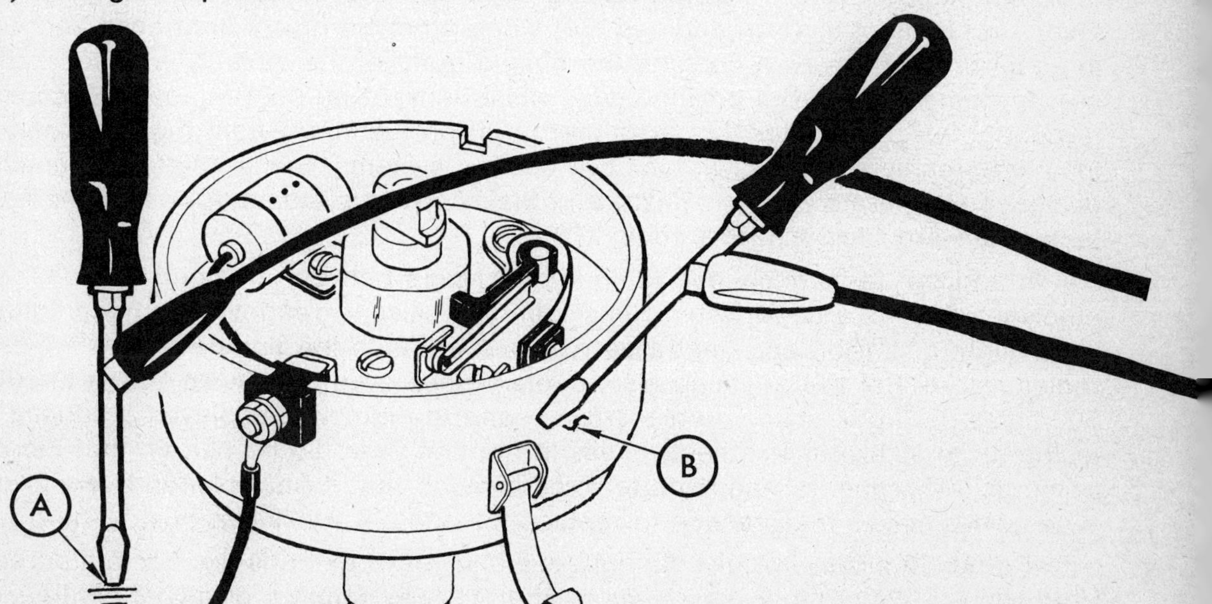

Finally, touch one probe to the engine block (A) while holding the other probe on the distributor housing (B). Voltage drop should be less than 0.05 volts.

Voltage drops above those specified indicate excessive resistance.

COOLANT HYDROMETER

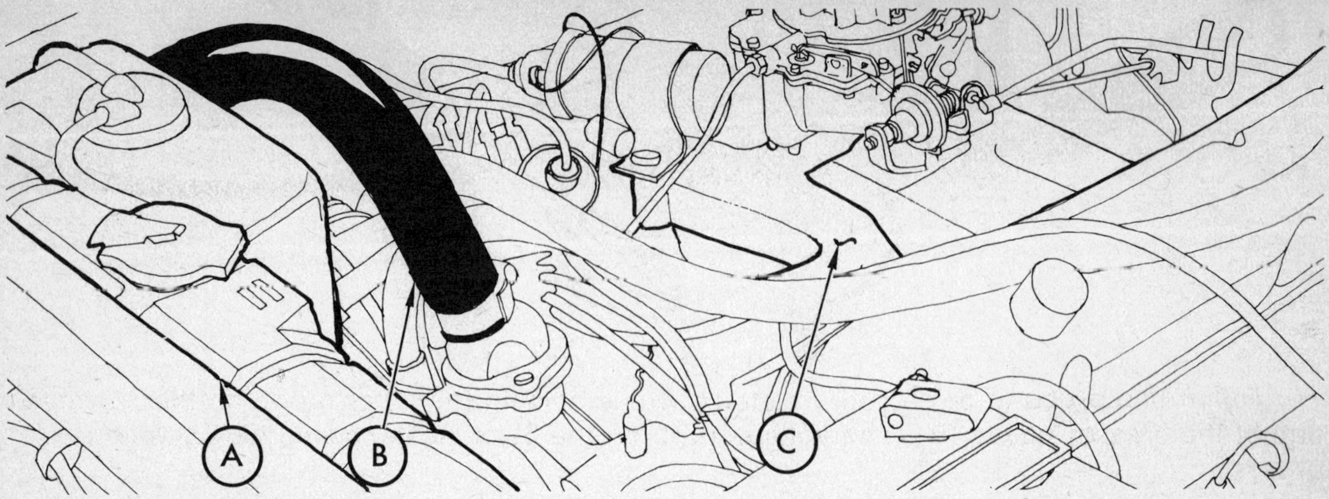

An automotive cooling system must be adequately protected from below-freezing temperatures or its coolant will freeze and expand, causing serious damage to the radiator (A), its connecting hoses (B) and even to the engine itself (C). While antifreeze has traditionally been thought of primarily in terms of this kind of protection, today's higher compression ratios and driving speeds combined with lower hood lines and limited radiator size make it necessary to use a coolant which protects against both overheating and freezing temperatures.

Ordinary water transfers heat somewhat better than an antifreeze/water mixture, but water boils away rapidly at the normal operating temperature of modern automotive engines. Thus it is no longer recommended or used as a coolant except in emergencies. For the same reason, methyl alcohol-based antifreeze solutions have practically disappeared. Because the alcohol boils at a lower temperature (180° F) than water (212° F), the alcohol evaporates before the water, leaving no protection against freezing while allowing the engine to run too hot.

Water boils at 212° F at sea level and freezes at 32° F. By pressurizing the cooling system, it is possible to raise the boiling point of water somewhat, but not enough to meet the needs of modern engines. But when ethelyne glycol antifreeze and water are used in equal proportions as a coolant, the boiling point of the mixture is raised sufficiently to prevent the solution from boiling away, while its freezing point is lowered considerably. Since glycol antifreeze provides this extra margin of protection at both high and very low temperatures, it is ideal for automotive cooling systems. Since 1962, automobiles have been designed to operate with this mixture in their cooling systems. All new cars leaving the factory are provided with this coolant.

In addition to its expanded temperature range between freezing and boiling points, ethelyne glycol is a good water pump lubricant. Also, it readily accepts inhibitor additives which fight corrosion and neutralize rust that forms in the engine before the rust can plug the radiator core. But like anything else, coolant does eventually wear out. It circulates 400 to 700 times per hour while you are driving—and the inhibitor additives gradually deteriorate, losing up to 50% of their effectiveness in the first year. To be certain that the coolant is doing its job properly, manufacturers recommend that it be replaced every 24 months; the wise owner makes it a practice to replace his coolant every year just to be on the safe side.

If a 50/50 mixture of glycol antifreeze and water extends the coolant's freezing protection to -34° F, then logic tells us that a 100% solution of glycol antifreeze should be even better. But in this case, logic is wrong, because pure glycol antifreeze freezes at 0° F. Thus it is important in cold-weather climates to know the *exact* amount of protection that the coolant in your car's cooling system will provide against freezing temperatures. This can be determined by the use of a good coolant hydrometer.

COOLANT HYDROMETER DESIGN

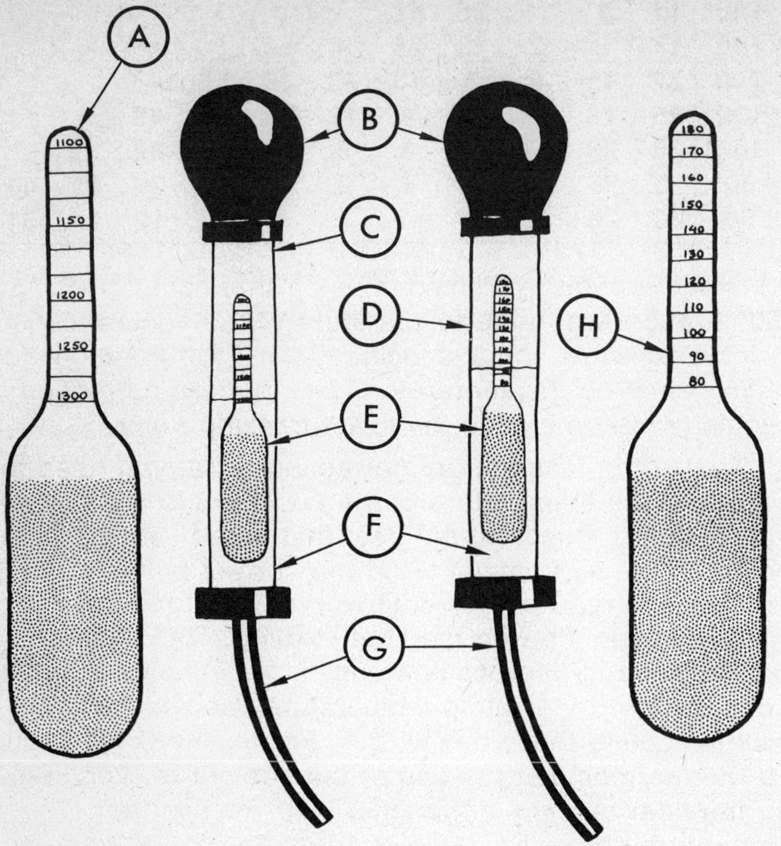

The coolant hydrometer, or antifreeze tester as it is sometimes called (C), looks exactly like the battery hydrometer (D) and is used in a similar manner. Both test the specific gravity of the solution in which they are used. Both consist of a weighted and calibrated float (E) inside a glass barrel (F). When the rubber bulb (B) at the top of the barrel is squeezed, coolant is drawn into the tester through a syringe tip or tubing (G) at its bottom.

The primary difference between the battery hydrometer and the coolant hydrometer is in the weighted float used and its calibration. The battery hydrometer scale reads the electrolyte specific gravity (H), but the coolant hydrometer scale is calibrated for degrees Fahrenheit (A).

COOLANT HYDROMETER SCALES

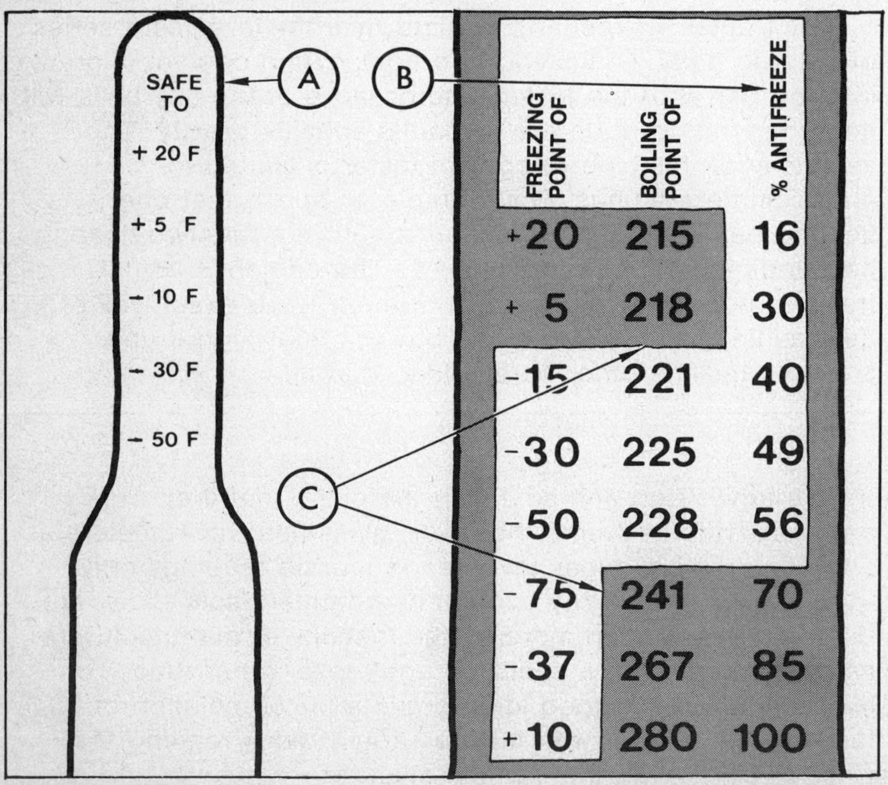

Coolant hydrometer scales vary considerably in design; it seems that every manufacturer uses one slightly different from that used by other manufacturers. Scale design ranges from the simple direct-reading type (A) to the complex scale that tells you both boiling and freezing points of the coolant as well as the percentage of glycol antifreeze in the mixture (B). These scales are usually color-coded (C) to warn of unsatisfactory protection levels.

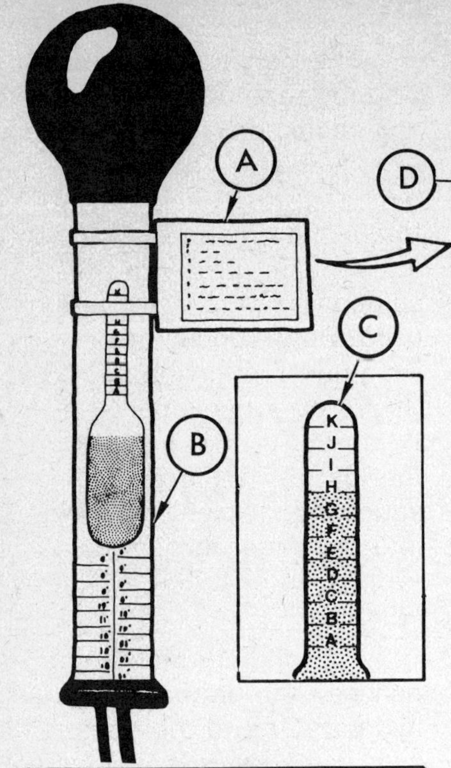

	A	B	C	D	E	F	G	H	I	J	K
180°	11	1	11	26	42						
160°	17	8	2	13	25	38					
140°	22	15	6	3	13	23	36	50			
120°	25	19	12	4	4	12	22	33	48		
100°	28	23	17	10	3	3	11	20	32	46	
80°	29	25	20	14	8	2	4	11	20	31	44
60°	30	26	22	17	12	7	1	5	12	21	31

Other coolant hydrometers use a thermometer (B). Since a 50° F decrease in the temperature of the coolant causes a 3% increase in its actual specific gravity, the thermometer reading must be transferred to a temperature correction scale or table (A) to determine the exact freezing point.

To use the temperature correction table, you need to know both the temperature of the coolant and the number or letter (C) on the neck of the float that is even with the coolant in the hydrometer. Find the correct temperature on the table and read across until you intersect the column which corresponds with the float reading. This number is the freezing point in degrees Fahrenheit. Most tables are color coded, with red indicating temperatures above 0° F and black indicating those below 0° F. For example, let's suppose that the thermometer reading of the coolant is 120° F and that the float reading is the letter "H." To use the temperature correction table and determine the exact amount of protection offered by the coolant, find 120° F on the table's temperature scale (D) and read across the table until your finger intersects the column marked "H" (E). This tells you that the coolant will protect against freezing down to 33° F below zero (F).

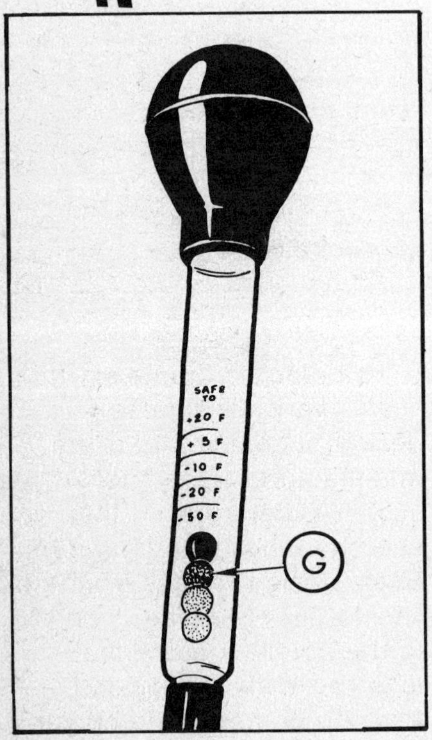

Still other inexpensive coolant hydrometers use a series of colored balls (G) instead of a float. When coolant is drawn into the barrel of the tester, one or more of the tiny balls will float, depending upon the coolant's specific gravity. The versatility of this type of coolant tester is limited to approximate readings. For example, suppose that one floating ball indicates protection to -30° F while two floating balls indicate protection to -50° F. The coolant's actual freezing point may be -42° F, but there is no exact way of determining this, because one ball will float high in the solution and the other barely float, if at all.

ACCURACY OF READINGS

The majority of coolant testers are advertised and sold with the claim that they work equally well in hot or cold solutions. As a rule of thumb, however, glass-float hydrometers are reliable only in hot coolant, while the floating-ball types give near-accurate readings only when used with cold coolant. Regardless of type, many coolant hydrometers sold today are simply not reliable. The problem is complicated even more by the inability of manufacturers to publish uniform boiling/freezing point ranges for a specified antifreeze/water ratio. To determine the accuracy of your hydrometer, it is a good idea to mix a 50/50 solution of glycol antifreeze and water and test the hydrometer with it. You should get a reading that indicates protection against freezing to -34° F with this sample mixture.

MEASURING COOLANT FREEZE POINT

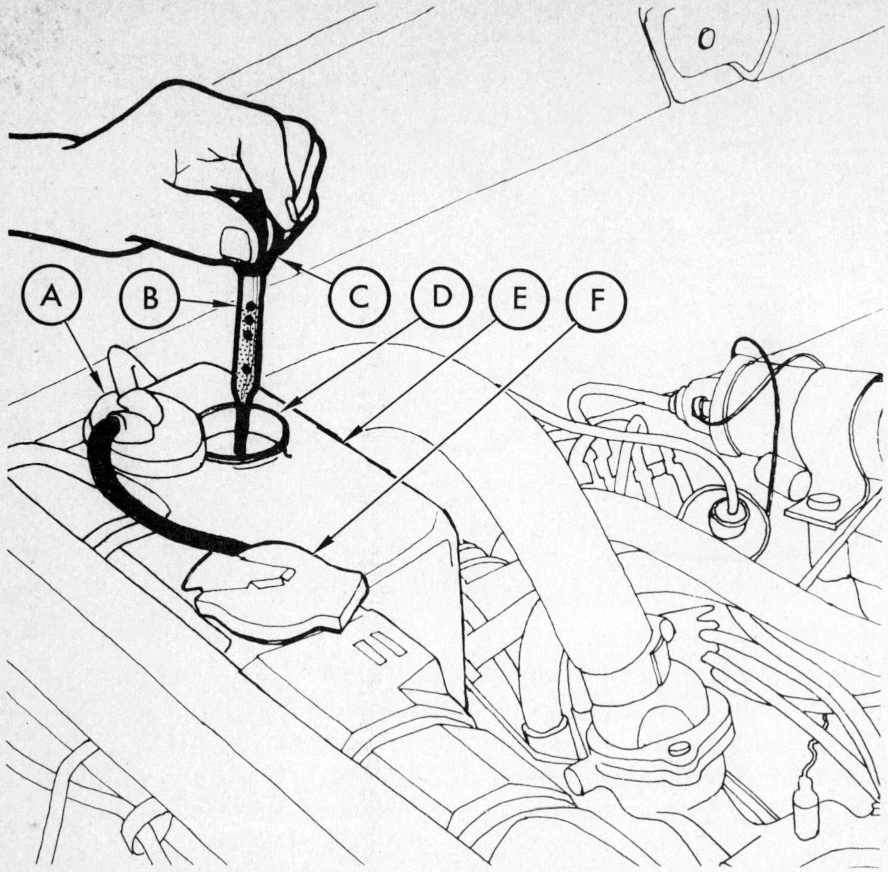

To use a coolant hydrometer, remove the radiator cap (F). If the cooling system to be checked uses a coolant reserve system (E), remove the reservoir tank cap (A) instead. *Do not* unnecessarily remove radiator caps on cars equipped with a coolant reserve system.

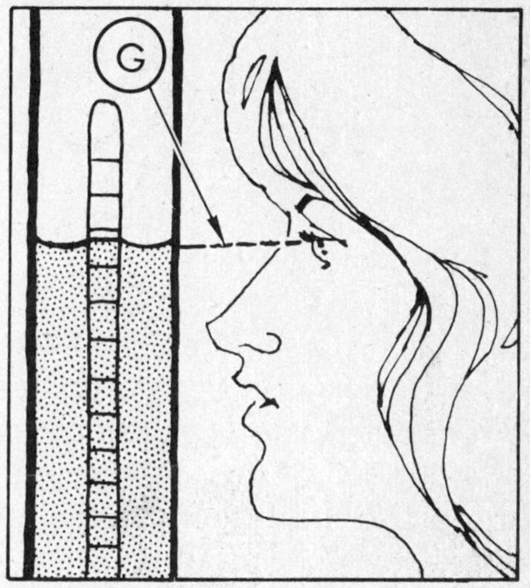

Insert the tube at the tester bottom into the filler neck (D) until it reaches the coolant inside. Holding the tester vertically, squeeze the rubber bulb (C) and release it. This should draw sufficient coolant into the tester barrel to freely float the hydrometer or tiny balls (B).

Withdraw the tester from the filler neck and hold it in a vertical position at eye level (G) so that the float or balls stand free of the barrel sides. Then read the scale. If your hydrometer uses a thermometer and temperature-correction scale, you will have to transfer the reading to the scale to arrive at the correct degree of protection.

PRECAUTIONS IN USE

Although some manufacturers claim that their coolant hydrometer can be used to test battery electrolyte, this is not advisable unless you have one of the special optical hydrometers used by some garages. Since these devices are very expensive and not generally used by nonprofessionals, they are not covered in detail here.

With the ordinary hydrometer, short of disassembling it and boiling it out with water, there is no way to satisfactorily prevent contaminating one solution with the residue of the other, which can lead to premature battery or coolant failure. It is much less expensive to buy each system its proper hydrometer than to replace a battery or repair a cooling system.

HYDROMETER FAILURE

Like battery hydrometers, the coolant hydrometer bulb and test tube are subject to premature failure due to the chemicals with which they come in contact. When buying a hydrometer for either purpose, avoid those made with genuine rubber. Look instead for one with a vinyl bulb and test tubing, as vinyl will outlast rubber. Also check the hydrometer in the store for airholes in the rubber/vinyl components. Holes render it useless in operation. Such manufacturing defects are not uncommon, especially when buying one of the lower-priced units.

REMOTE START SWITCH

Using a remote start switch lets you crank or start any starter-relay- (solenoid-) equipped engine from the engine compartment. This allows you to perform testing and tune-up procedures without the need for a second person inside the car to operate the ignition switch. To do this, the remote start switch must be properly connected to the car's ignition circuit so that it substitutes for the START function of the ignition switch or the pushbutton starter switch found on older cars.

Remote start switches come in a large variety of shapes, sizes and designs, from the simple pushbutton in a box or tube (A) to the rectangular squeeze-for-contact type (F). Regardless of the switch's outside appearance, its internal function is the same—to complete a simple electrical circuit when closed. Some are even incorporated into other tune-up and test equipment, such as in the handle of inexpensive neon timing lights (B) and in the more costly compression testers (E).

Whatever its shape, size or design, the remote start switch is equipped with twin hookup leads (C) that terminate in insulated spring-return or scissor-type clamps (D) for circuit connection.

CONNECTION PATTERNS

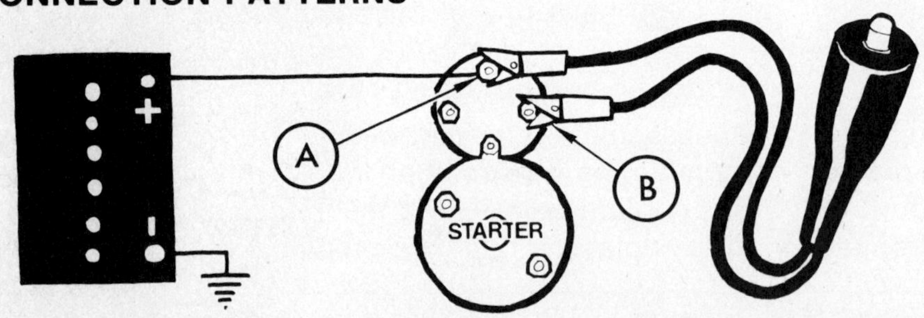

Although the remote start switch is a very simple piece of equipment, it is difficult to give an equally simple rule for connecting the switch into an ignition circuit, because of the great variety of circuit arrangements.

The simplest and most effective hookup is made by connecting one lead to the starter switch terminal (A) on the solenoid and the other lead to the starter solenoid battery terminal (B). But this hookup cannot always be used. Some cars do not use starter-mounted solenoids (Ford, Lincoln, Mercury, American Motors). Further, with most late-model cars (especially those with V-8 engines), access to the starter and solenoid from the engine compartment is severely limited or even prevented completely by the location of other engine components. Remote start switch connections to the solenoid in such cases would have to be made from beneath the car. That is impractical unless you are working with the vehicle on a hoist.

In general, then, if you can see and reach the solenoid easily, this connection pattern can be used efficiently. Older cars using a dash-mounted pushbutton, current six-cylinder Chrysler Corporation cars and many four-cylinder imports fall into this category.

CAUTION: BE CAREFUL WHEN ATTACHING LEAD CLIPS TO THE SOLENOID. THERE IS BATTERY VOLTAGE PRESENT.

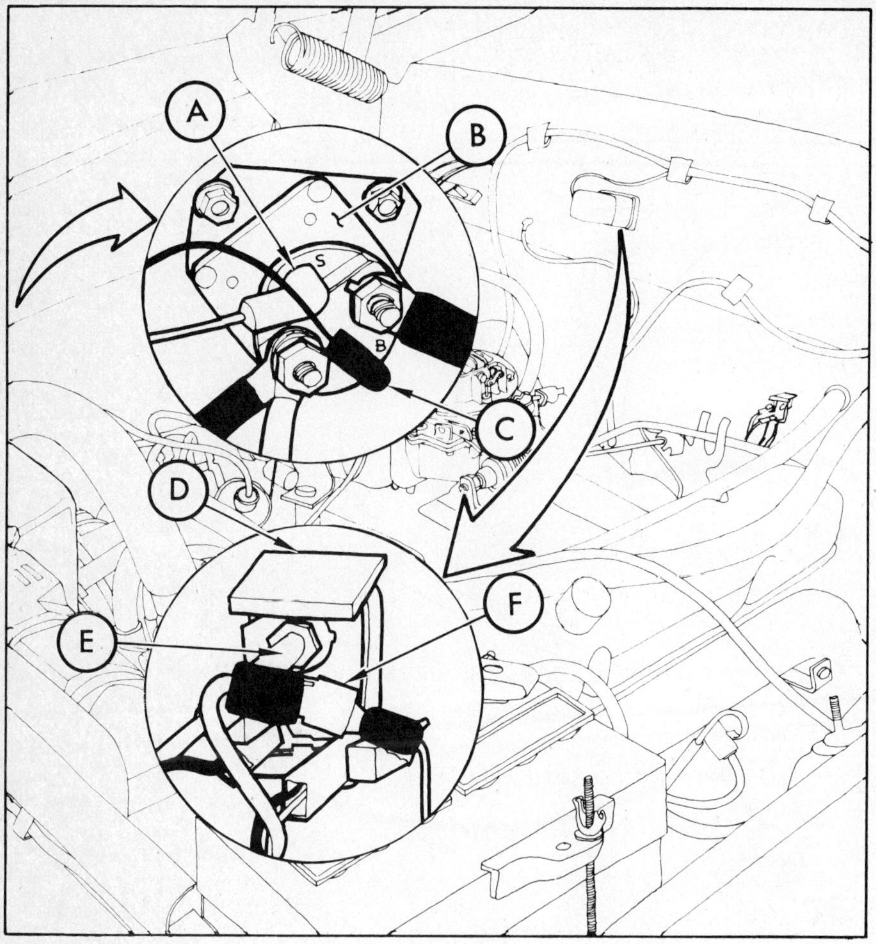

Solenoids (B,D) on AMC and Ford Motor Company cars are usually firewall- or fenderwell-mounted in the engine compartment and are plainly visible, making connection of the remote start switch quick and easy. Connect one of the switch leads to the battery side of the solenoid (C) and the other to the small terminal marked "S" (A). If an insulated quick-disconnect is used on the "S" terminal wire, remove the wire before connecting the leads.

Chrysler Corporation ignition systems incorporate a relay (D), usually mounted on the firewall, in the ignition switch-to-solenoid wiring. When the ignition switch is turned to its START position, the relay is activated, and this in turn activates the solenoid switch mounted on the starter. This relay makes an ideal hookup point for the remote start switch. Connect one lead to the solenoid side of the relay (F) and the other lead to the large terminal post (E) which is connected to the battery.

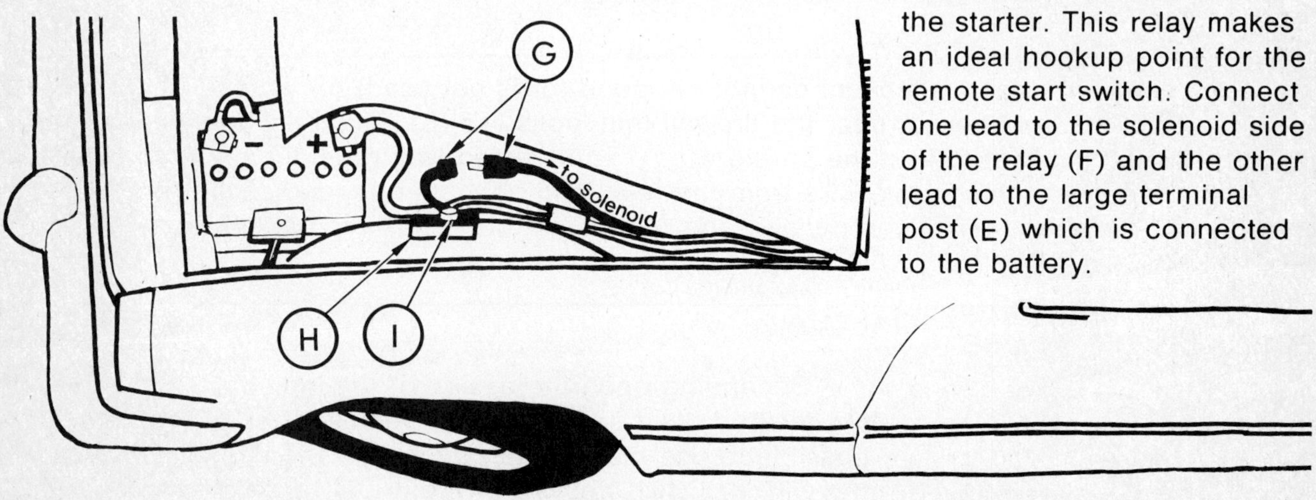

On General Motors cars, two connection patterns are possible. Look for a quick-disconnect in the primary solenoid side. This should be located close to the positive battery cable junction block (H) on the inside of the left fender. Connect one lead of the remote start switch to the battery side of the junction block (I). Pull the disconnect (G) apart and connect the other lead to the metal end of the wire leading to the solenoid.

REMOTE START SWITCH

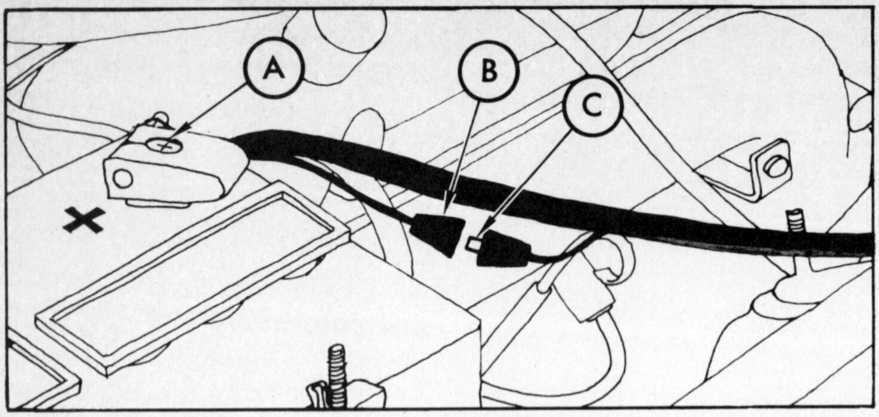

A similar quick-disconnect is provided near the positive battery terminal in the wiring of some foreign cars, especially Toyotas. After pulling the disconnect apart, connect one lead to the battery-side wire (B) or the positive battery terminal post (A) and the other lead to the wire leading to the solenoid (C).

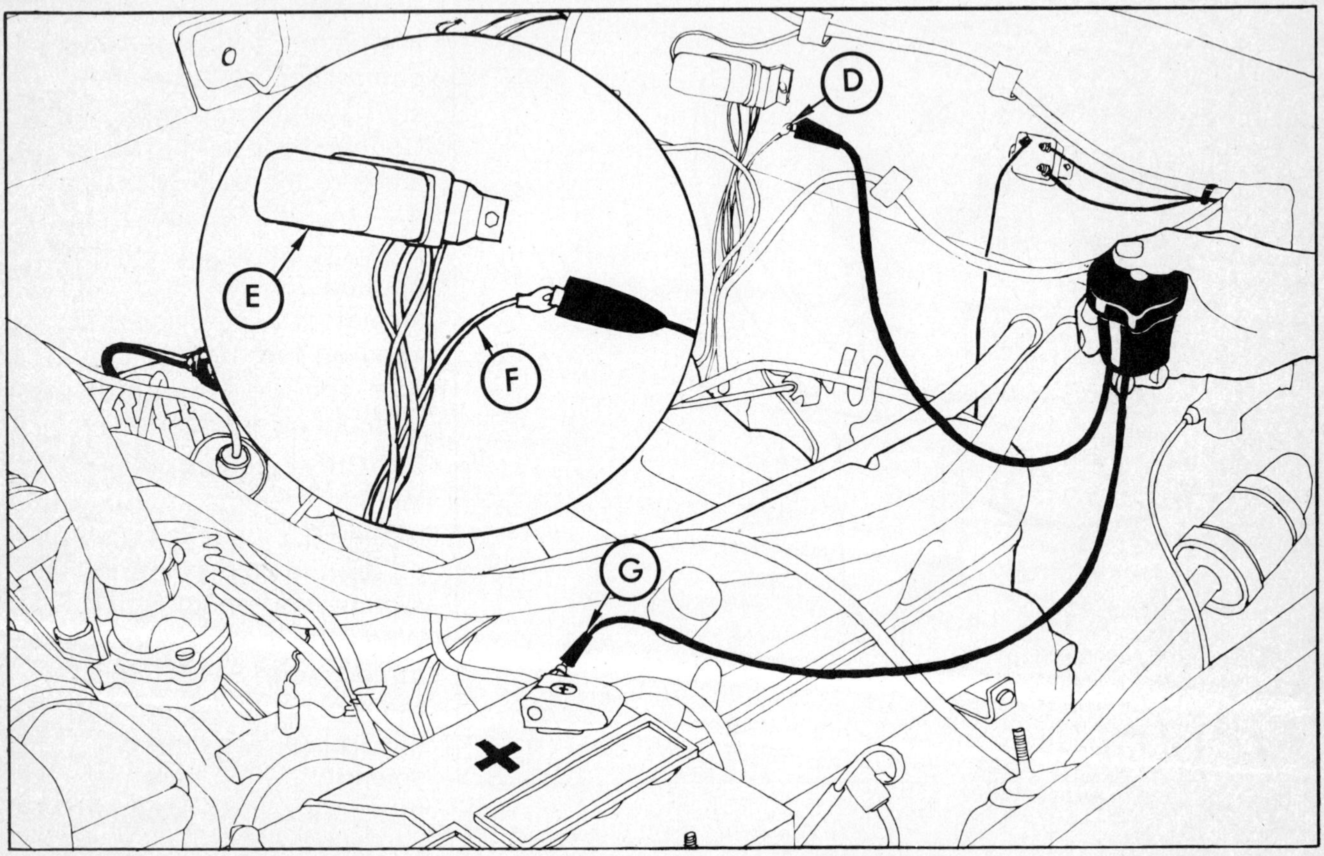

If the junction block disconnect cannot be found (or is not used) on your GM car, look for a relay box mounted on or near the firewall that looks like (E). This relay has nothing to do with the starting system, but the purple wire (F) connected to it runs directly to the starter solenoid. Disconnect this purple wire from the relay and connect one lead of the remote start switch to its metal end (D). Then connect the other lead to the positive battery terminal post (G).

PRECAUTIONS IN USE

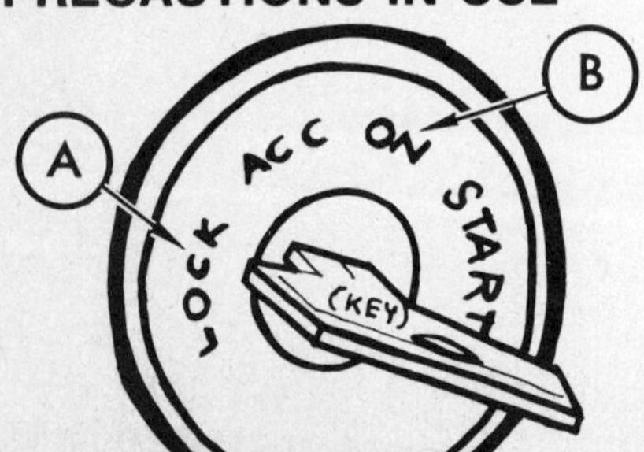

Depending upon the design of the ignition system and the method used to connect the remote start switch, it is possible to use the remote switch with the ignition switch in the OFF or LOCK position (A), but it is recommended practice to *always* turn the ignition switch to its ON position (B) before operating the remote start switch. Cars equipped with a HOT indicator lamp have the test circuit built into the ignition switch; on such cars, cranking the engine with the ignition switch OFF can damage the circuit.

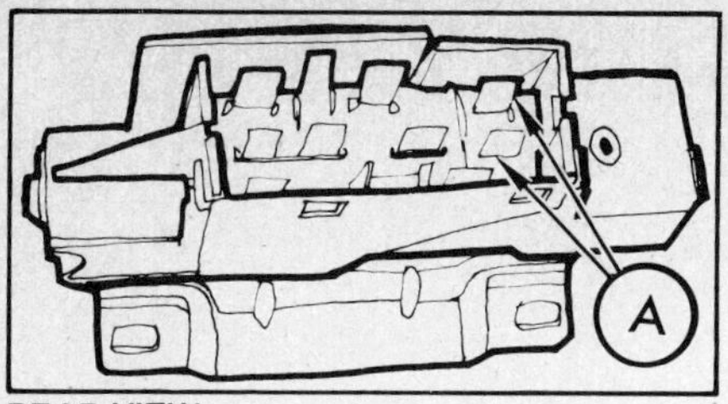

REAR VIEW

Also, many GM cars from 1961 on have a grounding terminal (A) in the ignition switch to prevent current flow with the switch OFF. Using a remote start switch to crank the engine can burn off the terminal, as it is a direct short to ground. The loose metal terminal then rattles around inside the ignition switch and can cause the ignition to short out while in use.

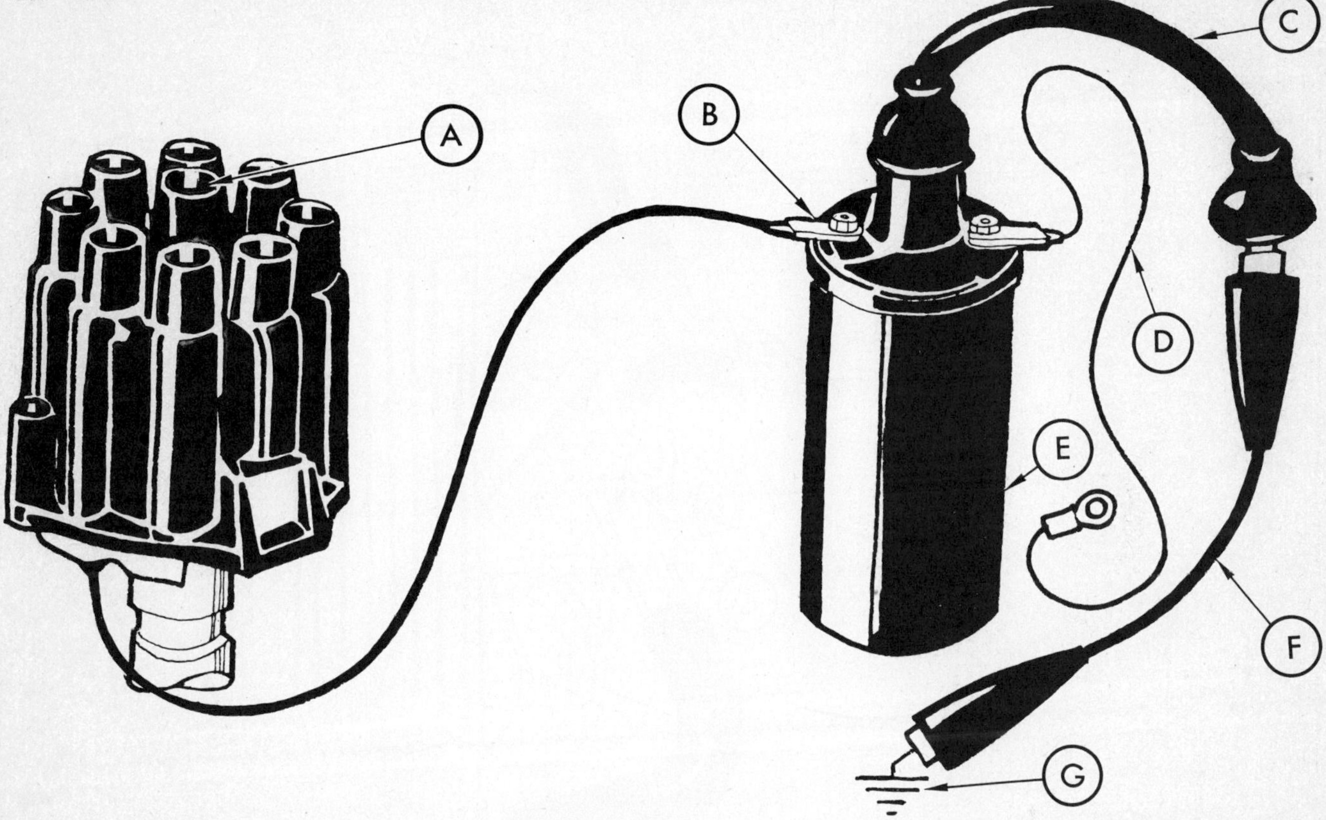

It is a good idea to remove the high-tension or secondary cable (C) from the center tower of the distributor cap (A) to prevent the engine from starting when cranked. To reduce the strain on the ignition coil (E), the wire should be connected to a good ground on the engine (G) by using a jumper lead (F). The ground connection *must* be a good one, especially on transistorized ignition systems, as they can be damaged by cranking the engine if the coil secondary is open (not grounded). If there is any doubt about the ground or the connection made to it, disconnect the coil battery lead (D) or its distributor lead (B).

CAUTION: WHEN REMOVING THE REMOTE START SWITCH, ALWAYS RECONNECT ANY IGNITION SYSTEM LEADS THAT YOU DISCONNECTED TO MAKE THE SWITCH HOOKUP.

REMOTE START SWITCH FAILURE

Although many remote start switches are sold with claims that they will not burn out, none are actually burnout-proof if subjected to constant use, as in a garage. Very few commercial remote start switches can take frequent use on a General Motors car, because the contacts inside the switch burn out. This is because the GM starter uses an electromagnet to pull the starter drive into engagement with the ring gear. It takes a great deal of current to operate the electromagnet, and of course this current goes through the remote start switch and gradually destroys the switch contacts. Thus, while you might consider such an elementary device to be a lifetime purchase, don't be surprised if its life is considerably shorter than yours, especially if you give it more than occasional use.

COOLING SYSTEM PRESSURE TESTER

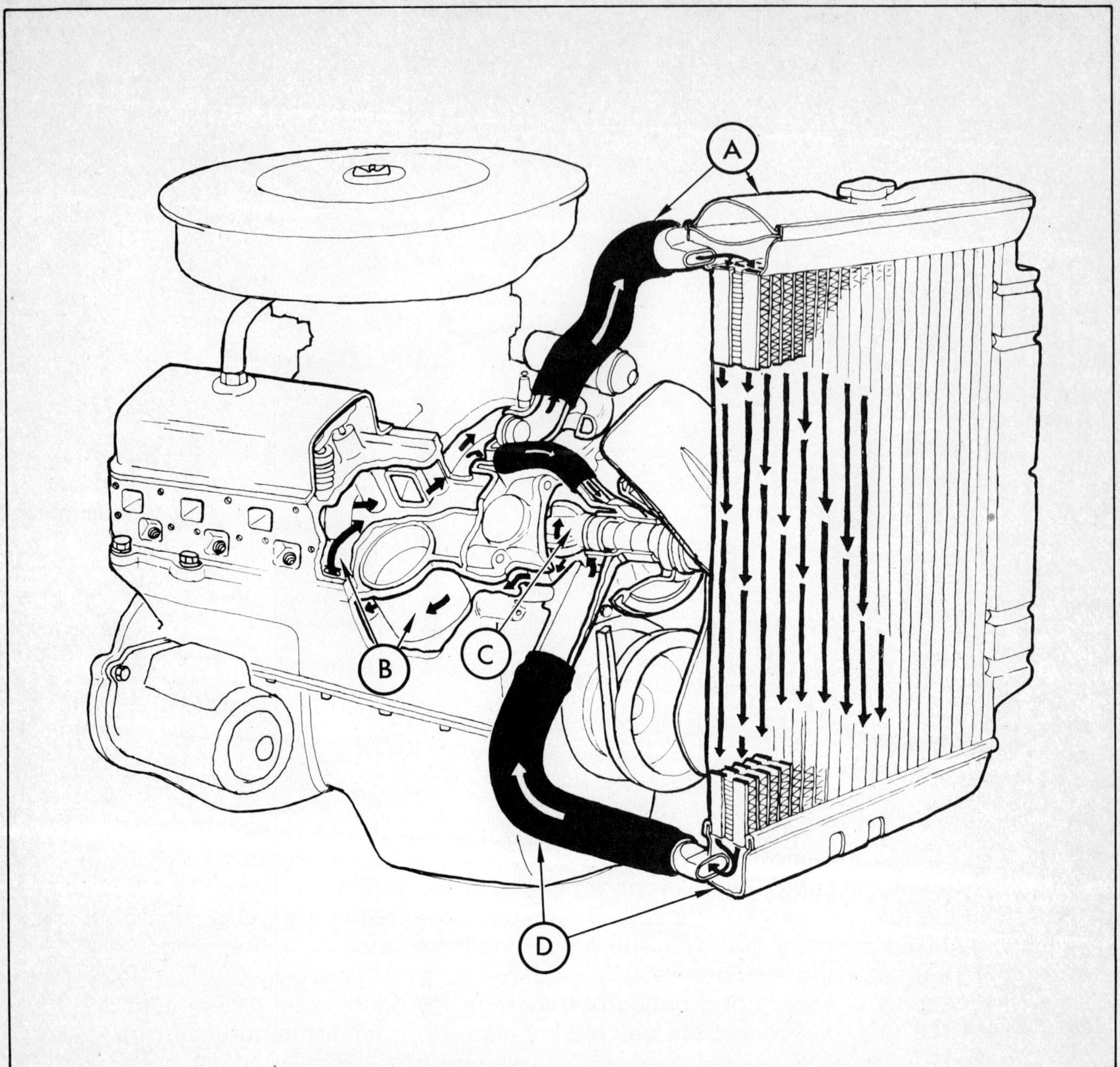

During operation, the automotive engine creates an excessive amount of heat by combustion of the air/fuel mixture. For greatest efficiency, engine temperature must be maintained at a specified level. This means that any excess heat must be removed as it is created or else the engine will "overheat" and be damaged. The cooling system is designed to transfer this excess heat to a point outside the engine where it can be dissipated (removed).

This is done by casting a "water jacket" or series of passages (B) into the engine block to allow fluid to circulate around the cylinder walls, valve seats/guides, combustion chamber top and wherever else heat must be removed. A belt-driven pump mounted at the front of the engine block (C) forces this fluid into the passages from the outlet tank (D) of the radiator. Also known as coolant, the fluid absorbs heat as it travels through the engine passages and returns to the inlet tank (A) of the radiator. In the process of passing through tiny copper tubes (an excellent conductor of heat) in the radiator core, the heat is lost. Once the coolant reaches the outlet tank of the radiator, it is ready to pass through the engine again to remove more heat.

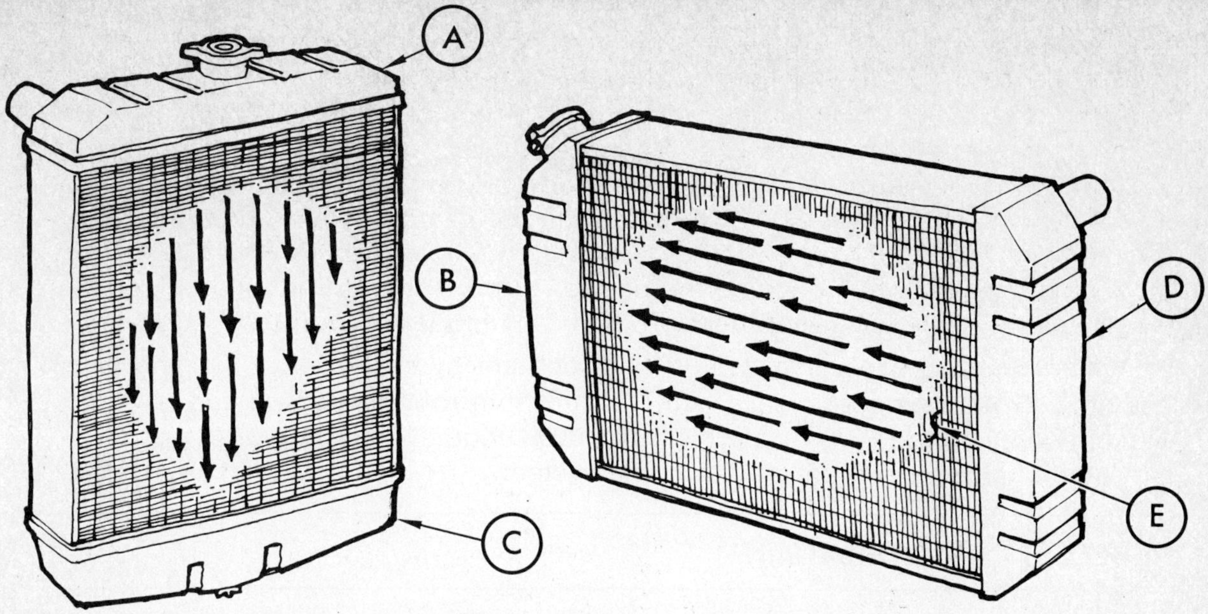

While the inlet tank is thought of as being at the top (A) of the radiator and the outlet tank (C) at the bottom, this design has been superseded. Most cooling systems today place the tanks (B & D) on the sides of the radiator, with the core tubes (E) running horizontally instead of up and down.

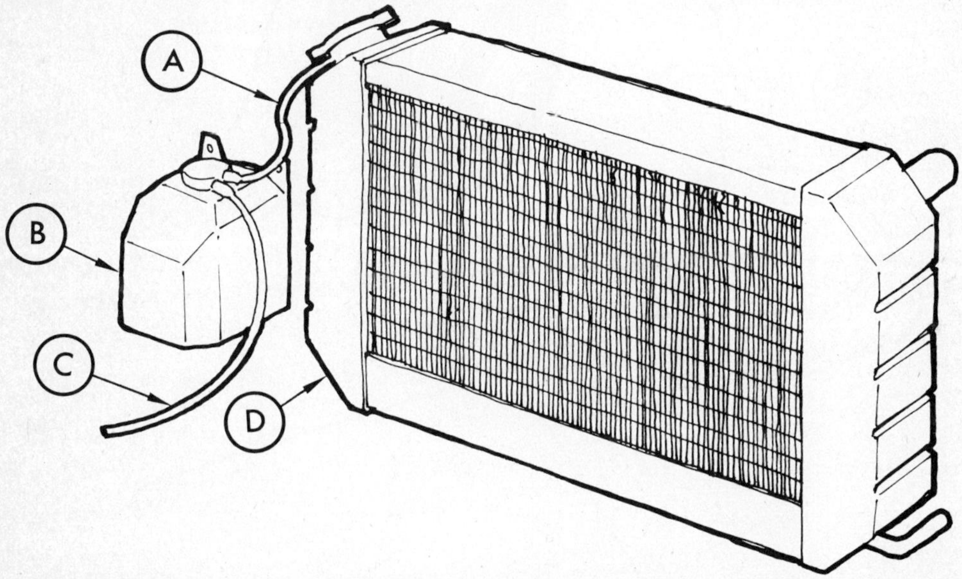

These newer radiator designs are also equipped with a reservoir (B) connected to the radiator filler neck by a hose (A). This prevents any excess coolant (which expands when hot) from being lost by overflow. When the cooling system begins to return to normal temperature, the coolant contracts and the excess coolant that was passed into the reservoir is drawn back into the radiator as the cooling system temperature decreases. This keeps the radiator full at all times. An overflow hose (C) is connected to the reservoir. If the system is overfull to begin with or if a blockage occurs in the cooling system, the hose allows the excess coolant to pass out of the system permanently.

Cars equipped with an automatic transmission use a transmission intercooler, usually located inside and at the bottom of the radiator (D). This intercooler removes excess heat from the automatic transmission fluid circulated through it. If the transmission fluid is hotter than the radiator coolant, then heat will pass from the fluid to the coolant. If the coolant is hotter, then heat will pass *into* the automatic transmission fluid. When the car is first started, the engine heats up faster than the transmission, which warms the transmission fluid so that the transmission operates better. Under heavy loads, the transmission runs hot, and the intercooler then tends to cool down the transmission fluid.

COOLING SYSTEM PRESSURE TESTER

PRESSURIZED COOLING SYSTEMS

With the advent of higher horsepower and compression ratios, beginning around 1950, it was found that cooling system efficiency could be improved considerably by pressurizing it. For every four pounds of pressure in the system, the cooling efficiency increases approximately 10%. Thus, while coolant in a non-pressurized or atmospheric system (such as that just described) boils at 212° F at sea level, the same coolant under 15 pounds of pressure boils at 257° F. As radiator size has gradually grown smaller due to cost and styling factors, pressurization has become the answer to providing the additional cooling capacity needed to handle the increased heat produced by modern automotive engines.

In a pressurized cooling system, loss of coolant by evaporation, surging or boiling is eliminated. Water pump efficiency is increased from roughly 85% at sea level in an atmospheric or non-pressurized system to 100% under 14 pounds of pressure. The coolant will not boil unless there is a defect in the system.

PRESSURE CAP OPERATION

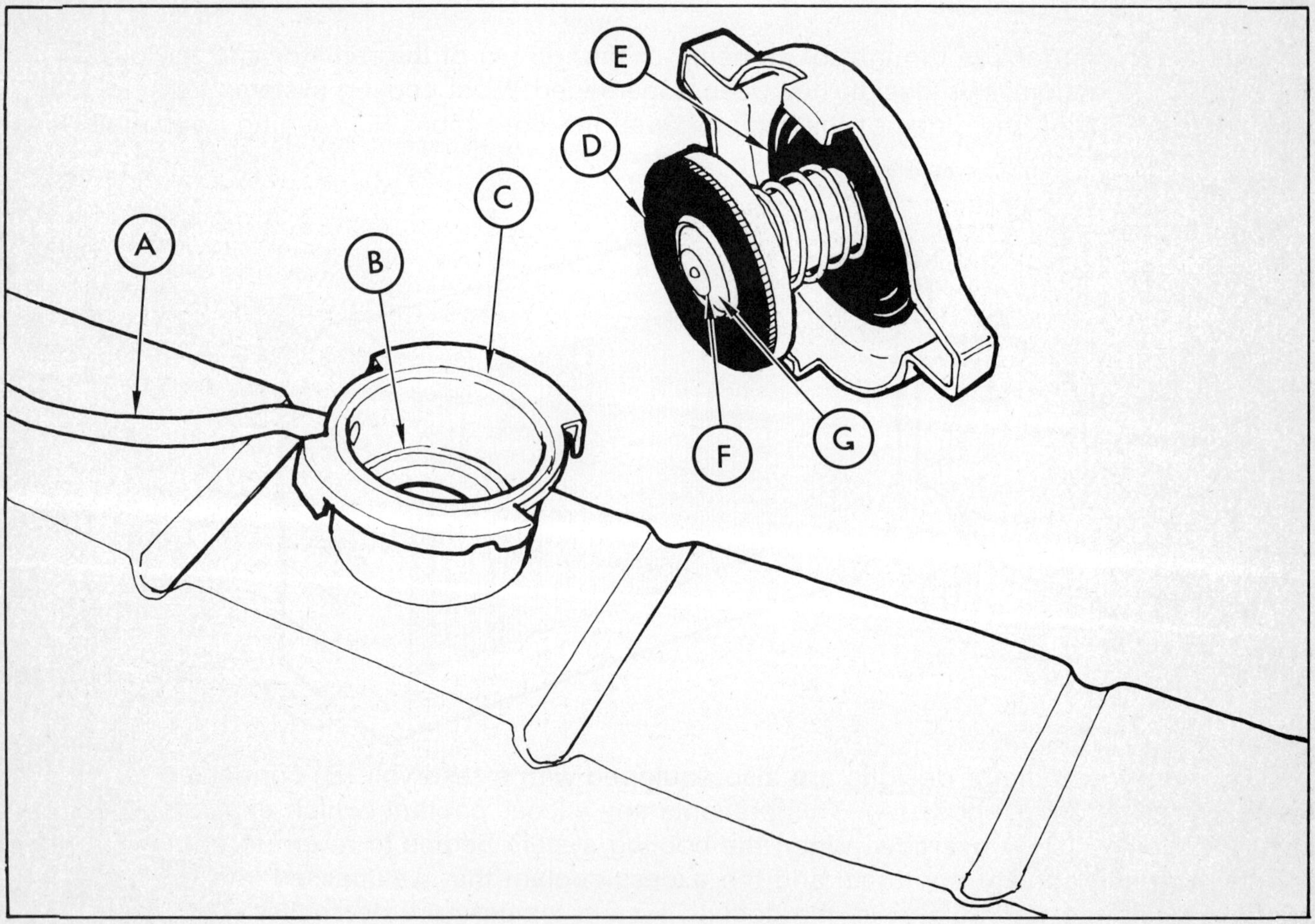

Pressurized cooling systems use a specially designed filler neck on the radiator and a pressure cap that is spring-loaded to maintain a specific system pressure. The filler neck has a top sealing seat (C) and a lower or inside sealing seat (B). The overflow tube (A) is placed at a point between the two seats.

The radiator pressure cap has a corresponding top sealing seat (E) that mates to the one on the filler neck (C) and an inside sealing seat (D) to mate with the one in the filler neck (B). It also has a pressure control or relief valve (F) that seals off the cooling system from the overflow tube, allowing the system to automatically pressurize itself as the coolant expands. To prevent a vacuum from forming in the cooling system that would otherwise collapse the radiator and/or hoses, a vacuum relief valve (G) opens automatically as the engine cools off. The cap thus functions as a safety valve to keep system pressure within safe limits.

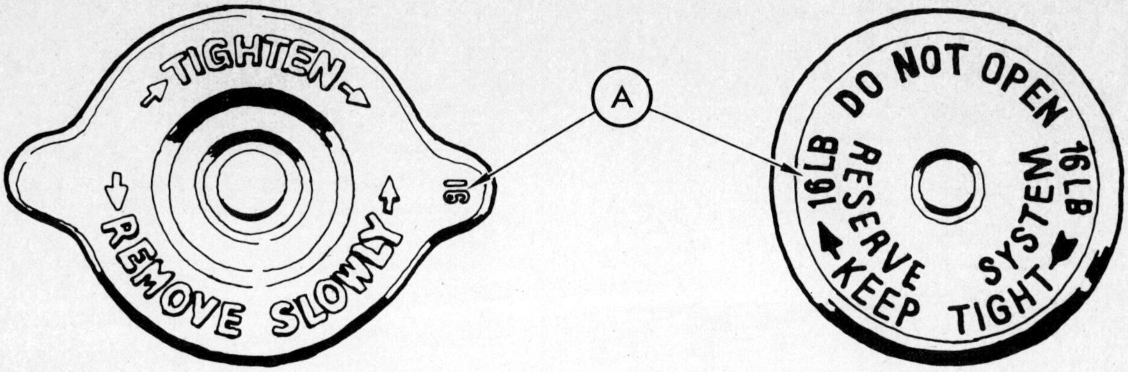

System pressure is determined by the rating of the pressure cap, and maximum cooling efficiency is reached at this pressure. The pressure cap is marked with a number (A) indicating the specified operating pressure.

Over the years, cooling systems have been designed to use one of four rated pressures. Each must operate within a predetermined range, as shown.

SYSTEM PRESSURE (LBS.)	OPERATING RANGE (LBS.)
4	3¼-4½
7	6½-8
13-14	12-15
15-17	14-17

When operating as specified, the radiator cap's pressure valve *must open* at or below the upper limit of the range, but *should not open* below the lower limit of the range. Thus a 7-lb. cap must open at 8 lbs., but not below 6½ lbs.

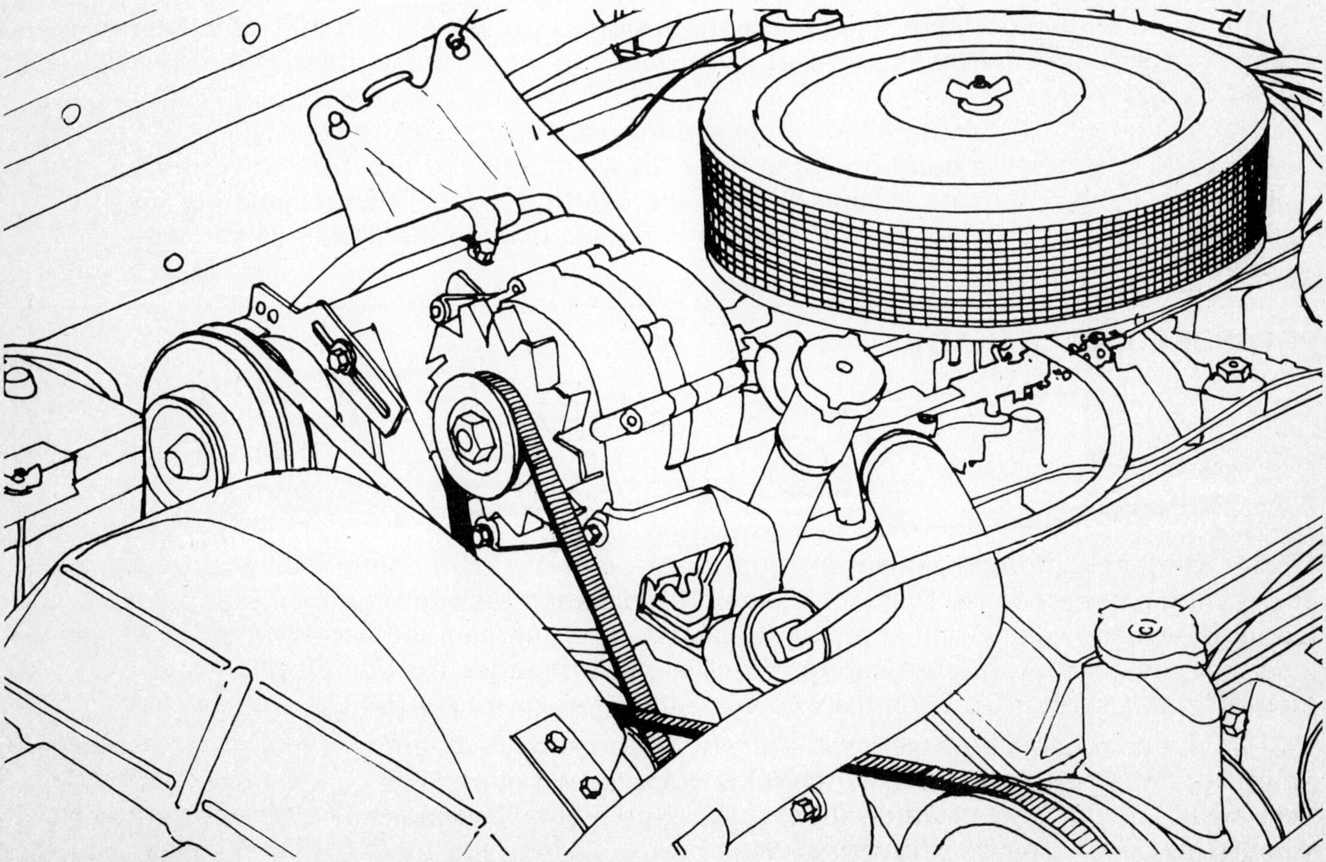

COOLING SYSTEM PRESSURE TESTER

PRESSURE CAP DESIGN

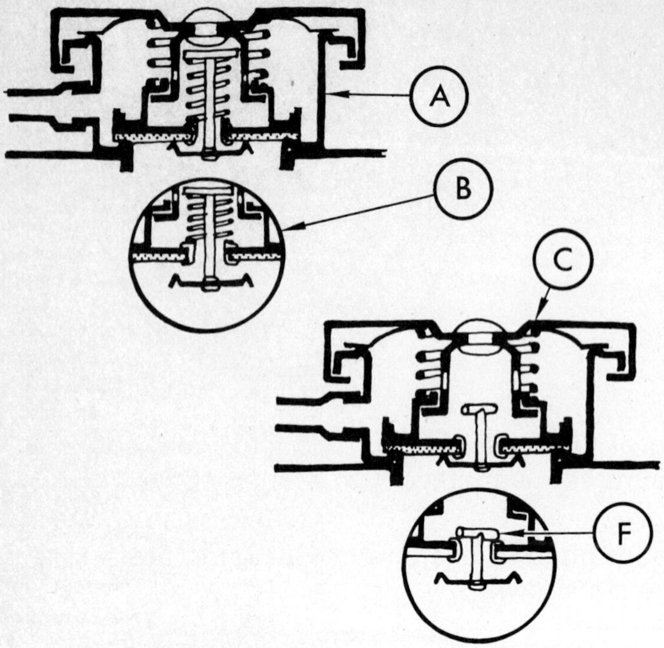

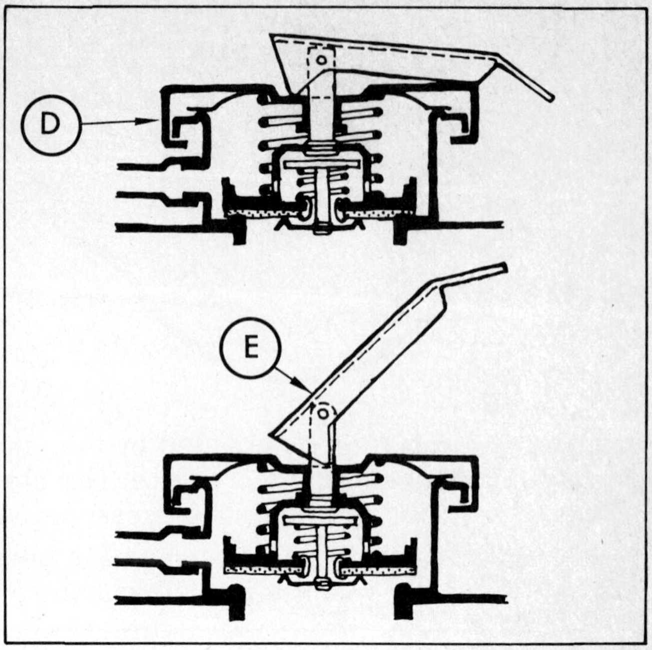

There are three types of pressure caps in use today: the constant-pressure cap (A), pressure-vent cap (C) and safety cap (D).

The constant-pressure cap (A) uses a light spring (B) to hold the vacuum valve in a closed position, sealing the cap.

The pressure-vent cap (C) has a loose vacuum valve with a small weight (F) to hold the valve open at atmospheric pressure. When the coolant temperature rises or when the coolant expands and touches the valve, it automatically closes. The cap acts as a constant-pressure type until the pressure drops to zero after the engine has been shut off and cools.

If the coolant temperature rises too high, the pressure will exceed that specified for the system. At this point, the pressure relief valve opens to vent excess pressure out through the overflow tube.

The safety cap (D) can be either a constant-pressure or pressure-vent type; it differs from the others in that it has a hand-operated lever (E) which lifts the pressure valve from its sealing seat, letting pressure escape through the overflow tube. This prevents hot coolant from spraying out of the filler neck if the cap is removed while the system is still under pressure.

PRESSURE CAP LEAKAGE

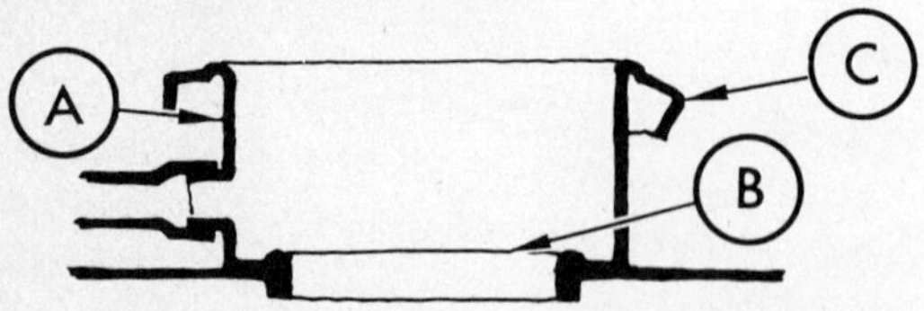

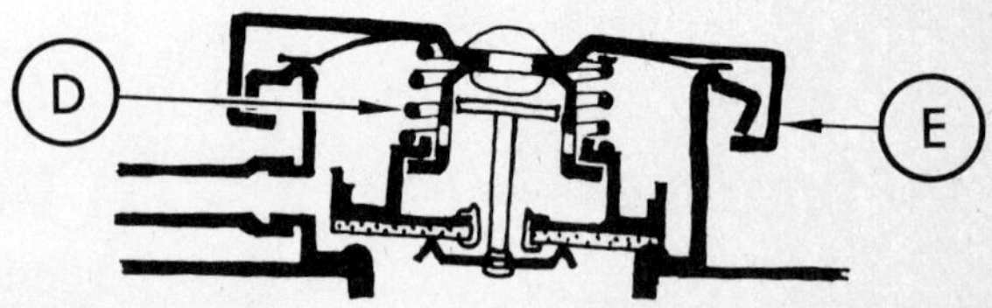

As filler neck and cap seats must both mate correctly, bent cams in either the filler neck or pressure cap can cause system leakage. The distance between the cam rest point (A) and the inside sealing seat (B) must remain undistorted. If one cam is bent down (C), the incorrect cock of the pressure cap on the filler neck changes the compression of the pressure valve spring (D). Then the cap will either leak or refuse to hold any pressure at all.

Should both cams be bent down equally, the cap will seal, but at a higher-than-specified pressure, since the pressure valve spring is compressed more. The cap will also seal when both cams are distorted upward, but at a lower-than-specified pressure. Bent pressure cap ears (E) cause the same conditions.

THE COOLING SYSTEM TESTER

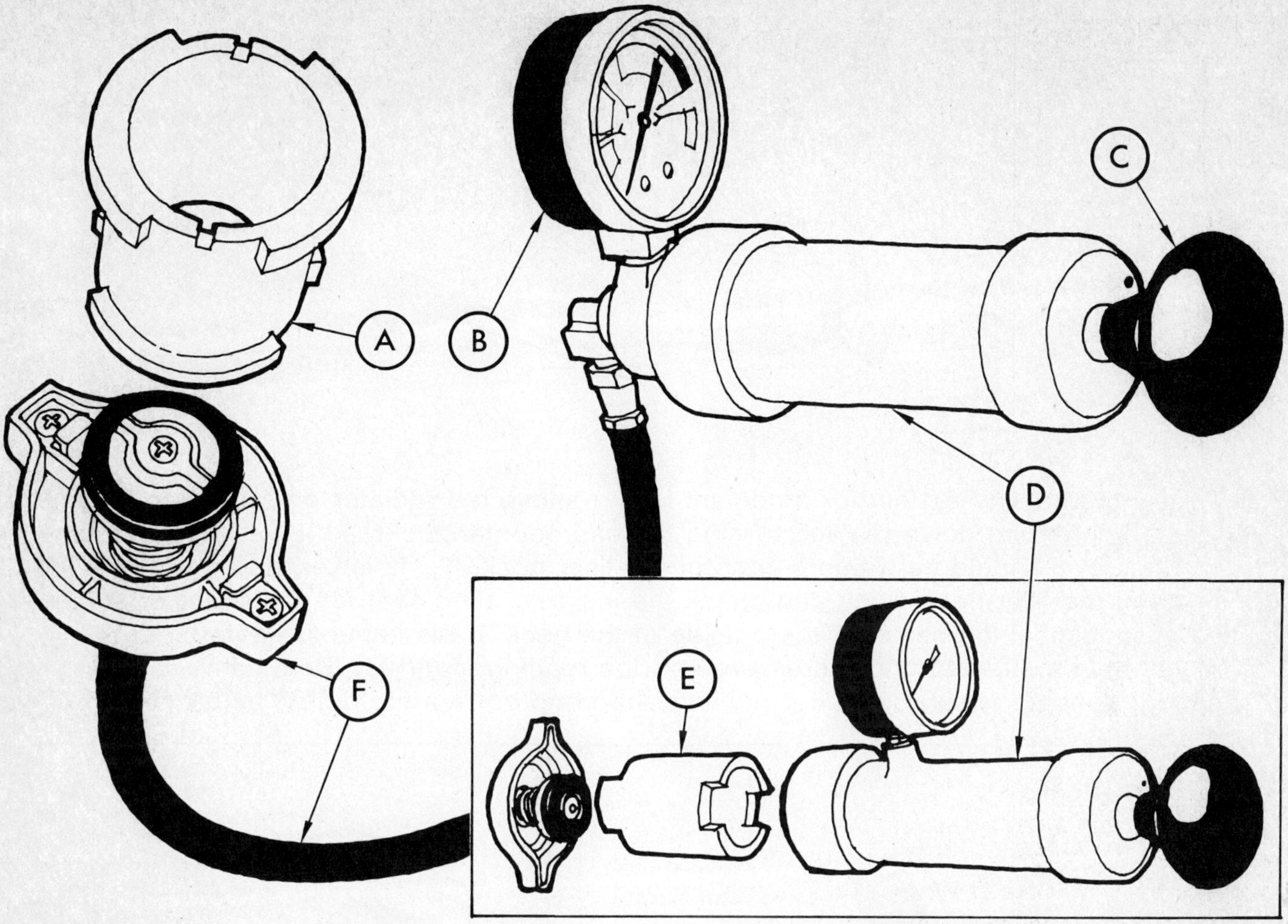

Leaks in the pressurized cooling system cause a loss of pressure (as well as a loss of coolant) and can lead to serious problems. Coolant leaks and system operation can be checked with the use of a cooling system pressure tester (D). This tester will check both the cooling system and the radiator pressure cap.

The cooling system pressure tester is a small hand pump with a plunger (C) at one end and a pressure gauge (B) at the other. Some use a flexible hose and attaching pressure cap (F), while other models contain the attaching cap as an integral part of the pump body (E). Radiator pressure caps are tested with an accessory cap adapter (A).

TESTER GAUGE

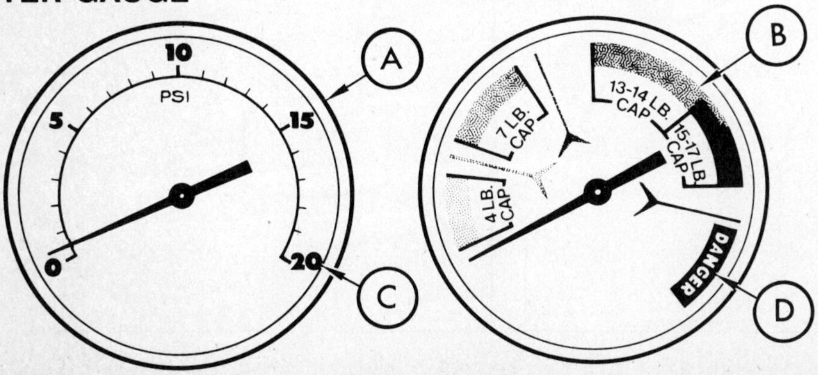

Cooling system tester gauges may be marked with a standard psi (pounds per square inch) scale (A) or a color-coded pressure range scale (B). If the gauge face does not have a red danger area (D), the 20 psi setting (C) should be considered a danger point.

WARNING: NEVER EXCEED 20 PSI IN THE COOLING SYSTEM WHEN USING THE COOLING SYSTEM PRESSURE TESTER.

COOLING SYSTEM PRESSURE TESTER

TESTING THE COOLING SYSTEM

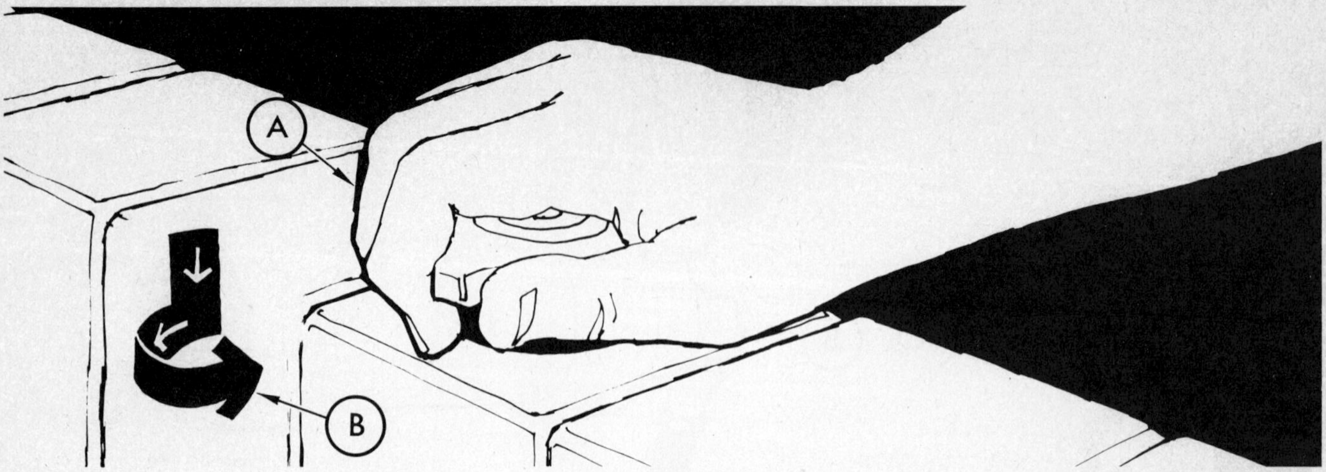

To test the cooling system for a coolant leak, remove the radiator pressure cap from the filler neck by pressing down (A) and turning the cap counterclockwise (B) at the same time. The coolant level should be within ½-inch of the filler neck. If it is lower, add coolant. Wipe the inside of the filler neck clean and check the inside sealing seat for nicks, dirt or solder bumps. Also inspect the cams on the outside of the neck. If the cams are deformed, the tester seal will be affected under pressure and the reading incorrect. Bent cams can be reshaped if done carefully. Consult a radiator shop and correct the defect before any testing.

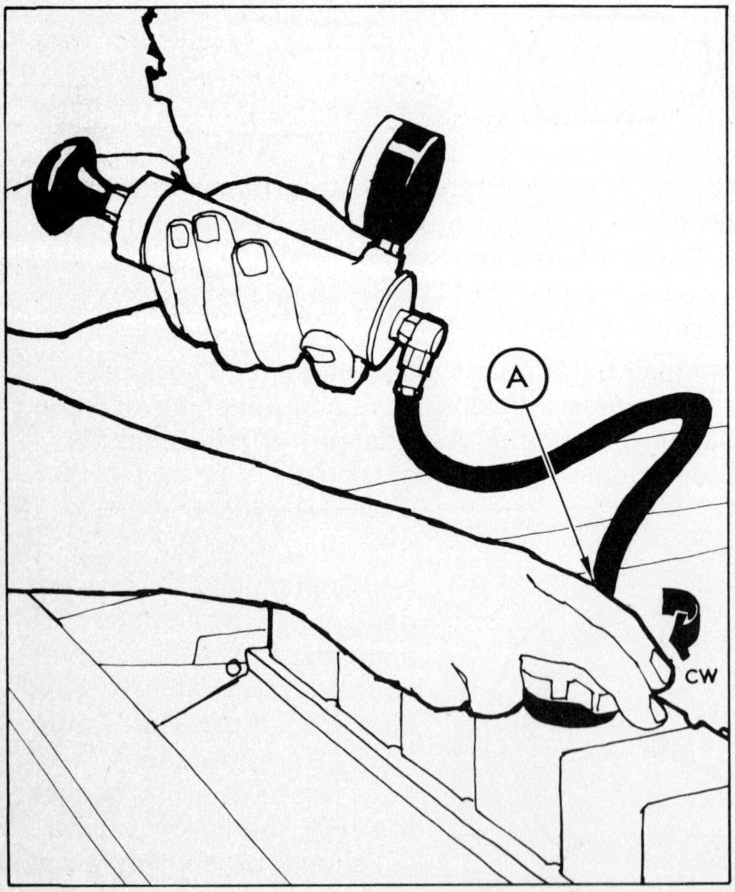

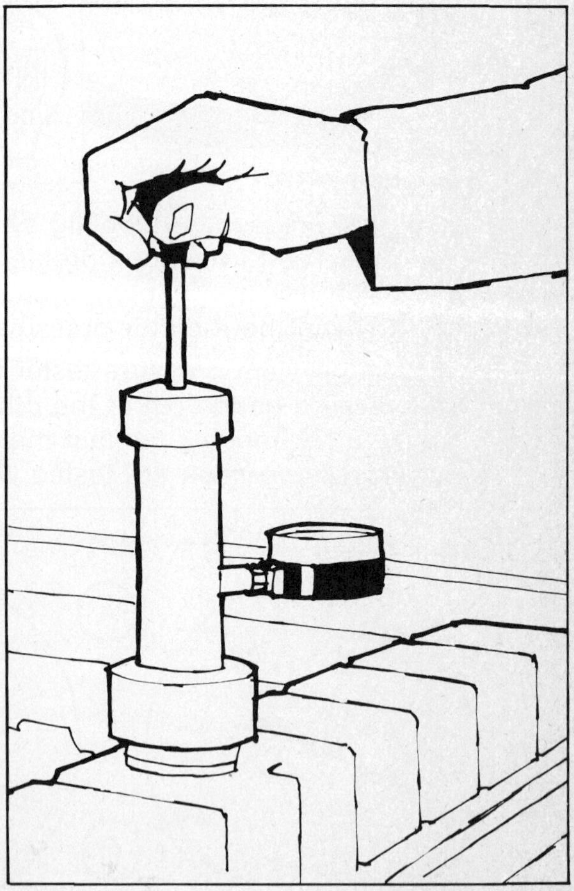

Run the engine until it is at operating temperature, then shut it off and attach the pressure tester (A) in place of the radiator pressure cap. Set the locking ears of the tester on the filler neck in line with the cutout notches, press down slightly and turn clockwise until the locking ears rest against the stop lugs on the filler neck. If this sounds complicated, remember that this is exactly how the pressure cap is placed on the filler neck.

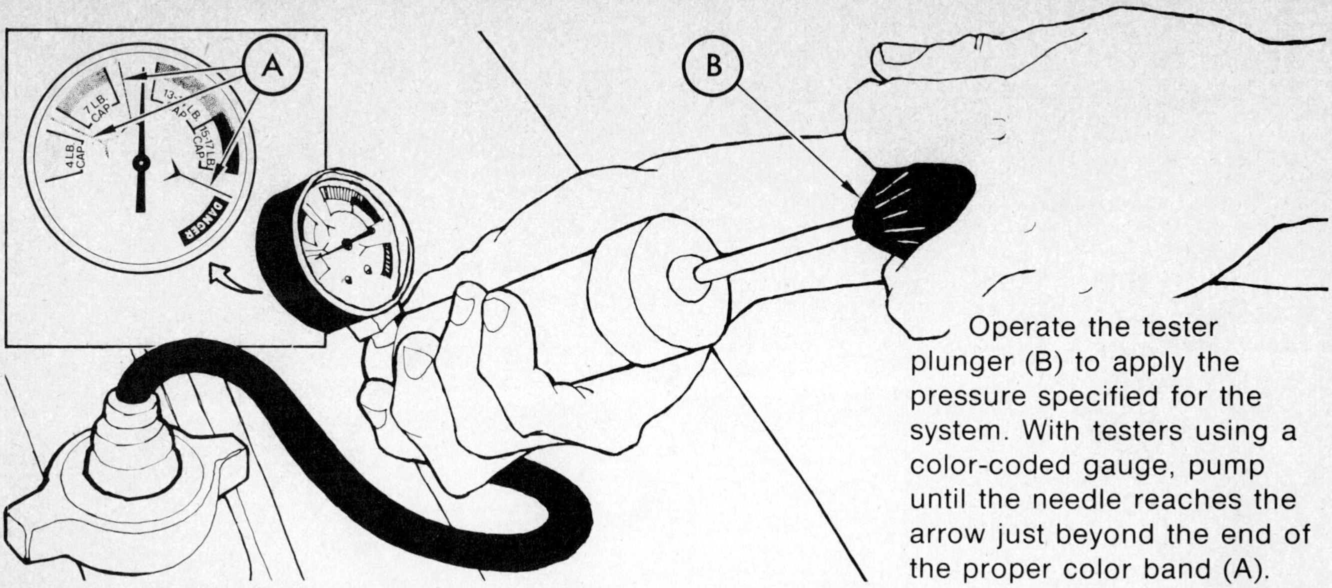

Operate the tester plunger (B) to apply the pressure specified for the system. With testers using a color-coded gauge, pump until the needle reaches the arrow just beyond the end of the proper color band (A).

PRESSURE TESTER READINGS

If the system is intact with no leakage present, the gauge needle should hold steady for at least two minutes.

A gauge needle that drops slowly instead of holding steady indicates that a small leak or seepage of some kind is present. Check the water pump, radiator and heater hoses, connections, clamps and gaskets. Hoses that swell excessively while the system is under test are weak and should be replaced. If no external leakage is found, there may be one or more tiny leaks in the radiator core.

A fluctuating needle indicates a combustion leak, usually a defective head gasket. This can also cause the needle to drop quickly after pressure is pumped into the system. In this case, check for external leaks. If none are found, detach the tester and run the engine to open the thermostat and let the coolant expand.

Reconnect the tester and pump to 7 pounds pressure with the engine running. Then race the engine and watch the tester needle. If it fluctuates at this point, you have a defective head gasket that is allowing exhaust gas to mix with the coolant.

CAUTION: PRESSURE BUILDS UP QUICKLY WHEN CONTINUOUS ENGINE OPERATION IS COMBINED WITH A GASKET LEAK TEST. RELEASE PRESSURE BY SHUTTING OFF THE ENGINE AND REMOVING THE TESTER. IN ANY CASE, *DO NOT* LET THE PRESSURE EXCEED THE MAXIMUM SPECIFIED FOR THE SYSTEM.

A combustion leak can be located by shorting out each spark plug (removing its cable) one at a time. The gauge needle will stop fluctuating when the leaking cylinder's plug is shorted out.

If the needle does not fluctuate but an abnormal amount of water comes out of the tailpipe during the test, the leak could be a defective head gasket, a cracked engine block or a cylinder head that is cracked near the exhaust ports.

When the tester gauge shows a pressure drop but no external leakage is present, run the engine briefly and check for an internal leak by removing the oil dipstick and inspecting for water mixed with the oil. Since transmission intercoolers are also a possible source of such leakage, check the transmission dipstick too on cars equipped with an automatic transmission.

If water appears on the dipstick along with the transmission fluid, there is a serious internal leak and the engine or transmission must be removed, disassembled and the leak located and corrected.

PRESSURE-TESTING RADIATOR CAPS

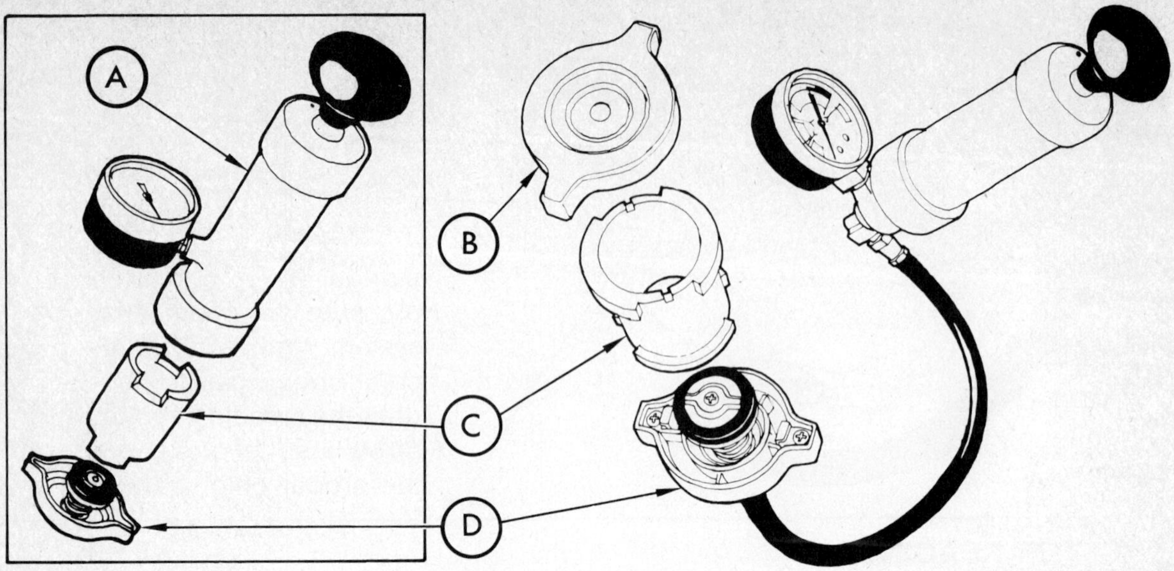

While the cooling system itself may test as good, the radiator pressure cap can be defective. To determine if the cap is working properly, attach the cap adapter (C) to the tester. With some models, the adapter and a neoprene seal fit directly onto the tester body (A); on other models, the adapter attaches to the tester cap (D).

Now remove the pressure cap from the radiator and check the pressure rating stamped on the top. If it is a replacement cap, it might be well to check this pressure rating against that specified by the manufacturer to make certain that it is the correct cap for the system. Clean the cap's seating surfaces and inspect for any distortions. Then dip the cap in water and attach the cap (B) to the open end of the adapter.

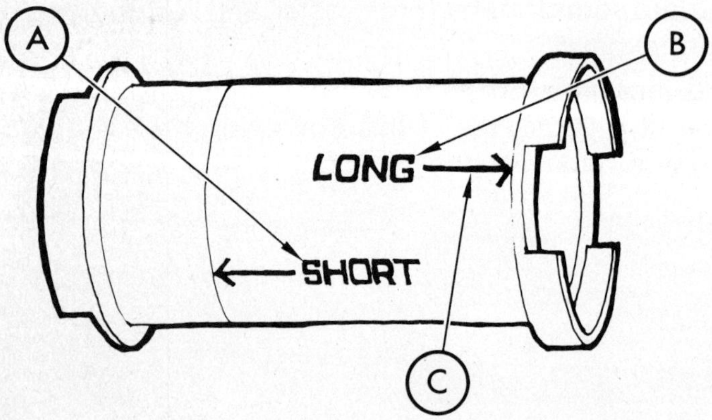

Some adapters are marked LONG (B) and SHORT (A), with corresponding directional arrows (C). With these adapters, the radiator cap must be attached to the proper end. If the cap has a long stem, attach it to the adapter end marked LONG; if it is a short-stem cap, attach to the end marked SHORT.

Operate the pump plunger until the specified pressure shows on the gauge face. If the pressure holds or falls very slowly but stays within the specified range for 30 seconds or longer, the cap is good.

If the needle falls comparatively fast, the cap is defective and must be replaced with a new cap of the same type and rating.

CAUTION: ***DO NOT*** **SUBSTITUTE A CONSTANT-PRESSURE CAP FOR A PRESSURE-VENT TYPE, OR VICE-VERSA.**

A pressure cap that has been in use on a radiator filler neck for some time forms a seat impression in the gasket. If the cap leaks on the first test, remove and reapply it to the adapter several times to see if the leakage is caused by the old impression seating incorrectly on the adapter. Dipping the cap in water as previously recommended helps overcome the effects of such seat impressions, as it tends to soften the gasket.

PRESSURE TESTER MAINTENANCE

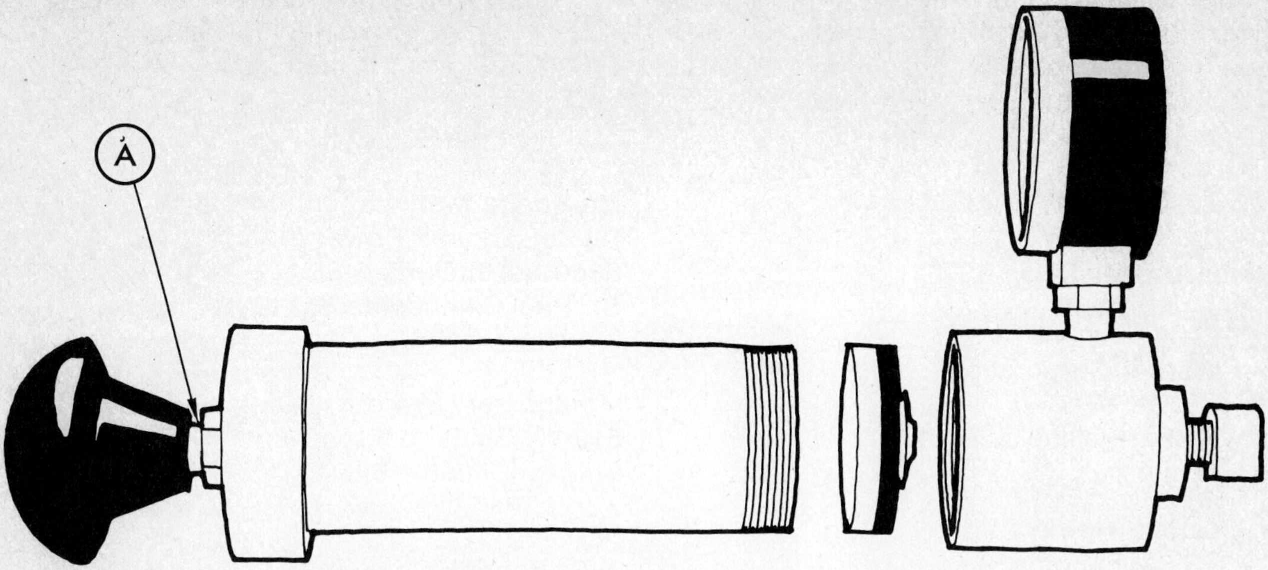

Like the compression gauge, the cooling system pressure tester is a simple instrument, but it should not be mistreated. Dropping or rough handling affect its accuracy and can damage the connecting flanges. To keep the leather piston inside the tester body flexible, occasionally put three or four drops of light oil in the pump air inlet (A).

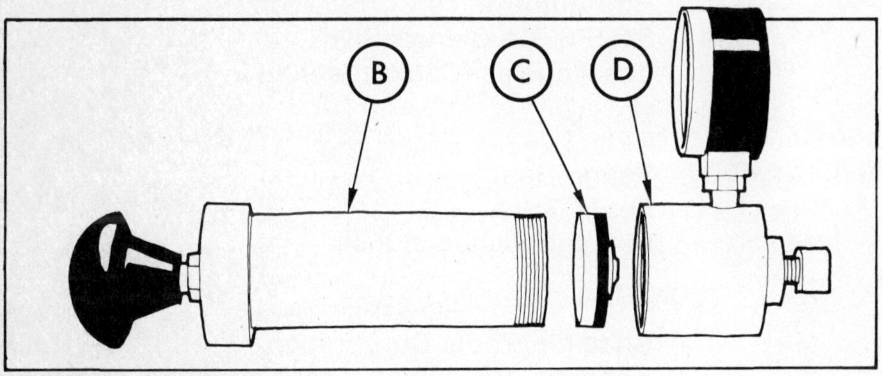

If the tester does not operate correctly, the check valve or gasket inside the tester body may be worn or damaged. Unscrew the pump tube (D) from the tester body (B) and lift the check valve and gasket (C) out.

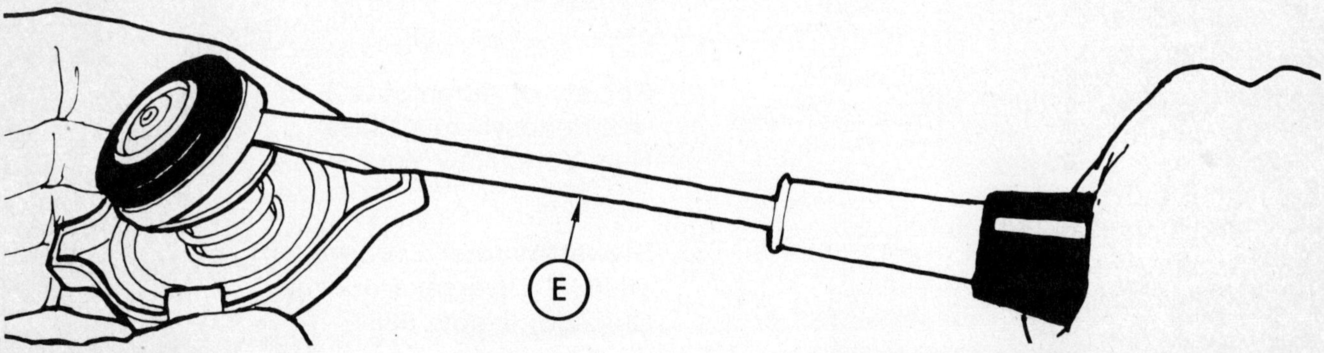

If check valve replacement is indicated and one is not at hand, the bottom part of a pressure cap can be pried off carefully with a screwdriver (E) and used as a replacement. Be sure to check the tester manual to determine the exact cap to be used for this purpose, however—not all spring-loaded valves will operate correctly as a replacement check valve.

Drop the replacement check valve into position, screw the pump tube on tightly by hand and tighten it securely.

Damage to the flexible hose on testers so equipped can be corrected with a length of universal oil filter line and the appropriate male connectors. Again, check your instruction manual for specified replacements and use a thread sealing compound when attaching the new hose to the fittings.

ACKNOWLEDGMENTS

Further information on automotive tune-up and test equipment and how to use it is available from the companies listed below. All of them have given valuable assistance in preparing this book. The editors wish to thank these companies for their generous contributions to the success of this volume of the Tool Book series.

Allen Testproducts Division
2101 N. Pitcher St.
Kalamazoo, Michigan 49007

Borg-Warner Corporation
11045 Gage Ave.
Franklin Park, Illinois 60131

Cal-Custom (Hawk)
23011 S. Wilmington
Carson, California 90745

Clayton Manufacturing Company
P.O. Box 550
El Monte, California 91734

Delco-Remy Division
General Motors Corporation
Anderson, Indiana 46000

Fox Products Company
4720 N. 18th St.
Philadelphia, Pennsylvania 19141

Heath Company
Benton Harbor, Michigan 49022

Ingersoll-Rand/Proto
Div. Pendleton Tool Ind. Inc.
2209 S. Santa Fe Ave.
Los Angeles, California 90058

Kal-Equip Company
P.O. Box 188
411 Washington St.
Otsego, Michigan 49087

Kar Check-Kal Equip
Otsego, Michigan 49072

Marquette Manufacturing Company
Div. of Applied Power Ind., Inc.
3800 North Dunlap St.
St. Paul, Minnesota 55112

Montgomery Ward & Company
619 W. Chicago Ave.
Chicago, Illinois 60607

J. C. Penney Company
1301 Avenue of the Americas
New York, New York 10019

Rite Autotronics Corporation
3485 S. La Cienga Blvd.
Los Angeles, California 90016

Sears, Roebuck and Company
Sears Tower
Chicago, Illinois 60684

Snap-On Tools Corporation
Kenosha, Wisconsin 53140

Society of Automotive Engineering
#2 Pennsylvania Plaza
New York, New York 10001

Stewart-Warner Corporation
1826 W. Diversey Parkway
Chicago, Illinois 60614

Sun Electric Corporation
3011 E. Route 176
Crystal Lake, Illinois 60014

Triplett Corporation
Bluffton, Ohio 45817